The U.N. Framework Convention on Climate Change Activities Implemented Jointly (AIJ) Pilot: Experiences and Lessons Learned

Institute for Global Environmental Strategies

VOLUME 1

The U.N. Framework Convention on Climate Change Activities Implemented Jointly (AIJ) Pilot: Experiences and Lessons Learned

Edited by

ROBERT K. DIXON

Office of Energy Efficiency and Renewable Energy,
U.S. Department of Energy,
Washington, DC, U.S.A.

Contributing Co-Authors (alphabetical order): L. Abron, K. Begg, R. Bradley, E. Brenes, K. Chatterjee, K. Danish, R. Dixon, M. Dutschke, M. Eisma, C. Figueres, W. van der Gaast, R. Gibbons, A. Hambleton, P. Hassing, J. Heister, C. Jepma, J. Jones, P. Karani, L. Kosloff, A. Leonard, W. Makundi, N. Matsuo, M. Mendis, A. Michaelowa, D. Michel, I. Mintzer, S. Parkinson, K. Poore, J. Rotter, J. Sathaye, R. Selrod, C. Sinha, M. Trexler, E. Vine

SPRINGER-SCIENCE+BUSINESS MEDIA, B.V.

A C.I.P. Catalogue record for this book is available from the Library of Congress.

ISBN 978-0-7923-6056-8 ISBN 978-94-011-4287-8 (eBook)
DOI 10.1007/978-94-011-4287-8

Printed on acid-free paper

Photo credits: Robert Dixon and Roger Taylor

Contents

Preface

José María Figueres Olsen
Former President Republic of Costa Rica

The heated debate about global climate change continues. Some say it is the gravest calamity our species has ever encountered. Others deny its existence altogether. As with most cases of human decision making, the truth is most likely somewhere in the middle.

The challenge of this particular set of decisions is the overwhelming sense of uncertainty. Science cannot fully attribute the climatic catastrophes occurring before our eyes to increasing levels of greenhouse gas concentrations. Neither can science prove that extreme events and warming trends are unrelated to human behavior. Economic models, sophisticated as they are, cannot agree on the costs of reducing carbon dioxide (CO_2) emissions in industrialized countries. International negotiations are thus mired in the morass of scientific and economic uncertainty.

The are only two elements of certainty in the whole debate. The first is the need for precaution. The potential impacts are such, that the risk of inaction is unaffordable to the human race. Under the current state of knowledge, mankind must take cautious but unequivocal steps to reverse current patterns.

The second element is that of constant learning. If only mildly true, the challenge ahead is one for which we have little previous relevant experience. This is a clear case in which the community of nations must take a step, review and learn. Then take another step, review and learn. It is only this commitment to continuous learning that will prepare us to move faster, and more comprehensively, should we need to do so in the near future.

Therein lies the importance and timeliness of this book. As the Parties to the Framework Convention on Climate Change (FCCC) forge emergent Clean Development Mechanism (CDM) and Joint Implementation (JI) instruments as conceived under the Kyoto Protocol, it is wise to turn our gaze to the experience with the FCCC Activities Implemented Jointly (AIJ) pilot. While we all agree that the experience could have been broader and more inclusive, we can also agree that this is the most relevant experience we have for those two forthcoming mechanisms. The seed of success of the CDM and of JI could already have been planted during the AIJ pilot phase. Only by searching for it will we know how to prepare the soil of our future endeavors.

I congratulate Robert Dixon, Editor, for his initiative in collecting lessons learned. I thank the individual authors for their thoughtful retracing of steps, which were sometimes fruitful, sometimes challenging, always innovative.

Foreword

TOWARD COUNTER MEASURES AGAINST GLOBAL WARMING IN THE 21ST CENTURY

HIROSHI OHKI
President of the Third Session of the Conference of the Parties to the UNFCCC; Former Minister, Environment Agency, Government of Japan

Urged by the common recognition that global warming must be tackled in order to preserve all living things on the Earth and to secure the welfare of future generations, ministers, officials, diplomats and technical experts representing many countries with different national interests and backgrounds worked together at the UN Framework Convention on Climate Change (FCCC) 3rd Conference of the Parties (COP-3), Kyoto, Japan, to find appropriate solutions. Opinions differed between developed countries and developing countries regarding greenhouse gas (GHG) emission targets and timetables, as well as, between the European Union (EU), USA and Japan regarding various components of the Protocol, including the flexibility mechanisms. This difference of opinion led to prolonged Protocol negotiations that could only be resolved at the crack of dawn of the last day of COP-3.

A primary controversy behind the tough COP-3 negotiations is the problem of how to harmonize the measures against global warming without damaging economic development of each Party (country). However, delegates from developing countries, EU, Japan and USA pooled their intellectual talents to address global warming while supporting the sustainable development goals for all Parties. In brief, negotiation of the Kyoto Protocol was a concrete demonstration of the commitment to sustainable development advocated at the UN Conference on Environment and Development (UNCED) in 1992.

As a result of our collective efforts, many new ideas and concepts have been incorporated into the Kyoto Protocol. Among the most important concepts contained within the Protocol include the so-called flexibility mechanisms, the Clean Development Mechanism (CDM), Joint

Implementation (JI), and Emissions Trading. Considerable effort is now being invested to formulate and implement the flexibility mechanisms at international, national and regional levels. In this context, the FCCC Activities Implemented Jointly (AIJ) pilot is very noteworthy as a precedent to design the Kyoto Protocol flexibility mechanisms.

I hope that this lessons learned book, an authoritative evaluation and summary of the AIJ pilot, will contribute to the development and implementation of the Kyoto Protocol flexibility mechanisms. Lessons learned from the AIJ pilot will help foster significant progress towards addressing global warming by all FCCC Parties.

Editor's Statement and Acknowledgements

Prominent players in the UN Framework Convention on Climate Change (FCCC) activities implemented jointly (AIJ) pilot phase prepared the chapters in this book. My 33 co-authors are drawn from 12 nations, representing the interests and experiences of developing, transition and developed countries. The authors include policy analysts, diplomats, scientists, engineers, project developers, attorneys, economists, and other professionals. Each co-author contributed a range of AIJ pilot phase experiences and in-depth expertise to this text. I offer my sincere thanks to this very special group of co-authors for their commitment to excellence in the completion of a book that attempts to summarize lessons learned from the technically and politically complex AIJ pilot phase. Each co-author volunteered their contribution(s) to this book and all toiled in this effort without any extra compensation.

The book attempts to cover a broad range of topics associated with the AIJ pilot phase. The three broad objectives of the book include:

- review, interpret and compile the experiences of participants and observers of the AIJ pilot phase;
- based on empirical data and skilled observations, identify and document lessons learned from the AIJ pilot; and,
- interpret and summarize lessons learned from the AIJ pilot and translate these experiences for future considerations by the FCCC Parties.

We endeavored to prepare a balanced text that treats cogent topics with sufficient technical and policy depth. Naturally some topics have received more emphasis than others have. It is recognized that many other AIJ pilot phase studies and reports are underway or have been completed. Hopefully, the chapters in this book complement the efforts and contributions of other workers in this field.

Each chapter in this book was subject to technical review by 3-5 technical and policy experts drawn from developing, transition and developed countries that are familiar with the topic. The chapters were revised based on the peer-review comments received by chapter co-authors. Each chapter was also subject to editorial review and revision. I offer my thanks to the 40+ referees from 20+ countries that volunteered their time to review and correct the chapters in this book. We have made every effort to minimize errors and omissions. Ultimately, the final text is the responsibility of the co-authors and the editor. This book has not been subject to technical or policy review by any FCCC Party and does not necessarily reflect the official views of any firm, government, intergovernmental body, or international organization.

Many professionals contributed to the preparation and publication of this book. While text space does not permit me to recognize all of those contributions I specifically commend the following individuals for their contributions to this book: K. Chow, J. Figueres, A. Hoffman, A. Haspel, M. Iwase, M. deJong, N. Matsuo, M. Moriya, A. Morishima, S. Nishioka, H. Oki, L. Perez, R. Pomerance, D. Reicher, D. Reifsnyder, D. Rose, A. Yu, A. Noordermeer-Zandee, and K. Zwally. The Institute for Global Environmental Strategies, Hayama, Japan, provided support to me during a brief period in 1998-99 and I thank them for their kind assistance.

Robert K. Dixon, PhD
Washington, DC, USA
August, 1999

Book Editor and Chapter Co-Authors

(in alphabetical order)

Dr. Lilia Abron
PEER Consultants
12300 Twinbrook Parkway, Suite 410
Rockville, MD 20852
USA
TEL: 1-301-816-0700
FAX: 1-301-816-9291
E-mail: peer1@ix.netcom.com
E-mail: peercpc@hotmail.com

Dr. Abron is President, PEER Consultants, an environment and engineering firm with offices throughout the USA. She is also President, PEER Africa, environmental engineering firm that developed the first Activities Implemented Jointly project in South Africa. Dr. Abron was a Professor of Civil Engineering for 13 years. She earned a Ph.D. in Civil Engineering from the University of Iowa in 1972, MS in Environmental Engineering from Washington University, St. Louis, MO in 1968 and BS in Chemistry from LeMoyne College, Memphis, TN, USA.

Dr. Richard Bradley
Office of Policy
U.S. Department of Energy
1000 Independence Avenue, SW
Washington, DC 20585
USA
TEL: 1-202-586-0154
FAX: 1-202-586-2062
E-mail: EconoNut@aol.com
E-mail: Richard.Bradley@hq.doe.gov

Dr. Bradley is an economist and has served in a variety of academic and government positions over the course of his career. He was a member of the USA delegations to the 1st, 2nd, 3rd and 4th FCCC Conference of the Parties (COP). Dr. Bradley earned his terminal degree at the University of California, Riverside, California, USA.

Dr. Katherine Begg
Centre for Environmental Strategy
University of Surrey
Guildford
Surrey GU2 5XH
United Kingdom
TEL: 44-1483-876687
FAX: 44-1483-879521
E-mail: K.Begg@surrey.ac.uk

Dr. Begg is coordinating co-editor of a major new text on the UN FCCC AIJ Pilot, *Accounting and Accreditation of Activities Implemented Jointly,* sponsored by the European Commission. Currently, she is with the Centre for Environmental Strategy at the University of Surrey. Prior to working on joint implementation and Clean Development Mechanism issues, Dr. Begg was with the U.K. Central Electricity Generating Board and engaged in acid deposition analysis.

Mr. Esteban Brenes, Esq.
Environment Department
The World Bank
1818 H Street, NW
Washington, DC 20433
USA
TEL: 1-202-483-4683
FAX: 1-202-522-0262
E-mail: ebrenes@worldbank.org

Mr. Brenes is with the World Bank. He has been an active participant in the World Bank Activities Implemented Jointly program offering technical assistance to joint implementation offices in Costa Rica, Guatemala, El Salvador, Argentina and Mexico. Mr. Brenes earned a JD from the University of Costa Rica and an LLM in Environmental Law from George Washington University Law School, Washington, DC, USA.

Dr. Kalipada Chatterjee
Global Environment Systems Group
Development Alternatives
B-32 Tara Crescent, Quatab Institutional Area
New Delhi 110016
India
FAX: 91-11-686-6031
E-mail: kc@sdalt.ernet.in
E-mail: c-kalipada@hotmail.com
E-mail: tara@sdalt.ernet.in

Dr. Chatterjee is Manager, Global Environment Systems Group, Environment Systems Branch, Development Alternatives. He is the Editor of the 1997 book, *Activities Implemented Jointly to Mitigate Climate Change: Developing Country Perspectives.*

Mr. Kyle Danish, Esq.
Hunton & Williams
1900 K Street, NW
Washington, DC 20006-1109
USA
TEL: 1-202-955-1567
FAX: 1-202-778-2201
E-mail: kdanish@hunton.com

Mr. Danish is an Attorney with the international law firm of Hunton and Williams, Washington, DC, USA. He has consulted with clients on the development of joint implementation projects and matters related to the UN FCCC and Kyoto Protocol. Mr. Danish is Vice-Chair of the American Bar Association Committee on climate change and sustainable development. He earned a JD from Temple University School of Law and MPA from Princeton University, Princeton, NJ, USA.

Dr. Robert Dixon
Office of Energy Efficiency and Renewable Energy
U.S. Department of Energy
1000 Independence Avenue, SW
Washington, DC 20585
USA
TEL: 1-202-586-9220
FAX: 1-202-586-9260
E-mail: robert.dixon@ee.doe.gov

Dr. Dixon is currently Director, International Programs, Office of Energy Efficiency and Renewable Energy, U.S. Department of Energy, Washington, DC, USA. He was a Visiting Research Fellow at IGES in Japan when this book was prepared and edited. Prior to this assignment he was Director of two U.S. Presidential Initiatives: U.S. Initiative on Joint Implementation and the U.S. Country Studies Program. Dr. Dixon earned his terminal degree in Biochemistry from the University of Missouri, Columbia, MO, USA.

Dr. Michael Dutschke
ESALQ
Universidae de Sao Paulo
BR 05630-130
Sao Paulo
Brazil
TEL: 55-11-3741-0992
FAX: 55-11-3741-0992
E-mail: dutschke@greco.com.br

Dr. Dutschke is Economist working on climate change issues at the Hamburg Institute for Economic Research. He has published a number of papers on joint implementation and the Clean Development Mechanism. Dr. Dutschke is Visiting Fellow at the Universidae de Sao Paulo.

Mr. Maarten J. Eisma
Joint Implementation Quarterly
Meerkoetlaan 30A
9765 TD Paterswolde
The Netherlands
TEL: 31-50-309-6815
FAX: 31-50-309-6815
E-mail: jiq@northsea.nl

Mr. Eisma is with the Joint Implementation Network, Paterswolde, The Netherlands. He has been involved in several Joint Implementation Network research projects and is Assistant Editor of the magazine, *Joint Implementation Quarterly*.

Ms. Christiana Figueres
Center for Sustainable Development in the America's
1700 Connecticut Avenue, NW
Washington, DC 20009
USA
TEL: 1-202-588-0155
FAX: 1-202-588-0756
E-mail: christiana@csdanet.org
E-mail: csda@wizard.net

Ms. Figueres is Executive Director, Center for Sustainable Development (CSDA) of the America's, Washington, DC, USA. Since founding CSDA in 1994, she has been active in the development of Activities Implemented Jointly in Latin America. Ms. Figueres has been in UN FCCC negotiations since 1995. She earned an MS in Economics from the London School of Economics, London, U.K.

Mr. Wytze van der Gaast
Joint Implementation Quarterly
Meerkoetlaan 30A
9765 TD Paterswolde
The Netherlands
TEL: 31-50-309-6815
FAX: 31-50-309-6815
E-mail: jiq@northsea.nl

Mr. van der Gaast is Research Fellow, Joint Implementation Network, Paterswolde, The Netherlands. He is Executive Director, of the magazine *Joint Implementation Quarterly*. He has conducted research and published reports on the Activities Implemented Jointly pilot and the Clean Development Mechanism.

Ms. Rebecca Gibbons
Trexler and Associates, Inc.
1131 SW River Forest Road
Portland, OR 97267
USA
TEL: 1-503-786-0559
FAX: 1-503-786-9859
E-mail: rgibbons@climateservices.com

Ms. Gibbons is a Policy Associate with Trexler and Associates, Portland, Oregon, USA. She conducts analysis on climate change mitigation with a focus on land-use change and forest sector projects. Ms. Gibbons holds a BA in Political Science from the University of Connecticut, USA.

Ms. Anne Hambleton
Center for Sustainable Development of the America's
1700 Connecticut Avenue, NW
Washington, DC 20009
USA
TEL: 1-202-588-0155
FAX: 1-202-588-0756
E-mail: a_hambleton@yahoo.com

Ms. Hambleton was formerly Program Director, Center for Sustainable Development of the America's, Washington, DC, USA. She has been active in the development of joint implementation activities in Latin America.

Mr. Paul Hassing
Directorate General for Development Cooperation
Ministry of Foreign Affairs
Bezuidenhoutseweg 67
P.O. Box 20061
2500 EB The Hague
The Netherlands
TEL: 31-70-348-4306
FAX: 31-70-348-4303
E-mail: p.hassing@dml.minbuza.nl

Mr. Hassing is currently Head, Climate, Energy and Environmental Technologies Division, Directorate for Development Cooperation, Ministry of Foreign Affairs. He is a regular member of the Dutch delegation to the UN FCCC Conference of the Parties. Mr. Hassing has been active participant in the UN FCCC Activities Implemented Jointly pilot and has published several significant reports on joint implementation and the Clean Development Mechanism.

Dr. Johannes Heister
Environment Department
The World Bank
1818 H Street NW
Washington, DC 20433
USA
TEL: 1-202-458-4280
FAX: 1-202-522-2130
E-mail: Jheister@worldbank.org

Dr. Heister is an Economist and was with the Kiel Institute of World Economics in Germany before joining the World Bank. He is the co-author of significant reports on tradable emission permits and published a book on the design of an international climate protection treaty. Dr. Heister has an MS in Economics and received his terminal degree in Political Science from Kiel University.

Dr. Catrinus Jepma
Department of Economics
Open University
University of Groningen
The Netherlands
TEL: 31-50-309-6815
FAX: 31-50-309-6815
E-mail: jiq@northsea.nl
E-mail: c.j.jepma@eco.rug.nl

Dr. Jepma is a Professor International Economics at the University of Groningen, University of Amsterdam and the Open University in the Netherlands. He has been active in the Intergovernmental Panel on Climate Change Second and Third Assessment Reports. Dr. Jepma is the editor or co-author of several major books on joint implementation and the Clean Development Mechanism, including, *On the Compatibility of Flexible Instruments*.

Mr. Jed Jones
Lloyd's Register Industry Division
Lloyd's Register House
29 Wellesey Road
Croydon CRO 2AJ
United Kingdom
TEL: 44-181-681-4727
FAX: 44-181-681-4839
E-mail: jed.jones@lr.org

Mr. Jones is Manager, Special Product Development, Lloyd's Register Industry Division, London, United Kingdom. He contributed to the development of monitoring, verification and certification protocols for the Activities Implemented Jointly pilot program. Mr. Jones earned a BS and MIEE.

Mr. Patrick Karani
Environment Department
The World Bank
1818 H Street, NW
Washington, DC 20433
USA
TEL: 1-202-473-4279
FAX: 1-202-522-2130
E-mail: pkarani@worldbank.org

Mr. Karani works on the World Bank AIJ Program and is formerly a Research Associate at the African Centre for Technology Studies. He is co-author of *Joint Implementation Under the Climate Change Convention: Development Opportunities in Africa* and a contributor to *A Climate for Development: Climate Change Policy Options for Africa.* Mr. Karani is a Ph.D. candidate at the University of Amsterdam, the Netherlands.

Ms. Laura Kosloff, Esq.
Trexler and Associates, Inc.
1131 SW River Forest Road
Portland, OR 97267
USA
TEL: 1-503-786-0559
FAX: 1-503-786-9859
E-mail: Lkosloff@climateservices.com

Ms. Kosloff is Vice-President, Trexler and Associates, Portland, Oregon, USA. She is senior counsel and directs negotiation and contracting for carbon offset projects in the forestry and energy sectors. She has developed legal agreements among private parties, governments and non-government organizations addressing legal issues associated with carbon offset projects. Ms. Kosloff holds a JD degree from the University of California, Davis, and BA from Antioch College, Yellow Springs, Ohio, USA.

Ms. J. Amber Leonard
Pacific Institute for Studies in Development, Environment and Security
9514 Garwood Street
Silver Spring, MD 20901
USA
TEL: 1-301-587-8714
FAX: 1-301-587-8716
E-mail: amber@igc.org

Ms. Leonard is Managing Editor, *Global Change* magazine and a Senior Associate of the Pacific Institute for Studies in Development, Environment and Security. She is a Project Director of the New Initiative for a North-South Dialogue on Climate Change and on behalf of the Global Business Network convenes roundtables for USA business leaders on the Clean Development Mechanism and joint implementation. Ms. Leonard was with the Stockholm Environment Institute prior to her current position(s). She earned a MBA from California State University, San Francisco, CA, USA.

Dr. Willy Makundi
Lawrence Berkeley National Laboratory
University of California, Building 90, Room 4000
Berkeley, CA 94720
USA
TEL: 1-510-486- 6852
FAX: 1-510-486-6996
E-mail: wrmldc@dante.lbl.gov

Dr. Makundi earned a Ph.D. in Resource Economics from the University of California-Berkeley, USA. He is currently a Scientist with Environmental Technologies Division, Lawrence Berkeley National Laboratory, Berkeley, CA, USA. Dr. Makundi was co-Editor of the book, *Sustainable Forest Management for Climate Change Mitigation: Monitoring and Verification of Greenhouse Gases.*

Dr. Naoki Matsuo
Institute for Global Environmental Strategies
1560-39 Kamiyamaguchi
Hayama, Kanagawa 240-0198
Japan
FAX: 81-468-55-3809
TEL: 81-468-55-3812
E-mail: n_matsuo@iges.or.jp

Dr. Matsuo is a Senior Research Fellow at the Institute for Global Environmental Strategies and Senior Researcher at the Global Industrial and Social Progress Research Institute, Tokyo, Japan. He is the author of many reports on joint implementation, Clean Development Mechanism and emissions trading. Dr. Matsuo is also a contributor to the Intergovernmental Panel on Climate Change (IPCC). He earned his Ph.D. in Theoretical Physics from Osaka University, Osaka, Japan in 1988.

Mr. Matthew Mendis
Alternative Energy Development
8455 Colesville Road, Suite 1225
Silver Spring, MD 20910
USA
TEL: 1-301-608-3666
FAX: 1-301-608-3667
E-mail: Mendis@CompuServe.com

Mr. Mendis is President and Founder, Alternative Energy Development, an energy and environment consulting firm in Silver Spring, Maryland, USA. He was formerly with the World Bank and has been active in the identification and development of greenhouse gas mitigation options in developing countries. Mr. Mendis helped lead the implementation of the Asia Least-Cost Greenhouse Gas Analysis for UNDP and the Asian Development Bank. He has published a number of reports on the joint implementation was a co-contributor to the UNDP report: *The Clean Development Mechanism: Issues and Options*. Mr. Mendis earned a BS in Mechanical Engineering and MA in Natural Resource Economics from the University of Maryland, USA.

Dr. Axel Michaelowa
177 Bd. de la Republique
92210 St.-Cloud
France
TEL: 33-1-477-12680
FAX: 33-1-477-12680
E-mail: michaelo@easynet.fr

Dr. Michaelowa has been working in association with the Hamburg Institute for Economic Research (HWWA), Hamburg, Germany on climate change issues since 1994. He specializes in the economic analysis of the Kyoto Protocol flexibility mechanisms and frequently publishes his work in journal articles and books. Dr. Michaelowa is a reviewer and helped design the World Bank Prototype Carbon Fund and associated joint implementation activities.

Mr. David Michel
School of Advanced and International Studies
The Johns Hopkins University
Washington, DC 20036
USA
TEL: 1-202-745-1377
E-mail: admichel_cop4@hotmail.com
E-mail: minfante@sais-jhu.edu

Mr. Michael current interests include the politics of the international climate change negotiations. He earned degrees in Political Science from Yale University and Sociology from Ecole des Hautes Etudes, Paris, France. He is Ph.D. candidate at The Johns Hopkins University.

Dr. Irving Mintzer
Pacific Institute for Development, Environment and Security
9514 Garwood Street
Silver Spring, MD 20901
USA
TEL: 1-301-587-8714
FAX: 1-301-587-8716
E-mail: imintzer@igc.apc.org

Dr. Mintzer is Executive Director, *Global Change* Magazine and a Senior Associate of the Pacific Institute for Studies in Development, Environment and Security. He is also a Senior Consultant with the Global Business Network. Dr. Mintzer was a lead author and co-author in the 1995-96 Intergovernmental Panel on Climate Change (IPCC) Assessment. He has testified on climate change issues before the USA Congress, the British Parliament, the German Bundestag, and the European Commission. Dr. Mintzer holds a Ph.D. in Energy and Resources and Masters in Business Administration from the University of California, Berkeley, CA, USA.

Dr. Stuart Parkinson
Centre for Environmental Strategy
University of Surrey
Guildford
Surrey GU2 5XH
U.K.
TEL: 44-1483-300-800
FAX: 44-1483-259-394
E-mail: s.parkinson@surrey.ac.uk

Dr. Parkinson was a contributor to a major new text on the UN FCCC Activities Implemented Jointly pilot, *Accounting and Accreditation of Activities Implemented jointly,* sponsored by the European Commission. He is Research Fellow with the Centre for Environmental Strategy at the University of Surrey. Dr. Parkinson has been employed with private industry, academia and non-government organizations.

Ms. Kerri Poore
Environment Department
The World Bank
1818 H Street, NW
Washington, DC 20433
USA
TEL: 1-202-473-2898
FAX: 1-202-522-2130
E-mail: kpoore@worldbank.org

Ms. Poore works on the World Bank Activities Implemented Jointly Program. She earned an MA in Comparative and Regional Politics from the American University, Washington, DC, USA. Ms. Poore contributed to the development of the World Bank Prototype Carbon Fund.

Mr. Jonathan E. Rotter, Esq.
General Counsel Division
The Nature Conservancy
4245 N. Fairfax Dr., Suite 100
Arlington, VA 22203-1606
USA
TEL: 1-703-841-4593
FAX: 1-703-841-0128
E-mail: jrotter@tnc.org

Mr. Rotter is Chief Divisional Counsel for Latin America and the Caribbean, Office of International Counsel, International Headquarters, The Nature Conservancy, Arlington, VA, USA. He has been a leader and Chief Counsel for the development of joint implementation activities at The Nature Conservancy. Mr. Rotter serves as Vice-Chair for the American Bar Association committee on climate change and sustainable development. He earned a JD from the William and Mary School of Law, USA.

Dr. Jayant Sathaye
Lawrence Berkeley National Laboratory
University of California, Building 90, Room 4000
Berkeley, CA 94720
USA
TEL: 1-510-486-6294
FAX: 1-510-486-6996
E-mail: jasldc@dante.lbl.gov

Dr. Sathaye is a Senior Scientist and Leader of the International Energy Studies Group at the Lawrence Berkeley National Laboratory. He earned a PhD in environmental engineering. Dr. Sathaye is a Coordinating Lead Author of Intergovernmental Panel on Climate Change (IPCC) reports on Technology Transfer, and, Land-use Change and Forestry. He is also a Coordinating Lead Author for the IPCC Third Assessment Report.

Mr. Rolf Selrod
Norwegian Consortium for Development and Environment
Kirkegatan 12, N-0153 Oslo
Norway
TEL: 47-22-82-5200
FAX: 47-22-94-0581
E-mail: rselrod@online.no

Mr. Selrod is a Partner in the Bureau for Environmental Analysis and Director of the Norwegian Consortium for Development and Environment, Oslo, Norway. He has served as Program Director at the Centre for International Climate and Environment Research at the University of Oslo, Special Advisor on matters relating to Global Environmental Issues in the Norwegian Ministry of Foreign Affairs and a Program Officer with UNEP in Nairobi, Kenya. Mr. Selrod earned an M Phil. in Political Science from Oslo University, Oslo, Norway.

Dr. Chandra S. Sinha
Environment Department
The World Bank
1818 H Street, NW
Washington, DC 20433
USA
TEL: 1-202-458-7475
FAX: 1-202-522-2130
E-mail: Csinha@worldbank.org

Dr. Sinha provides technical advice on Global Environment Facility projects and was involved in the World Bank's Activities Implemented Jointly Program and the Prototype Carbon Fund. He has been involved in the design of the India Agricultural Demand Side Management Project and the Philippines Renewable Energy Electrification for Remote Island Applications. Dr. Sinha earned his terminal degree in Energy Studies from the Indian Institute of Technology, New Delhi, India and is on leave from the TATA Energy Research Institute.

Dr. Mark C. Trexler
Trexler and Associates, Inc.
1131 SW River Forest Road
Portland, OR 97267
USA
TEL: 1-503-786-0559
FAX: 1-503-786-9859
E-mail: mtrexler@climateservices.com

Dr. Mark Trexler is President, Trexler and Associates, Portland, Oregon, USA. He was an early leader in assessing forest sector potential in climate change mitigation while serving as a Senior Fellow at the World Resources Institute, Washington, DC, USA. Dr. Trexler participated in the development of the first forest sector carbon offset project, the CARE Guatemala Project funded by AES Corp. Dr. Trexler earned a PhD and MPP from the University of California, Berkeley, CA and BA from Antioch College, Yellow Springs, OH, USA.

Dr. Edward Vine
Lawrence Berkeley National Laboratory
University of California, Building 90, Room 4000
Berkeley, CA 94720
USA
TEL: 1-510-486-6294
FAX: 1-510-486-6996
E-mail: ELVine@lbl.gov

Dr. Vine is a Staff Scientist in the International Studies Group at Lawrence Berkeley national Laboratory. He is widely published on the topics of energy efficiency programs and energy policy. Dr. Vine recently published two major reports on the monitoring, evaluation, reporting, verification and certification of Activities Implemented Jointly (AIJ) projects. Dr. Vine is a member of the American Evaluation Association, Association of Energy Services Professionals, the California Demand-Side Management Measurement Advisory Committee, and an Affiliated Faculty member of Energy and Resources Group at the University of California, Berkeley, USA.

Chapter and Book Referees

(in alphabetical order)

Dr. Lilia Abron
Peer Consultants
12300 Twinbrook Parkway, Suite 410
Rockville, MD20852
USA
TEL: 1-301-816-0700
FAX: 1-301-816-9291
E-mail: peer1@ix.netcom.com

Mr. Kenneth Andrasko
Office of Reinvention and Policy
U.S. Environmental Protection Agency
401 M Street, SW
Washington, DC 20460
USA
TEL: 1-202-260-4497
FAX: 1-202-260-6405
E-mail: andrasko.ken@epa.gov

Dr. Michael Apps
Climate Change
Canadian Forest Service
Natural Resources Canada
5320 122nd Street
Edmonton, Alberta T6H 3S5
Canada
TEL: 1-780-435-7305
FAX: 1-780-435-7359
E-mail: mapps@NRCan.gc.ca

Dr. Joseph Asamoah
OMEGA Scientific Research
Moselata Park
0044 Pretoria
South Africa
E-mail: joasa@smartnet.co.sa

Ms. Suzanne Barnes
Environment Department
The World Bank
1818 H Street, NW
Washington, DC 20433
USA
TEL: 1-202458-8353
FAX: 1-202-477-0565
E-mail: Sbarnes@worldbank.org

Mr. Richard Baron
International Energy Agency
9 rue de la federation
75739 Paris Cedex 15
France
E-mail: richard.baron@iea.org

Dr. Katherine Begg
Centre for Environmental Strategy
University of Surrey
Guildford
Surrey GU2 5XH
United Kingdom
TEL: 44-1483-876687
FAX: 44-1483-879521
E-mail: K.Begg@surrey.ac.uk

Dr. Richard Bradley
Office of Policy
U. S. Department of Energy
1000 Independence Avenue, SW
Washington, DC 20585
USA
TEL: 1-202-586-0154
FAX: 1-202-586-2062
E-mail: Richard.Bradley@hq.doe.gov
E-mail: Econonut@aol.com

Dr. K. Chatterjee
Development Alternatives
B32 Tara Crescent
Qutub Institutional Area
New Delhi 110016
India
FAX: 91-11-686-6031
E-mail: tara@sdalt.ernet.in

Mr. Jeffrey Dunoff
Temple University School of Law
1719 N. Broad Street
Philadelphia, PA 19122
USA
TEL: 1-215-204-8233
FAX: 1-215-204-1185
E-mail: dunoffj@vm.temple.edu

Ms. Jane Ellis
Environment Directorate
Organization for Economic Cooperation and Development
2 rue Andre Pascal
75775 Paris Cedex 16
France
TEL: 33-1-4524-1598
FAX: 33-1-4524-7876
E-mail: jane.ellis@oecd.org

Dr. Tibor Farago
Former Chairman, Subsidiary Body on Scientific and Technical Advice, UN FCCC
504, UCTA 2
Budapest H-1173
Hungary
E-mail: bit@mail.matav.hu

Mr. Charles Feinstein
Environment Department
The World Bank
1818 H Street, NW
Washington, DC 20433
USA
TEL: 1-202-473-2896
FAX: 1-202522-2130
E-mail: Cfeinstein@worldbank.org

Ms. Christiana Figueres
Center for Sustainable Development in the America's
1700 Connecticut Avenue, NW
Washington, DC 20009
TEL: 1-202-588-0155
FAX: 1-202-588-0756
E-mail: christiana@csdanet.org

Mr. Frank Friedman
Elf Aquitane, Inc.
910 17th Street, NW
Washington, DC 20006
USA
TEL: 1-202-785-2955
FAX: 1-202-785-0522
E-mail: ffriedman@ix.netcom.com

Dr. Ibrahim Abdel Gelil
Egyptian Environmental Affairs Agency
P.O. Box 995
El Maadi Post Office
Cairo
Egypt
TEL: 202-375-3215
FAX: 202-378-4285
E-mail: iagelil@idsc.gov.eg

Mr. Donald Goldberg
Center for International Environmental Law
1367 Connecticut Ave., NW
Washington, DC 20036
USA
TEL: 1-202-785-8700
FAX: 1-202-785-8701
E-mail: cieldg@apc.ipc.org

Dr. Erik Haites
Margaree Consultants, Inc.
145 King Street West, Suite 1000
Toronto, Ontario M5H 3X6
Canada
TEL: 1-416-369-0900
FAX: 1-416-369-0922
E-mail: Ehaites@netcom.ca

Dr. Elmer Holt
Office of Policy
U.S. Department of Energy
1000 Independence Avenue, SW
Washington, DC 20585
USA
TEL: 1-202-586-0714
FAX: 1-202-5862062
E-mail: Elmer.Holt@hq.doe.gov

Dr. Joseph C.K. Huang
U.S. Country Studies Program
1000 Independence Avenue, SW
Washington, DC 20585
USA
TEL: 1-202-586-3090
FAX: 1-202-586-3485
E-mail: joseph.huang@ee.doe.gov

Dr. Maithili Iyer
Institute for Global Environmental Strategies
1560-39 Kamiyamaguchi, Hayama, Kanagawa
Japan 240-0198
TEL: 81-468-55-3810
FAX: 81-468-55-3809
E-mail: miyer@iges.or.jp

Dr. Catrinus Jepma
Department of Economics
University of Groningen
University of Amsterdam
Open University
The Netherlands
FAX: 31-50-309-6815
E-mail: jiq@northsea.nl
E-mail: c.j.jepma@eco.rug.nl

Dr. Gregory Kats
Office of Energy Efficiency and Renewable Energy
U.S. Department of Energy
1000 Independence Avenue, SW
Washington, DC 20585
USA
TEL: 1-202-586-1392
FAX: 1-202-586-9260
E-mail: greg.kats@ee.doe.gov

Dr. Alexy Kokorin
Institute of Global Climate and Ecology
Glebovskaya 20B, IGCE
Moscow 107258
Russia
TEL: 7-95-169-2198
FAX: 7-95-413-6263
E-mail: Alexey.Kokorin@relcom.ru

Ms. Duane Lakich
Global Environment Center
U.S. Agency for International Development
Ronald Reagan Bldg., Room 3.08
1300 Pennsylvania Avenue, NW
Washington, DC 20523-3800
USA
TEL: 1-202-712-5304
FAX: 1-202-216-3174
E-mail: dlakich@usaid.gov

Dr. Amin Aslam Malik
ENVO RK: Research and Development Organization
H#7, Street #30
F-7/1
Islamabad
Pakistan
E-mail: amin@isb.comsats.net.pk

Mr. Matthew Mendis
Alternative Energy Development
8455 Colesvilles Road, Suite 1225
Silver Spring, MD 20910
USA
TEL: 1-301-608-3666
FAX: 1-301-608-3667
E-mail: Mendis@CompuServe.com

Dr. Nandita Mongia
Regional Coordinator for Climate Change
United Nations Development Program
1 United Nations Plaza
New York, NY 10017
USA
TEL: 1-212-906-5833
FAX: 1-212-906-5825
E-mail: nandita.mongia@undp.org

Dr. Mark Mwandosya
Centre for Energy, Environment, Science and Technology
1372 karume Road, Oysterbay
P.O. Box 5511
Dar Es Salaam
Tanzania
TEL: 255-51-66-7569
FAX: 255-51-66-6079
E-mail: mjmwandosya@intafrica.com

Dr. Anne Arquit Niedeberger
SWAPP
Bundesamt fur AuBenwirtschaft
Effingerstrasse 1
3003 Bern
Switzerland
E-mail: anne.arquit-niederberger@bawi.admin.ch

Mr. Lubomir Nondek
DHV CR
Taboriska 23
13087 Prague
Czech Republic
E-mail: dhv@terminal.cz

Mr. Mark Perlis
Dickstein, Shapiro, Morin and Oshinsky
2102 L Street, NW
Washington, DC 20037
USA
TEL: 1-202775-4703
FAX: 1-202-887-0689
E-mail: perlism@dsmo.com

Dr. Jonathan Pershing
International Energy Agency
9 rue de la federation
75739 Paris Cedex 15
France
E-mail: jonathan.pershing@iea.org

Ms. Annie Roncerel
Energy and Atmosphere Program
United Nations Development Program
1 United Nations Plaza
New York, NY 10017
USA
TEL: 1-212-906-6616
FAX: 1-212-906-5148
E-mail: annie.concerel@undp.org

Dr. Jayant Sathaye
Lawrence Berkeley National Laboratory
University of California, Building 90, Room 4000
Berkeley, CA 94720
TEL: 1-510-486-6294
FAX: 1-510-486-6996
E-mail: jasldc@dante.lbl.gov

Mr. Youba Sokona
Environmental Development Action in the Third World
B.P. 3370, 54 rue Carnot
Dakar
Senegal
TEL: 221-822-2496
FAX: 221-823-5157
E-mail: ysokona@enda.sn

Mr. David South
Technology and Markets Group
Energy Resources International, Inc.
1015 18th Street, Suite 650
Washington, DC 20036
USA
TEL: 1-202-785-8833
FAX: 1-202-785-8834
E-mail: south@energyresources.com

Mr. Mark R. Stevens
Department of Primary Industries
Edmund Barton Building, Barton, ACT
Canberra 2601
Australia
TEL: 61-6-272-4791
FAX: 61-6-271-6599
E-mail: mark.stevens@isr.gov.au

Mr. Evind Tandberg
International Monetary Fund
Ministry of Finance
Rakovski Street 102
1000 Sofia
Bulgaria
TEL: 359-2-9874-583
FAX: 359-2-9874-583
E-mail: E.Tandberg@minfin.govrn.bg

Dr. Franz Tattenbach
Officina Costarricense de Implementation Conjunta
P.O. Box 7170-1000
San Jose
Costa Rica
TEL: 506-290-1283
FAX: 506-290-1283
E-mail: ocicgm@sol.racsa.co.cr

Ms. Eveline Trines
United Nations Climate Change Secretariat
Haus Carstanjen, Martin Luther-King-Strasse 8
D-53175 Bonn
Germany
TEL: 49-228-815-1525
FAX: 49-228-815-1999
E-mail: etrines@unfccc.de

Ms. Kristi Varangu
International Energy Agency
9 rue de la federation
75739 Paris Cedex 15
France
E-mail: kristi.varangu@iea.org

Dr. Ted Vinson
Department of Civil Engineering
Oregon State University, Apperson Hall
Corvallis, OR 97330
USA
TEL: 1-541-737-3494
FAX: 1-541-737-3052
E-mail: vinsont@ccmail.orst.edu

Ms. Pamela Wexler
The Cadmus Group, Inc.
1901 N. Ft. Meyer Drive
Arlington, VA 22209
USA
TEL: 1-703-558-0211
FAX: 1-703-558-0210
E-mail: pwexler@cadmusgroup.com

Abbreviations and Acronyms

AIJ	Activities Implemented Jointly
ADB	Asian Development Bank
AEP	American Electric Power
AOSIS	Alliance of Small Island States
APSEB	Andhra Pradesh State Electricity Board
ASEAN	Association of South East Asian Nations
BIODIVERSIFIX	Biodiversity and Carbon Fixation Project in Costa Rica
C	Carbon
CARFIX	Carbon Fixation Project in Costa Rica
CCAP	Center for Clean Air Policy
CDM	Clean Development Mechanism
CER	Certified Emission Reduction
CFE	Comision Federal de Electricidad
CFC	Chlorofluorocarbon
CFL	Compact Fluorescent Lamp
CHP	Combined Heat and Power
CIDA	Canadian International Development Agency
CINDE	Coalicion Costarricense de Iniciativas de Desarrollo
CJII	Canadian Joint Implementation Initiative
CNG	Compressed Natural Gas
COP	Conference of the Parties to the UN FCCC
CO_2	Carbon dioxide
CPA	Carbon Purchase Agreement
CVA	CDM Validation Assessment Report
CTO	Certifiable Tradable Offset
DNV	Det Norsk Veritas
DOE	U.S. Department of Energy
DSM	Demand Side Management
EAES	Environmentally Adapted Energy System
ECOLAND	Piedras Blancas National Park forest preservation project
EIA	Environmental Impact Assessment
EMR	Emission Reduction Monitoring Report
EPA	U.S. Environmental Protection Agency
EPRI	Electric Power Research Institute
ERPA	Emission Reduction Purchase Agreement
ER	Emission Reduction
ERU	Emission Reduction Unit
ESMAP	Energy Sector Management Program

EU	European Union
FACE	Forests Absorbing Carbon dioxide Emissions
FCCC	UN Framework Convention on Climate Change
FEMP	Federal Energy Management Program
GATT	General Agreement on Tariffs and Trade
GDP	Gross Domestic Product
GEF	Global Environment Facility
GHG	Greenhouse Gas
GWP	Global Warming Potential
G77	Group of 77 developing countries
ha	hectare
HFO	Heavy Fuel Oil
HV	High Voltage
HVDS	High Voltage Distribution System
IAB	International Accreditation Body
ICSU	International Council of Scientific Unions
IEA	International Energy Agency
IET	International Emissions Trading
IIEC	International Institute for Energy Conservation
IDB	Inter-American Development Bank
ILUMEX	High Efficiency Lighting Project
INC	Intergovernmental Negotiating Committee
INE	Instituto Nacional de Ecologia
IPCC	Intergovernmental Panel on Climate Change
IPMVP	International Performance Measurement and Verification Protocol
IPP	Independent Power Producer
IREDA	Indian Renewable Energy Development Agency
IRR	Internal Rate of Return
IUEP	International Utility Efficiency Partnerships
JI	Joint Implementation
JIRC	Joint Implementation Registration Centre
KP	Kyoto Protocol
LBNL	Lawrence Berkeley National Laboratory
LPG	Liquid Petroleum Gas
LUCF	Land-use Change and Forestry
LV	LowVoltage
LVDS	Low Voltage Distribution System
Mg	Megagram
MINAE	Ministry of Environment and Energy
MITI	Ministry of International Trade and Industry
MOP	Meeting of the Parties to the UN FCCC
MRV	Monitoring, Reporting and Verification
MVP	Monitoring and Verification Protocol

NEDO	New Energy and Industrial Technology Development Organization
NGO	Non-Government Organization
NO_x	Nitrogen Oxide(s)
NUTEK	National Board for Industrial and Technical Development
OCIC	Costa Rica Office for Joint Implementation
ODA	Official Development Assistance
OECD	Organization for Economic Cooperation and Development
OLADE	Organizacion Latinoamericana de Energia
O&M	Operation & Maintenance
OTC	Over The Counter
PAP	Protected Areas Project
PCF	Prototype Carbon Fund
PPA	Power Purchase Agreement
Petagram	Pg
PV	Photo Voltaic
R, D&D	Research, Development and Deployment
QELRC	Quantified Emission Reduction or Limitation Commitments
RDP	Reconstruction and Development Program
RIL	Reduced Impact Logging
ROR	Rate of Return
RPTS	Regional Program on Traditional Energy Sector
Rs	Indian Rupees (1US$=43Rs)
RUSAFOR	Saratov, Russia Afforestation Project
RUSAGAS	Russia Fugitive Gas Capture Project
SADC	Southern Africa Development and Cooperation
SBSTA	Subsidiary Body for Scientific and Technological Advice
SEB	State Electricity Board
SGS	Societe Generale de Surveillance
SO_2	Sulfur dioxide
STIBOR	Stockholm Interbank Offered Rates
TAA	Trexler and Associates
TNC	The Nature Conservancy
Teragram	Tg
ton	t
UN	United Nations
UNCED	United Nations Conference on Environment and Development
UNDP	United Nations Development Program
UNEP	United Nations Environment Program

UNFCCC	United Nations Framework Convention on Climate Change
URF	Uniform Reporting Format
USA	United States of America
USCSP	U.S. Country Studies Program
USIJI	United States Initiative on Joint Implementation
VCR	Voluntary Challenge and Registry
WBSCD	World Business Council for Sustainable Development
WCED	World Commission on Economic Development
WEC	World Energy Council

Chapter 1

INTRODUCTION TO THE FCCC ACTIVITIES IMPLEMENTED JOINTLY PILOT

R. DIXON[1], I. MINTZER[2]
[1]Institute for Global Environmental Strategies; [2]Pacific Institute for Studies in Development, Environment and Security

Key words: joint implementation, Activities Implemented Jointly, Clean Development Mechanism, Conference of the Parties, Kyoto Protocol, UN FCCC

Abstract: Increasing scientific evidence suggests there is a discernable human influence on the global climate system. The 1992 United Nations Framework Convention on Climate Change (FCCC) was conceived by representatives from over 170 countries to achieve stabilization of greenhouse gas (GHG) concentrations in the atmosphere at a level that would prevent dangerous anthropogenic interference with the climate system. Such a level should be achieved within a timeframe sufficient to enable economic development to proceed in a sustainable manner. Parties to the FCCC are now considering policies and measures to mitigate GHG emissions. The concept of joint implementation (JI) was incorporated into the FCCC text as part of Article 4.2(a). This paragraph states that, developed country Parties and other Parties included in Annex I may implement... policies and measures jointly with other Parties and may assist other Parties in contributing to the objective of the FCCC. Given the absence of any practical experience with JI, the Parties established the Activities Implemented Jointly (AIJ) pilot phase in 1995. A wide range of questions regarding the potential costs and benefits of JI projects could best be answered in an international pilot program. The AIJ pilot provides an opportunity for testing and experimentation with the development, management, monitoring and reporting of JI activities. Experiences and lessons learned during the AIJ pilot will assist Parties with the development and potential implementation of the 1997 Kyoto Protocol and its new flexibility mechanisms: JI, emissions trading and the Clean Development Mechanism (CDM).

1. BACKGROUND

For more than a century, scientists raised concerns about the risks of accelerated climate change due to human activities (Arrhenius, 1896). In 1956, Roger Revelle and Hans Suess elevated the level of urgency about this issue with a warning that the continuing atmospheric buildup of GHGs constituted a grand global experiment on the only livable planet currently known to science (Carraro, 1999). In 1983, a group of scientists and economists held an assessment conference that was convened jointly by the UN Environment Program (UNEP), the World Meteorological Organization (WMO), and the International Council of Scientific Unions (ICSU) in Villach, Austria. At this meeting, for the first time, scientists, economists, and other experts concluded that addressing the risks of climate change might require international cooperation, an active dialog between scientists and decision-makers, and the implementation of specific policies to reduce the risks of climate change.

From 1985 to 1990, scientists and policy-makers engaged in an increasingly intense exchange of views. Formal discussions proceeded at a series of international meetings held in Toronto (Canada, June 1988), Tokyo (Japan, July 1989), Bergen (Norway, May 1990), and at the Second World Climate Conference in Geneva, (Switzerland, April 1990). In each, the issue of human influence on the atmosphere was debated at length.

In December 1990, the President of the Maldives urged the UN General Assembly to undertake global cooperative efforts to reduce the risks of climate change. He urged that these efforts begin immediately, so as to slow the rate of sea-level rise before his small island country was lost forever. The Intergovernmental Panel on Climate Change (IPCC), formed under the auspices of the WMO, concluded in their 1995 Assessment Report that increasing scientific evidence suggests that there is a discernible human influence on the global climate system (Watson et al. 1997).

The objectives of this chapter include:

- Consider the topic of global climate change and present an overview of the UN FCCC on climate Change;
- Introduce the AIJ pilot phase under the UN FCCC; and,
- Review the concepts and principles of JI.

1.1 Evolution of the UN FCCC and the concept of JI

In response to appeals from member countries, the UN General Assembly passed Resolution 89/212 and initiated the process of negotiating an international treaty to protect the climate system (United Nations, 1989). For nearly two years, representatives of more than 120 countries and numerous non-governmental organizations (NGOs) joined in the work of the Intergovernmental Negotiating Committee (INC). They worked diligently to hammer out an effective and equitable agreement that could be acceptable to all countries (Carraro 1999).

In 1991, the government of Norway introduced a new proposal to the negotiations. Norway observed that, because national circumstances differed, it might be more expensive to reduce GHG emissions in some countries than in others. This observation implied the idea that it might be cost-effective for two countries to form a partnership in these activities. In principle, they could share the costs of implementing an emissions reducing project in the country where costs were lowest. This idea grew into the concept known today as JI.

The precedent for the concept of JI can be found in several international agreements and treaties: the 1985 Vienna Convention for the Protection of the Ozone Layer, the 1976 Convention Concerning the Protection of the Rhine River Against Pollution by Chlorides, and the 1979 European Commission Long-Range Transboundary Air Pollution Convention (Hanafi, 1998). All of these agreements identify specific obligations of Parties but provisions were also made for joint actions to mitigate subject pollutants. The UN Convention to Combat Desertification also allows Parties to implement their obligations... individually or jointly, either through existing or prospective bilateral or multilateral agreements of combinations thereof. While these examples reflect a narrower scope than the concept of JI in the FCCC, they serve as indicators of new forms of international cooperation among Parties to an international agreement or treaty (Jepma and van der Gaast 1999).

Negotiations on a climate agreement continued in the INC until May 1991 when the text of the new legal instrument was drafted. Representatives from 172 countries convened at the UN Conference on Environment and Development (UNCED also called the Earth Summit) in Rio de Janeiro, Brazil, June 1992. A major achievement of the Earth Summit was the adoption of the FCCC. The FCCC represented a shared commitment by nations around the world to reduce the potential risks of global climate change (United Nations, 1992a). The objective of the FCCC is to:

> Achieve...stabilization of GHG concentrations in the atmosphere at a level that would prevent dangerous anthropogenic (human) interference with the climate system. Such a level should be achieved within a time frame sufficient to allow ecosystems to adapt naturally to climate change, to ensure that food production is not threatened, and to enable economic development to proceed in a sustainable manner.

The FCCC is the first international legal instrument that deals directly with climate change. By June 1993, 165 states plus the European Union (EU) had signed the FCCC. The 50th ratification on March 21, 1994 triggered the FCCC's entry into force by the end of 1994. To achieve the ultimate objective of the FCCC, a series of general commitments and principles, as well as, specific emission reduction targets or aims were identified for particular countries. The FCCC distinguishes three primary categories of states: members of the Organization for Economic Cooperation and Development (OECD) industrial countries (referred to as Annex I countries), industrial countries excluding those countries with economies in transition (EIT) to a market system (Annex II countries), and developing countries (non-Annex I countries) (Carraro 1999; Barrett 1994).

This approach is consistent with the recognized principle of common but differentiated responsibility among states. The FCCC encourages OECD countries to take the strongest steps towards GHG emission reductions while EITs are allowed more flexibility. The FCCC recognizes that reductions by developing countries in their rate of emissions growth will depend on the provision of financial and technical assistance from developed countries.

In August 1993, developing country representatives raised a series of pointed questions in the INC about the operational aspects of JI (Parikh, 1994). Would credits for JI projects be made available before industrialized countries took on legally binding targets for domestic emissions reductions? Would credits be available for JI projects in developing countries or would the regime be limited to projects undertaken by two industrial countries? The remainder of this chapter highlights the principle issues, risks and concerns regarding JI raised by the Parties and analysts.

2. WHY JOINT IMPLEMENTATION?

Many developing and EIT countries are in the process of expanding or restructuring their energy or natural resource sector infrastructure. The occasion of development or reconstruction of this infrastructure offers some of the most cost-effective options for mitigation of GHG emissions (Dixon et al. 1998, Jepma 1995). Because costs of reducing or sequestering GHG emissions may vary among countries and all such emissions have the same effect on global climate regardless of location, JI offers the opportunity to reduce emissions at a lower cost than would be possible if each country acted alone. The FCCC Parties recognized that policies and measures to deal with global change should be cost-effective so as to ensure global benefits at the lowest possible cost. With this in mind, the FCCC stated that efforts to address climate change might be carried out cooperatively by interested Parties.

JI activities provide benefits for partner country participants and for the global community as a whole. The global benefits include reducing the costs of GHG emission reductions while promoting sustainable development (Chatterjee, 1997). JI Projects can contribute to technology transfer and help to meet the development objectives of host countries while also achieving the FCCC environmental objectives.

Benefits accruing to host country participants from JI projects could include the following:

a) *Expanding technology transfer*, to encourage commercial diffusion of innovative technologies that can help meet host country development priorities while reducing or sequestering GHG emissions;
b) *Increasing private and public investments* in technologies and projects that reduce GHG emissions while contributing to overall host country development objectives;
c) *Reducing local environmental damages and human health consequences of development*, by preventing or reducing air, water, or soil pollution, and contributing to more sustainable use of natural resources;
d) *Enhancing local economic benefits*, through training, construction of new or improved facilities, public participation in projects, and provision of new energy services;
e) *Promoting sustainable development* by encouraging additional private sector investment in the development and dissemination of environment-friendly technologies and practices while reducing or sequestering GHG emissions; and
f) *Influencing the future of JI* by providing participants with an opportunity to influence the direction and structure of the AIJ pilot, the Kyoto Protocol and associated flexibility mechanisms, and demonstrating international collaboration to resolve environmental problems.

There are also important benefits of JI to participants outside the host countries including:

a) Improving market access, by providing entrée into energy and environmental markets in host countries. Participants may also be eligible for host country assistance in terms of relaxed administrative regulations or removal of other impediments to project development, for example, import restrictions;
b) *Lowering the cost of green technologies* by enhancing the competitiveness of climate-friendly technologies and accelerating their worldwide application, thus further reducing production costs;
c) Enhancing prospects for financing by expanding partnership opportunities, providing greater visibility and credibility to the potential

project, all of which may increase the credit worthiness of a particular activity;

d) *Reducing risk* by increasing the security of investment in foreign countries;
e) *Expanding knowledge of the FCCC and flexibility mechanisms* by providing the opportunity to participate in the analysis and structure of international environment agreements;
f) *Enhancing recognition* by demonstrating participants' commitment to reduce the threat of global change and contribute to sustainable development;
g) *Establishing a public record of GHG emission reductions* including development of transparent and functional methodologies for monitoring and verification; and
h) *Strengthening international credibility* by creating a track record in international markets by working with governments, businesses and organizations in foreign countries.

2.1 Specific Concerns and Risks Associated with JI

The concept of JI is considered to be complex by some Parties, participants and observers. Some have argued that it brings special risks for developing countries. The primary concerns associated with JI can be grouped into three broad categories: equity among Parties, sovereignty of the Parties, and leakage of GHG emission reductions.

2.1.1 Equity is a concern to all UN FCCC Parties

The FCCC is based on the premise that Annex I countries should be leaders in mitigating GHG emissions since they have made the largest contributions to historical and current-year emissions. Concerns have been raised by some countries that JI is a means for industrialized countries (Annex I) to transfer their environmental problems to developing countries (Chatterjee 1997). Some non-Annex I countries have concerns that JI will allow Annex I Parties to purchase GHG emission reductions internationally while implementing minimal mitigation measures domestically. Furthermore, JI could be abused to subsidize Annex I country fossil fuel consumption and foster disproportionate GHG emission patterns.

Annex I countries have equity concerns because projections of future GHG emissions growth in non-Annex I Parties, such as India and China, are enormous. With improved understanding, most Parties recognize that the purpose of JI is not to provide a mechanism for developed countries to export GHG emissions. Instead, FCCC delegates recognized that JI is a potentially significant economic instrument for reducing GHG emissions while offering technology choices for developing countries and helping host countries to meet their sustainable development goals.

2.1.2 Sovereignty risks to Parties should be avoided in the AIJ Pilot Phase

Protection of national sovereignty is an issue for both Annex I and non-Annex I Parties. Many Parties have expressed concerns that JI could pose risks to national sovereignty. Potential sovereignty risks include alteration of economic development policies, improper ownership and mismanagement of natural resources, and implementation of JI projects that do not transfer technology or support sustainable development goals.

National economic development policies could be negatively influenced by JI activities in non-Annex I countries. JI projects, funded by Annex-I donors, could influence national, regional and local economic investment decisions, technology choices, and the socioeconomic fabric of developing countries. Investors without proper consideration of appropriate technology could unduly influence technology selection and deployment. Domestic industrial policy of non-Annex I countries could be altered by imported technology or manufacturing processes that are embodied in JI activities. Without appropriate human and institutional capacity building, differences in the ability of developing countries to properly evaluate, compete for, and implement JI projects could lead to uneven distribution of technology and financial resources (Jackson and Begg 1999).

In theory, JI projects could contribute to international conflict over the ownership and management of domestic natural resources (e.g., land, forest systems, agro-ecosystems, and minerals). For example, forest sector C conservation and sequestration projects are governed by long-term contracts over land use and as well as local land tenure arrangements. The flow of economic goods (fiber, food or fuel) and environmental services (protection of biological resources and watersheds) from forest and agro-ecosystem JI projects can be positively or negatively influenced by the long-term goals of C sequestration and conservation (Dixon et al. 1994).

Similarly, investor firms are justifiably concerned that their investments in developing country natural resource projects must be properly insured against damage or destruction (eg, forests can be destroyed by fire). JI projects involving natural resources that are considered to be national assets must carefully balance trade-offs (e.g., food production vs. C sequestration) and be responsive to national sustainable development goals.

Developing and EIT countries seek JI projects that transfer technology and meet their sustainable development goals. Disproportionate investment in low cost JI projects in developing countries may ultimately lead to a situation where only high cost GHG mitigation options remain available should non-Annex I countries take on quantitative emission reduction targets. Some Parties view limiting future choices of GHG mitigation options as a potential loss of sovereignty. This scenario presupposes that non-Annex I countries will eventually take on emission reduction targets, a subject of serious debate and negotiation.

2.1.3 JI Project Leakage

Many delegates and analysts have argued that JI projects should take precautions to minimize leakage of GHG emission reductions between and among countries. Some Parties have argued that JI projects could actually have a negative impact on global GHG emission reductions if project baselines are inflated or if carbon accounting methods become unreliable. Sources of GHG leakage that could be attributable to JI projects include:

- Joint implementation projects could reduce or remove incentives for non-Annex I countries to take on quantitative emission reduction targets because it is easier to attract finance and technology without having Annex I status. The net result could be net growth of non-Annex I GHG emissions.
- Establishment of credible emission baselines is complex and the methodologies are not fully refined. Annex I countries and non-Annex I countries could improperly manipulate GHG emissions baselines leading to inflation or double counting of emission reductions. For example, an investor country could claim credit for reducing GHG emissions at domestic facilities under their own National Climate Change Action Plan and also take credit for a JI project using the same or similar technologies or processes in another country. Investor and host countries share the incentive to inflate their emissions reduction credits by claiming JI reductions that would have occurred anyway.

JI projects offer Annex I countries a broad portfolio of cost-efficient GHG emission reduction options that can be realized using available technologies. Some Parties have argued that technical innovation and development of, improved renewable energy and energy efficiency technologies could be delayed or postponed if investment funds are diverted to JI projects. The overall result is slower or incomplete mitigation of GHG emissions.

The arguments for and against JI activities and projects have resulted in a politically charged working environment for participants in the FCCC AIJ pilot phase (Jepma 1995). Given that practical hands-on experience with JI projects was limited in the early 1990s, an institutional framework was needed for the FCCC Parties to work together to achieve practical experience, methodologies, operational criteria and procedures. The combined effects of these conditions made it difficult for the FCCC Parties to craft a pilot program capable of realizing the full potential of JI in an AIJ regime with no credits.

3. THE ACTIVITIES IMPLEMENTED JOINTLY PILOT PHASE

The concept of JI was formally adopted in the FCCC text in 1992 (UN FCCC 1992). Article 4.2 (a) of the UN FCCC states that developed country Parties and other Parties included in Annex I may implement... policies and measures jointly with other Parties and may assist other Parties in contributing to the objective of the Convention. Article 3.3 of the FCCC states that policies and measures to deal with climate change should be cost-effective so as to ensure global benefits at the lowest possible cost, and highlights the link to JI by stating that efforts to address climate change may be carried out cooperatively by interested parties. The FCCC definition of JI was intentionally broad recognizing the need for future analysis, development and negotiation. Criteria for selecting, monitoring and crediting appropriate JI projects were omitted from the original FCCC text.

Article 4 of the FCCC outlines the national commitments undertaken by Parties to protect the global climate system. Paragraph 4.1(a) summarizes the general commitments of all signatories and includes a specific commitment to promote and cooperate in the development, application and diffusion, including transfer of technologies that control, reduce or prevent emissions of GHGs (United Nations, 1992b). Paragraph 4.2 specifies that industrial countries, in particular, will implement policies and measures to reduce GHG emissions (United Nations, 1992c). Sub-paragraph 4.2(a) deals with the concept of JI. This sub-paragraph notes that OECD Parties to the FCCC may implement such policies and measures jointly with other Parties and may assist other Parties in efforts to achieve the objective of the FCCC.

The wording of paragraph 4.2(a) reflected the original Norwegian concept of JI. The underlying idea behind JI was to encourage industrial countries (or private entities within industrial countries) to invest in projects to exploit low-cost reduction opportunities in developing countries and in EITs. In this way, the benefits of global environmental protection might be achieved in a more cost-effective manner than if all of the necessary reductions were achieved in countries with the highest historical emissions. The approach anticipated that credit for the resulting emissions reductions could be divided between the investing country and the host country for such projects.

At the First FCCC Conference of the Parties (COP-1), Berlin, Germany, March and April 1995, the AIJ pilot was initiated. The purpose of the AIJ pilot was to promote operational learning and to address methodological issues raised by this new class of joint ventures. To avoid confusion and promote cooperation, a suggestion by Malaysia was endorsed at COP-1 that the name of the new regime be changed from JI to AIJ. As part of this compromise, it was agreed that AIJ projects could be developed in either developing countries or in countries with economies in transition. In addition, Decision 5 stipulated that there would be no internationally fungible credits awarded for projects that either achieved emissions reductions or expanded the uptake of greenhouse gases by natural sinks during the AIJ pilot. This pre-operational period would last until an evaluation was completed, at a date identified as before the end of the decade (United Nations, 1995.

In the early stages, the Nordic countries, the Netherlands, and the USA were active leaders in supporting the goals and principles of the AIJ pilot. Many OECD countries invested financial, technological and material resources in human and institutional capacity building activities to support a thorough test of the AIJ pilot. For example, the USA established the U.S. Initiative on Joint Implementation (USIJI) as a part of President Clinton's Climate Change Action Plan. Similarly, Japan, Germany and the Netherlands also developed robust national JI programs modeled after the AIJ pilot. Parties reported their activities on an annual basis and the FCCC Secretariat summarized these national reports (Jackson and Begg 1999).

Between 1995 and 1997, many governments and analysts expressed dissatisfaction with the results of the AIJ pilot. For some advocates of technology cooperation and GHG emissions reductions, the principal problem was that there were simply not enough projects under development (< 100 worldwide). A small proportion of the projects were actually operational and visible. As a consequence, it was perceived that little learning by doing was actually occurring.

For others Parties, the main issue was not so much the number of projects but the geographic and sectoral distribution of projects. Almost half of the AIJ projects reported to the FCCC Secretariat by the end of 1997 were in the Baltic States. The second largest concentration occurred in Latin America. There were relatively few projects in Asia and only one under development in all of Africa. Nearly all the reported projects involved either the energy production sector or the forestry sector. Among analysts, there were sharp concerns that agreement on fundamental methodological issues was nowhere in sight. Consensus among Parties about common approaches to technical issues including the development of project baselines, performance monitoring, evaluation, verification, certification of emissions reductions remained elusive during the early stages of the AIJ pilot.

Governments, the private sector and non-government organizations made significant contributions to advance the establishment and implementation of the FCCC AIJ pilot phase. For example, the USIJI accepted over 33 JI projects in four years. These projects represented estimated private sector investments of approximately U.S.$500 million (Dixon 1997). The Governments of Japan, Norway and the Netherlands established special JI project development funds to help implement the AIJ pilot and stimulate project development (1998). An important information sharing activity, the establishment of *Joint Implementation Quarterly* and web-site, was sponsored by the Dutch government. The World Business Council for Sustainable Development (WBCSD), the Nature Conservancy (TNC), the Center for Clean Air Policy (CCAP) and other NGOs were leaders in the development, management and evaluation of JI projects and activities. The FCCC AIJ pilot phase provided an opportunity to test and evaluate methodologies for the design, implementation, monitoring and verification of GHG mitigation projects. The practical experience gained by many partners and Parties offers valuable insight into the potential benefits of JI and the challenges that must be addressed in order to achieve those benefits.

4. THE KYOTO PROTOCOL

In December 1997, Kyoto, Japan, COP-3 Parties agreed to a historic Protocol to the UN FCCC. This new instrument, called the Kyoto Protocol is designed to reduce GHG emissions by harnessing the economic forces of the global market place, in order to protect the global environment. Key aspects of the Protocol include emission reduction targets, timetables for industrialized nations, and market-based measures (e.g., flexibility mechanisms) for meeting the targets (Carraro 1999).

The Kyoto Protocol was opened for signature in March 1998. Over 50 countries have signed the Protocol but only a small number have ratified it. To enter into force, the Protocol must be ratified by at least 55 countries, accounting for 55 percent of the total 1990 carbon dioxide (CO_2) emissions of developed countries.

A central feature of the Kyoto Protocol is a set of binding GHG emission reduction targets that it establishes for developed countries. The specific limits vary from country to country, although the limits for the European Union (EU), Japan and the U.S. are similar, 8 percent below 1990 emission from the EU, 7 percent for the U.S. and 6 percent for Japan. Emission targets are to be reached as the average rate over a five-year period rather than in a single year. The first budget period will be 2008-2012. Allowing emissions to be averaged across a budget period increases flexibility by helping to smooth out short-term fluctuations due to economic or climatic factors. The emissions targets include all six major GHGs and provide for sequestration of CO_2 by forest systems (Carraro 1999).

Article 17 of the Kyoto Protocol allows nations with emissions targets to trade GHG allowances (Anonymous 1995). Using this mechanism, countries can cooperate with others to identify the most efficient and low cost emission reductions available. The emissions trading concept was developed in the USA and by Nordic countries to dramatically reduce sulfur dioxide (SO_2) emissions (Hanafi 1998; Jepma and van der Gaast 1999). Rules and guidelines for emissions trading verification, reporting and accountability were discussed at the 4th Conference of the Parties, Buenos Aires, Argentina, November 1998. While some progress has been made in developing international emissions trading guidelines a robust structure for the international emissions trading regime is yet to finalized. Some Parties to the UN FCCC (eg, Australia, Canada, Japan, New Zealand, Russia, Ukraine and the USA) have reached a conceptual agreement to pursue, through an informal umbrella group, the implementation of a trading regime.

Article 12 of the Kyoto Protocol outlines the CDM, another market-based component that embraces the concept of JI for credit between Annex I (industrial countries and countries with economies in transition) and non-Annex I (developing countries). Under the proposed CDM, industrial countries will be able to use certified emission reductions (CER) from project activities in developing countries to contribute to their compliance with GHG emission reduction targets (Goldemberg, 1998). CERs achieved through CDM projects can count toward compliance with national emissions reduction targets during the first budget period. Credits can be earned by such projects starting in 2000. The CDM will facilitate entry by firms in the industrial world into cooperative projects to reduce emissions in the developing world. Candidate projects might include energy efficiency, renewable energy or forest establishment activities. Firms will be able to reduce emissions at lower costs than they could accomplish domestically, while developing countries would receive sorely needed transfers of technology and project finance. The CDM will help developing countries advance projects that meet their sustainable development goals, as well as to certify and monitor project activities (Anonymous 1998). The CDM also ensures that some portion of the resources will be used to help highly vulnerable developing countries adapt to the environmental consequences of global change.

The Kyoto Protocol also addressed JI among developed countries. Article 6 states that countries with emissions reduction targets may obtain credit toward their targets through project-based emission reductions in other countries. Although not yet fully defined by the Parties, the operational modalities of this flexibility mechanism will likely be based on the experiences gained in the AIJ pilot. The Parties hope to agree upon final details at COP-6.

5. CURRENT STATUS OF AIJ PILOT AND OBJECTIVES OF THIS BOOK

The AIJ pilot under the FCCC has been operational for almost five years. Approximately 130 joint implementation projects have been reported to the FCCC Secretariat. The FCCC Secretariat has published three AIJ Pilot summary reports. Over 70 countries (Parties) have a government office that administers the complex technical, logistical and financial matters associated with JI. National JI programs have been subject to analysis and assessment, with several major reports published in recent months (Goldemberg 1998; Hanafi 1998; Lile et al. 1998). A cottage industry of JI project developers, JI project monitors and allied institutions have mushroomed over the past five years to support the AIJ pilot and national JI programs. The collective experiences and lessons learned by participants and observers of the AIJ Pilot have not been interpreted or systematically compiled (Jepma and van der Gaast 1999). This book is a contribution towards that goal.

In this context, the broad objectives of this book include:

- Review, interpret and compile the experiences of participants and observers of the AIJ pilot phase;
- Based on empirical data and skilled observations, identify and document lessons learned from the AIJ pilot; and
- Interpret and summarize lessons learned from the AIJ pilot and translate these experiences for future consideration by FCCC Parties.

REFERENCES:

Anonymous (1995) *controlling carbon dioxide emissions: the tradable permit system*, UN Conference on Trade and Development, UNCTAD/GID/11, Geneva.

Anonymous (1998) *Mitigation and Adaptation Cost Assessment Concepts, Methods, and Appropriate Use*, UNEP Collaborating Centre on Energy and Environment, Roskilde.

Arrhenius, S. (1896) on the influence of carbonic acid in the air upon the temperature of the ground. *Phil. Magazine* 41: 237.

Barrett, S. (1994) *The Strategy of Joint Implementation in the Framework Convention on Climate Change*, UN Conference on Trade and Development, UNCTAD/GID/10, Geneva.

Carraro, C. (Ed.) (1999) *International Environmental Agreements on Climate Change*, Kluwer Academic Publishers, Dordrecht, in press.

Chatterjee, K. (Ed.) (1997) *Activities Implemented Jointly to Mitigate Climate Change: Developing Country Perspectives*, Development Alternatives, Delhi.

Dixon, R.K. (1998) The U.S. Initiative on Joint Implementation: An Asia-Pacific Perspective. *Asian Perspective* 22:5-19.

Dixon, R.K. (1997) The U.S. Initiative on Joint Implementation. *Intern. Journal of Environment and Pollution* 8:1-18.

Dixon, R.K., Brown, S., Houghton, R.A., Solomon, A.M., Trexler, M.C. and Wisniewski, J. (1994) Carbon pools and flux of global forest systems. *Science* 263: 185-190.

Goldemberg, J., ed. (1998) *Issues and Options: The Clean Development Mechanism*, UNDP, New York.

Hanafi, A.G. (1998) Joint Implementation: Legal and Institutional Issues for an Effective International Program to Combat Climate Change. *The Harvard Environmental Law Review* 22:441-508.

Jackson, T. and Begg, K. (1999) Accounting and Accreditation of Activities Implemented Jointly, European Commission, Brussels.

Jepma, C.J. and van der Gaast, W. (1999) *On the Compatibility of Flexible Instruments*, Kluwer Academic Publishers, in press.

Jepma, C., ed. (1995) The *Feasibility of Joint Implementation*, Kluwer Academic Publishers, 212 p.

Lile, R., Powell, M. and Toman, M. (1998) *Implementing the Clean Development Mechanism: Lessons from U.S. Private Sector Participation in Activities Implemented Jointly*, Resources For the Future, Washington, DC.

Parikh, J. (1994) *North-South Cooperation in Climate Change through Joint Implementation*, Indira Ghandi Institute for Development Research, Bombay.

Sathaye, J. et. al. (Eds.)(1997) Sustainable *Forest Management for Climate Change Mitigation: Monitoring and Verification of Greenhouse Gases*, Kluwer Academic Publishers, Dordrecht.

United Nations (1989) UN General Assembly Resolution 89/212.

United Nations (1992a) *Framework Convention on Climate Change*, Article 2, United Nations, Geneva.

United Nations (1992b) *Framework Convention on Climate Change*, Article 4.1(c), United Nations, Geneva.

United Nations, (1992c) *Framework Convention on Climate Change*, Article 4.2(a), United Nations, Geneva.

United Nations (1995) *Decision 5/CP.1*, Climate Change Secretariat, United Nations, New York.

UN FCCC (1998) The *Kyoto Protocol: to the Convention on Climate Change*, UN FCCC Secretariat, Berlin.

Watson, R.T., Zinyowera, M.C., Moss, R.H. and D.J. Dokken (1997) The *Regional Impacts of Climate Change: An Assessment of Vulnerability*, Intergovernmental Panel on Climate Change, UN Environment Program, Nairobi.

Chapter 2

OVERVIEW OF THE UN FCCC ACTIVITIES IMPLEMENTED JOINTLY PILOT: COP-1 DECISION 5, REPORTING GUIDELINES AND CASE STUDIES

C. JEPMA[1], R. DIXON[2] and M. EISMA[1]
[1]*Foundation Joint Implementation Network;* [2]*Institute for Global Environmental Strategies*

Key words: activities implemented jointly, learning, baselines, reporting, cost-effectiveness, land-use change and forestry, energy efficiency, renewable energy

Abstract: The activities implemented jointly (AIJ) pilot was established by Decision 5 of the UN Framework Convention on Climate Change (FCCC) First Conference of the Parties (COP-1) in 1995. Decision 5 provides a broad framework for implementation of the pilot and identifies miminum criteria and guidelines for Parties to follow. With the advent of the AIJ pilot, only a hand-ful of countries were prepared to fully engage in this activity. Today, approximately 80 FCCC Parties are participating or have plans to participate in the pilot. Information from the pilot is reported to the FCCC Secretariat annually. AIJ project reporting guidelines have evolved over the past four years and standard guidelines have been developed by the Parties and are being used by most project developers. Project baselines and additionality of projects remain two AIJ pilot concepts which remain very controversial and difficult to report uniformly. Analysts continue to define additionality and develop methodologies for for project baseline development. Three AIJ projects are examined in detail: Burkina Faso Energy Management, Indonesia Reduced Impact Logging (RIL) and RUSAGAS fugitive gas capture. Lessons learned from each project are presented which are relevant to development of mature joint implementation (JI) and Clean Development Mechanism (CDM) regimes.

1. INTRODUCTION

Parties to the UN FCCC recognized in the mid-1990s that practical experience with JI is very limited. Some Parties had specific experience with JI activities and projects, while others only had a theoretical understanding of the broad principles (Jepma and van der Gaast 1999; Dixon, 1998). Thus, it was important to learn from the Parties with JI expertise and experience and share it with others. Practical experience could help provide a proper foundation for development of a mature JI scheme and an evolving CDM (Goldemberg, 1998). The objectives of this chapter include:

- Review Decision 5 of the FCCC COP-1;
- review the development and evolution of AIJ pilot reporting guidelines; and,
- Consider three AIJ project case studies.

2. DECISION 5, COP-1, THE BASIS OF THE AIJ PILOT

The FCCC AIJ pilot was defined at COP-1, Berlin, Germany in 1995. The Parties formally addressed JI and provided further definition of this developing concept. The result was Decision 5 of COP-1, activities implemented jointly under the pilot phase, which states:

1. Decides:
    a. To establish a pilot phase for activities implemented jointly among Annex I Parties and, on a voluntary basis, with non-Annex I Parties that so request;
    b. That activities implemented jointly should be compatible with and supportive of national environment and development priorities and strategies, contribute to cost-effectiveness in achieving global benefits and could be conducted in a comprehensive manner covering all relevant sources, sinks and reservoirs of GHGs;
    c. That all activities implemented jointly under this pilot phase require prior acceptance, approval or endorsement by the Governments of the Parties participating in these activities;
    d. That activities implemented jointly should bring about real, measurable and long-term environmental benefits related to the mitigation of climate change that would not have occurred in the absence of such activities;
    e. That the financing of activities implemented jointly shall be additional to the financial obligations of Parties included in Annex II to the

FCCC within the framework of the financial mechanism as well as to current official development assistance (ODA) flows;

f. That no credits shall accrue to any Party as a result of GHG emissions reduced or sequestered during the pilot phase from activities implemented jointly;

2. Further decides that during the pilot phase:
   a. The Subsidiary Body for Scientific and Technical Advice (SBSTA) will, in coordination with the Subsidiary Body for Implementation (SBI), establish a framework for reporting, in a transparent, well-defined and credible fashion, on the possible global benefits and the national economic, social and environmental impacts as well as any practical experience gained or technical difficulties encountered in activities implemented jointly under the pilot phase;
   b. The Parties involved are encouraged to report to the COP through the Secretariat using the framework thus established. This reporting shall be distinct from the national communications of Parties.
   c. The SBSTA and SBI, with the assistance of the Secretariat are requested to prepare a synthesis report for consideration by the COP;
3. Further decides:
   a. That the COP shall, as its annual session, review the progress of the pilot phase on the basis of the synthesis report with a view to taking appropriate decisions on the continuation of the pilot phase;
   b. In doing so, the COP shall take into consideration the need for a comprehensive review of the pilot phase in order to take a conclusive decision on the pilot phase and the progression beyond that, no later than the end of the decade.

3. PARTICIPATING COUNTRIES AND ORGANIZATION OF THE PILOT PHASE

When the AIJ pilot was approved by the Parties only a handful of countries were poised to participate (Dixon, 1998; Dixon, 1997). Among Annex I countries, Norway, the Netherlands and the USA had announced or were on the verge of announcing national AIJ pilot activities (Begg and Jackson 1999). Costa Rica was a leader among non-Annex I countries establishing a fully staffed office to help launch the AIJ pilot. Other countries active in developing the AIJ pilot include: Australia, Belgium, Canada, Chile, the Czech Republic, France, Germany, Guatemala, Japan, Mexico, the Nordic countries (Denmark, Finland, Iceland, and Sweden), Poland, South Africa, Sri Lanka and Vietnam. Today, over 80 countries have established or plan to establish national offices to manage participation in the FCCC AIJ pilot (Jepma and van der Gaast 1999). AIJ projects that have been accepted, approved and/or endorsed by the designated national authorities for AIJ of the host country and the investing (donor) countries can be reviewed on the FCCC Secretariat web site: *http://www.unfccc.de/fccc/ccinfo/aijproj.htm*. Many of the countries now participating in the AIJ pilot have established national rules of procedure for evaluation, selection, establishment and monitoring of projects which complement FCCC AIJ operational criteria (Carraro 1999). These criteria will be thoroughly discussed in other chapters of this book.

4. EVOLUTION OF FCCC AIJ REPORTING GUIDELINES

The AIJ pilot phase is a learning opportunity. Therefore, AIJ reporting can also be seen as a learning experiment. The primary question arising from AIJ reporting include:

- what should be learned from the reporting; and,
- what can be learned from the actual reporting;

In following this approach the reader should keep in mind the context that the AIJ reporting system evolved. The AIJ reporting format, as used, has not been set up by independent experts to serve their needs for controlled, reliable information. Rather, the reporting system should be viewed as the outcome of FCCC negotiations. Compromises were reached in these negotiations. Today, the FCCC Parties have adopted a uniform reporting format (URF) for AIJ projects. The reporting system balances the need for as much and as detailed information on the various projects as possible, and, the reluctance of some projects' participants to disclose information or invest resources in collecting and arranging project information. Project participants may have a right to be reluctant, because not only do they have to undergo the difficult process of the project acceptance, they also face the obligation to report once the project is underway (Dixon, 1998). Unfortunately, no credits are yet attached to the GHG mitigation contributions of AIJ projects (Carraro 1999). As far as the host country participants are concerned, especially in countries with a limited capacity to handle AIJ projects' administration, they also have a point when arguing that they have difficulty with the requirement to provide input in overly detailed reporting formats (Jackson and Begg, 1999). Thus, it cannot be over emphasized that the flexibility provided to the project participants as to the degree of specification in their reporting is not only a necessary but also a logical consequence of the fundamental character of the reporting process.

4.1 WHAT SHOULD BE LEARNED FROM THE AIJ REPORTING SYSTEM?

Expectations as to what information should be derived from AIJ reporting will differ, depending on what purpose the reporting is considered to play. Various reporting levels can be distinguished such as:

- (1) broad indication of: project participants, mitigation impact, cost-effectiveness ($ Mg CO_2), externalities and funding;
- (1) + detailed quantitative assessment of expected abatement costs ($ Mg CO_2), and externalities (at least specified qualitatively);
- (2) proper description of how the baseline has been arrived at + proper indication
- of additionality; and,
- (3) description of how the mitigation can or will be achieved and sufficiently elaborate description of baseline determination, so that report is suitable for verification.

The primary distinction between 1-2 and 3-4, is that levels 1-2 reporting provides a broad picture of what is going on, but does not allow for much insight into how the data presented have been derived. Level 3-4 reports increasingly provide the information needed for a critical assessment and possible framework of verification. The overall picture of the present state of reporting is that all reports can now be categorized in levels 1-2. In other words, if a verification process needed to start right now for the purpose of GHG crediting, the actual reports would provide insufficient information. The reports are deficient because the baseline employed commonly lacks clear additional information as to how it has been derived (Jackson and Begg 1999). Moreover, assumptions regarding project boundaries (eg, leakage) are seldom specified. Externalities are usually only provided in a qualitative manner. Alternative technologies are not considered or explained. Financial additionality is quite often not discussed at all.

Therefore, the present AIJ reporting and reporting format are on the whole, notwithstanding some exceptions, is not yet suitable for verification of GHG emissions. A major deficiency is uniform methods for deriving baselines. Project's participants can not be blamed for this situation, as long as concrete guidelines as to how baselines should be derived, is not generally available. Since the Kyoto Protocol calls for activating crediting in the year 2000, it seems quite necessary to inform project participants that in the nearest future they may need to provide more information than is currently the case.

In order to assess project cost-effectiveness, it is important to make a clear distinction between the various cost-concepts: transaction costs, direct costs-indirect costs, capital costs, marginal-average costs (Jepma and van der

Gaast, 1999). Similarly, various mitigation concepts: net vs. gross mitigation, with or without GHG implications of secondary effects, requires definition. Precise definitions of these concepts for the reporting system will not be provided here because this is beyond the scope of this chapter. It is important to consider that only after very precise definitions have actually been specified on both costs and mitigation, can project cost-effectiveness data be appropriately compared.

A relevant question is, should more than broad superficial information about cost-effectiveness should be provided through the AIJ reporting system? This question relates to the fundamental character of the additionality element of AIJ pilot. In essence, two approaches can theoretically be used to determine whether a particular project brings about environmental benefits related to the mitigation of climate change that would not have occurred in the absence of such activities, or not. One approach would be to look at the financial aspects of a project, and, the other to look at the technical GHG emission's aspects of a project. In the first case, a project would be additional if the AIJ subsidy, or in the JI phase, the credits that would accrue to the investor through the project, would make the difference between being sufficiently commercially attractive to get it started or not. For example, the minimum rate of return (ROR) required by the investor to start a particular AIJ project would be 15% and that without AIJ subsidy or JI credits the return would be 14% only, but including subsidy/credits it is 15%. The AIJ subsidy or JI credits would then turn the ROR balance, the criterion for additionality, and would cause the project investment to proceed. The net-abatement impact of the project could then fully be ascribed to the subsidy. Additionality assessment would, in this system, be based on knowledge about a project's ROR and financial information would need to be provided by the investor, because no clear indication of additionality could be achieved.

The other system for additionality assessment focuses not on financial but on technical data. In this system, the same output of something, products, services, or a particular type of energy, can be produced with the help of various technologies, each having different implications in terms of net GHG emissions (Jackson and Begg, 1999). Switching from one technology to another, therefore, will most likely have net GHG implications. If such a switch is not in line with the usual or normal technology development (eg, the technology change itself is not included in the baseline), and if the overall impact is that less GHGs are emitted per unit of output, then there would be additionality. In this approach, the information to be provided by the AIJ project's participants is just about the technology employed and about the savings in terms of GHG emissions avoided. Thus, financial information would simply be irrelevant.

In comparing the two approaches, it seems quite clear that the AIJ additionality concept is based on the second approach, not on the first one. Collecting reliable financial information is difficult and unlikely for assessing additionality and/or crediting. This conclusion raises the issue why the AIJ reporting system requires AIJ participants to report on financial costs and cost-effectiveness if the ultimate purpose would be to collect the data required for crediting. This is not to say that systematic and comparable data on mitigation cost-effectiveness via AIJ could not be highly informative for potential investors, sponsors and for the research community. For the purpose of being able to start and verify crediting financial data may not be essential.

One of the elements, covered in the reporting system, deals with financial additionality (ie, additionality to the financial obligation of Annex II Parties under the finance mechanism of the FCCC as well as current ODA flow). In order for financial additionality to be assessed, project participants are requested to indicate their source(s) of project funding (including the pre-feasibility phase). In this respect, the question arises what the real meaning of this information is in terms of indicators of additionality. Suppose a donor government reduces its ODA commitments with an amount x, which is subsequently channeled to developing countries via a climate assistance program. If there is no clear linkage between the reduction and the alternative allocation, one could claim financial additionality since there is no future norm vis-à-vis the size of ODA. However, if the reverse were true, the climate program channels part of its JI budget to the agency involved with ODA distribution, it would be difficult to successfully claim financial additionality.

In summary, since official government resources are fungible, financial additionality will be hard to prove generally. Financial additionality with regard to a particular AIJ/JI project even more difficult to determine. In fact, one could argue that the financial additionality criterion, unlike the environmental additionality criterion, should primarily be viewed in its national, macro-economic context. In other words, the test of the financial additionality should not be carried out at the level of the investor in a particular AIJ/JI project, but at the level of the donor-country. Testing financial additionality would then probably become easier insofar as the easy-to-measure overall national ODA performance and its development in the course of time would be the dominating criterion for the determination of additionality of AIJ/JI initiated from the donor country.

4.2 LESSONS LEARNED FROM BASELINES

In order to determine an AIJ project GHG ER it is necessary to determine a reference scenario to estimate what the emissions would have been in absence of the project. The main difficulty in determining such a reference scenario (referred to as the *baseline*) is that it is counterfactual (scenario describes a situation where the project never exits). As a result, analysts have argued that because of this difficulty the additionality issue is the Achilles heal of the AIJ pilot. Both the host partner and the investing partner have incentives to inflate the baseline so that a higher ER can be claimed. Careful (third party) verification is required to judge whether a project's baseline is factual and equitable. Discussion of additionality and baselines by FCCC Parties is not finished and will be continued at the future sessions of COPs or Meeting of the Parties (MOPs).

Several options for baseline determination have been proposed (Anonymous 1999; Ellis, 1999; Jackson and Begg 1999; Dixon 1997). The fundamental point in this debate seems to be characterization of the baseline. On the one hand, the baseline needs to be an as technically precise description of the counterfactual scenario of a specific AIJ or JI project. This approach requires detailed technical information about the conditions under which a particular project is undertaken. In contrast, irrespective of the amount of detailed technical information gathered to construct a baseline, bold assumptions are required to devise the counterfactual scenario. In other words, baseline determination is to a certain extent an arbitrary and subjective process. Thus, the specific technical and logistical circumstances of a specific project should lead to the development of a mutually reasonable baseline which is acceptable to all project partners. Four different approaches to baseline assessment will be presented in the following sections.

4.2.1 Baseline Option 1

The first option for baseline assessment deals with an *ex ante* best estimate of GHG emissions on the site would have amounted to in absence of the project. This estimate can be completed in several ways depending on the characteristics of the project, as well as, the project host country characteristics. The detailed complexities of this option (eg, project boundaries, incorporation of externalities) will not be presented because this literature is available elsewhere (Jackson and Begg, 1999). It is illuminating to illustrate how difficult it is to determine what the *ex ante* baseline is for a simple or straightforward AIJ project.

For example, a number of pilot energy efficiency improvement projects have been carried out in the Baltic States region under the auspices of the Nordic Council of Ministers (1997). These energy efficiency projects have stimulated investment in the region. Without the projects, the host countries probably would have made the investment anyway, but with a delay of three to five years. If these projects were JI projects under the Kyoto Protocol, the projects' baseline would only deviate from the actual emissions (e.g. result in credits) during the first three to five years of the projects.

This example illustrates the complexity of baseline determination. Some of the potential host countries are undergoing a process of a rapid economic transition. In these countries, several JI (or CDM) projects probably only accelerate investments that would have been carried out by the countries themselves in the mid- or long- term (Anonymous 1999). In developing countries where investment in energy sector projects has been historically slow (eg, sub-Saharan Africa) it is less likely that the JI (or CDM) project investment would have been carried out anyway in the short- or mid-term (Anonymous, 1997). For some developing countries, the period for which the JI (or CDM) project is additional is often a longer period of time.

Determining the length of the period during which a JI (or CDM) project is additional is not the only uncertainty surrounding the baseline determination. Other factors such as economic growth, energy prices, currency prices and political risks are also important considerations. For example, suppose a project developer has determined a 10-year ER baseline for a JI (or CDM) project. However, after five years it turns out that the host country invests in several similar projects. The reported ER is larger than what has actually been achieved. Such a case is obviously beneficial for the investing and host country Parties but global GHG ERs are not achieved.

4.2.2 Baseline Option 2

A second option to develop baselines is to follow the approach in option #1 but allow for *ex post* corrections of the baseline. Such corrections may be required if it turns out that the underlying assumptions for the baseline were erroneous. For investors and host country parties, this option may increase the risk to invest in JI (or CDM) projects, as it is not clear beforehand how many credits will be generated. *Ex post* corrections of the baseline has the advantage that generated credits are most likely more based on real ERs. Under this option the global environment will more likely be protected but with potentially greater risks for the project partners.

Were the Parties to decide, on *ex post* corrections of baselines, project developers would probably only select those projects highly unlikely to be implemented by host countries themselves in the mid- or long-term. A detailed analysis of 30 pilot projects in Central and Eastern Europe revealed three primary factors that hamper systematic improvements in the energy production sector in the region:

- funding required for emission reduction investments in power plants and district-heating plants is often insufficiently available;
- in several Central and Eastern European countries the legislation prescribing energy efficiency improvements is often lacking; and,
- technical and management skills to implement and maintain new energy efficient technologies are often insufficient.

Based on this analysis, the Nordic Council of Ministers drew a distinction between demand side (eg, district heating system improvements) and supply projects (development of power plants). Energy supply side investments are often much larger than demand side investments which makes it more financially attractive to invest in district heating improvements. Moreover, as a result of the gradual reduction of energy subsidies during the economic transition process in Eastern Europe and the former Soviet Union there is greater pressure on governments to improve the energy efficiency at the energy demand side. Finally, consumers in Central and Eastern European countries are becoming more and more eager to have comfortable living conditions, including a comfortable domestic heating system. Based on this analysis, the Nordic Council of Ministers concludes that there is growing pressure on local and central governments to invest in demand side rather than supply side projects. As a result, *ex post* corrections of baselines, energy supply side projects in Central and Eastern Europe are probably less risky AIJ projects than demand side projects, since the baseline for supply side projects is probably more stable.

The Costa Rica Protected Areas Project (PAP) is an example of a project for which a methodology has been developed to deal with *ex post* baseline corrections (Dixon, 1998). This project aims to sequester 15.6 million Mg C on an area of 530,000 ha. Through an international verification and certification procedure, the government of Costa Rica has been able to issue certified tradable offsets (CTOs) for the first one million Mg C sequestered via the project. In order to minimize risk for the buyers of CTOs, 700,000 Mg C have been retained in buffer zones. According to the project developers, the buffer zone helps reduce uncertainty regarding the project baseline. It is estimated this uncertainty corresponds with 16.1% of the total amount of C sequestered.

4.2.3 Baseline Option 3

A third approach for baseline determination was developed by the USA Center for Clean Air Policy (CCAP). This option is based on development of top-down baselines by governments (Anonymous 1999). This methodology is based on the concept that national governments of JI/AIJ host countries would use their overall ER commitment as a basis to calculate commitments from various economic sectors or technologies. For example, the quantified emission limitation and reduction commitment (QELRC) of a Party can only be achieved if the CO_2 emissions per unit of energy produced in the power sector would be 20% less than the average under current conditions. In that hypothetical scenario, the 20% figure would then determine the baseline for JI/AIJ projects in that particular sector.

With respect to CDM projects this methodology may not be easily applied because non-Annex I Parties will be the host countries and they have not accepted QELRCs. To solve this dilemma it has been suggested to construct baselines on the basis of acceptable simulated QELRCs for non-Annex I countries. However, non-Annex countries find the top-down approach potentially contentious.

4.2.4 Baseline Option 4

A fourth option is to adopt default project or technology specific baseline with possible differentiation for specific countries or regions. A panel of experts could determine a baseline for a number of project types and establish a benchmark for all FCCC Parties (Anonymous 1999). This project categorization could then be expended to regions or countries resulting in a region-by-project matrix. So, a matrix of baselines can be constructed which can be consulted and employed by project developers. If an investing and a host country agree on a project, they can locate the project baseline in the matrix and calculate the credits.

Some advantages of this option include:

- transaction costs for the project developers will be lower and technical consultants will be required to a lesser degree; a visit to the FCCC web-site may be sufficient;
- third party check for each individually determined baseline is no longer necessary, which may also result in a significant cost saving.; and,
- categorization system provides a uniform basis for choosing the correct baseline out of several ones each of which can equally well be defended as being correct.

One could argue that the matrix approach is too crude, because in particular circumstances the matrix elements are so clearly unfair to the project participants and an ad hoc adjustment are needed to achieve fairness. Therefore, as an additional element of this matrix approach it has been suggested to enter the possible for the projects' participants to appeal for an adjustment of the baseline used in their particular case. This opportunity would be optional. Project participants could decide for themselves if they take the risk to lose the appeal, by making an investment in data gathering in order to apply for an exemption. The extra costs associated with this procedure, as well as possible extra third party verification will have to be borne by them.

With respect to the procedure to set up the matrix system just mentioned, it has been suggested to let FCCC authorized international third parties participate in the process of determining the aggregate sector/technology set of baselines (Anonymous 1999). If the actual developments would go in this direction an appeal procedure as described above may introduce an asymmetry into the system. Project participants will never complain if the matrix-based baseline turns out to be rather favorable for them in the particular circumstances of the project, and will only do so if the matrix-based baseline is considered unfair. Insofar as a part of the appeals will be honored, the overall implication of this asymmetry will be a certain degree of baseline inflation. A periodical international verification of aggregate baselines would be necessary insofar as technological progress would require this. A particular point in this respect is the risk of leakage between sectors. Setting a target for one sector in a non-Annex I Party may affect the appropriate target for other sectors.

The reporting information on approximately 100 AIJ projects does not give a clear guidance as to what approach with regard to the baseline should be preferred. However, it is not very realistic to assume that the detailed information necessary for an advanced technical baseline determination will or can be reported in great detail. This approach could lead to a normative, possibly matrix-based baseline determination procedure. Or lead to a process where general criteria rather than detailed project host country data would provide the basis for the baseline.

A final point, which can be derived from the actual reporting relevant for the process of baseline determination, is the role of best professional judgement elements in the baseline process. In fact some participants reporting on their AIJ projects have indicated that they are not be prepared to provide a baseline for their particular project, because they are convinced that doing this would be meaningless. The argument used is that quite often so many arbitrary assumptions have to be used that, if left to the project participants, the baseline is hardly to be taken seriously by the project participants themselves. To determine what would have happened without the project is a task that goes far beyond the competence and skills of the project partners. This is especially true if the assumptions that are acceptable should be consistent across the whole range of comparable projects. Individual project's participants are often not even aware of other, comparable AIJ projects. The large number of sometimes rather arbitrary assumptions to be used only would add to this feeling of insecurity among project participants.

4.3 HOW TO CHOOSE A BASELINE FOR CDM AND AIJ?

This example illustrates the complexity of choosing the correct baseline for a JI or CDM project (Table 1). Suppose that a company from an OECD country wants to invest in a coal gasification project in India. The GHG emissions per unit of energy of the technology the company wants to implement in the project amount to 400 Mg. Technologies applied by comparable multinationals from the OECD in a similar investment would cause 450 Mg GHG emission per unit of energy (let us call this the OECD average). In India the GHG emissions per unit of energy from the most modern power plant available amount to 500 Mg and those from an average power plant to 600 Mg. Finally, the GHG emissions from an average power plant in South Asia amount to 700 Mg per unit of energy.

What would be the most appropriate baseline if this investment would apply if the project developers submitted an application for AIJ/JI recognition? Unfortunately, the answer is not very straightforward. The FCCC could argue, for example, that since it might be very complicated to determine a baseline for each AIJ/JI power plant project in India and other countries in the region, the best way to calculate a baseline would be to take the South Asian average as a reference. In that case, the baseline would be 700 Mg per unit of energy, so that the credits would amount to 300 Mg (700 minus the new plant's emissions of 400 Mg).

Others could also argue that the differences between India and several other countries in South Asia are too big to use an average South Asian baseline for AIJ/JI power plant investments in India. In that case, only the average for India should be taken, which is 600 Mg per unit of energy. Using this figure as the baseline would result in 600 - 400 = 200 Mg per unit of energy to be credited to the AIJ/JI investment.

But even this baseline could overestimate the GHG mitigation over the project lifetime. After all, why should the current average emissions from a power plant in India as a reference point, as it may be fair to assume that as a result of India's economic development better and more efficient power plants will be established anyway? Then the GHG emissions from the most efficient power plant that is currently used in India is the best baseline. In that case the credits would only amount to 500 - 400 = 100 Mg per unit of energy.

A fourth possibility to determine a baseline on the basis of the information given is not to evaluate the emissions from Indian power plants but to take the current average GHG emissions from power plants in OECD countries into account. After all, a successful economic development in India may result in a commercially driven transfer of the current average OECD power plant techniques to India. Calculating with this possible development would only result in 50 credits (the current OECD average of 450 Mg per unit of energy minus 400) for the JI investors.

Finally, it could be argued that investing in this project is commercially feasible anyhow, so that no AIJ/JI credits should be given because the project would not satisfy the additionality criterion (Ellis, 1999).

TABLE 1. A hypothetical example of how to choose a baseline for an AIJ/JI gasification project in India.

Baseline (Mg GHG emission per unit of energy)	**GHG emissions AIJ/JI project (Mg per unit of energy)**	**Credits**
South Asian average (700)	400	700-400 = 300
Indian average (600)	400	600-400 = 200
Most efficient Indian power plant (500)	400	500-400 = 100
OECD average power plant (450)	400	450-400 = 50
Project commercially feasible (400)		400-400 = 0

This example, which is summarized in the Table 1, reveals five different baselines for this AIJ/JI gasification investment could be factually derived. It is therefore a very important task for the FCCC decision-makers to formulate criteria or at least decision rules for JI (and CDM) baseline determination. The fundamental trade-off in the procedure to determine the baseline is between precision, fairness and transparency and transaction costs and loss of time and momentum.

5. WHAT CAN BE LEARNED FROM AIJ REPORTING?

Currently, the general impression of the quality of AIJ project reporting is rather favorable. After the inspection of 74 project reports in great detail, it could be concluded that reporting has improved considerably, compared to the overall quality of reporting of a comparable subset submitted in 1997 (Jackson and Begg 1999; Dixon 1997). Current project reports are more complete and the quality of information (detail and precision) is improved.

Unfortunately, project developers are using various formats for reporting and uniformity has not yet been fully achieved.

The quality of reporting among AIJ project developers and Parties varies to a considerable degree. The primary differences are in the degree to which the mitigation claimed could be explained by the actual investment in terms of technology. Moreover, the specification of mitigation costs per unit of CO_2 is quite often absent. A considerable part of the differences in reporting can be explained by the maturity of the project. Projects that are mature report more information of better quality than projects that are in their infancy (Dixon, 1998).

Comparing earlier reports to more recent ones information regarding emissions of gases other than CO_2 is incomplete. Only few reports reveal the project impact on the others GHGs mentioned in the Kyoto Protocol.

The following conclusions can be drawn from an examination of 74 project reports submitted to the FCCC Secretariat in 1998:

- most of the projects involve 3-6 participants; a few projects have only two participants, some more than 6;
- about one-third of project investments comes from private sources (especially from the USA investors); the remainder are financed with the help of government resources, although several of these projects are partly based on private sources with the government financing the AIJ study component;
- most projects started in 1995 (9), or 1996/7 (14); some before 1995 (3) or in 1998 (6);
- project duration varies widely; for forestry and agricultural projects almost always 25 years or much longer (up to 60 years); for energy efficiency ranging between 1.5 and 15 years (av. some 7 yr.); for renewable energy projects from 10-30 years;
- costs of emissions avoided are not always specified (in less than half of the number of projects); project participants quite often (broadly half of the cases reported) claim a no-regrets character of their project with negative costs ranging between some US$ 5-20 Mg CO2, mostly in the sphere of energy efficiency and renewable energy; in other cases the (positive) costs range between almost zero and about $US 25 Mg CO_2;
- amount invested (cumulative) is not always clear because most reports indicate the overall project investment (including elements that are not at all related to AIJ); others only report on the size of the AIJ investment; about a quarter of the projects can be considered small (< $500,000); another quarter of the projects are intermediate size ($500,000> < $2m); size of the remaining projects on which the investment size has been reported (roughly another quarter) varies between some $5m - $60m; a

quarter of the projects' participants (mainly private investors) does not provide information about the amounts invested in the project.
- about one-third at most of the project reports contains quantitative information on environmental, social or economic benefits; there are no consistent patterns in this element of the reporting process, except there is no reporting on one of the externality elements;
- information regarding baselines is sometimes provided; sometimes extensively; however the relative (i.e. vis-a-vis the baseline) ER of the AIJ project can seldom be derived from the reports; information on the absolute size of the ER is often incorporated in the reports.

6. CASE STUDIES

One of the best ways to analyze what can be learned from the AIJ reporting is to examine actual projects in detail. Two projects are presented as case studies. In selecting these project case studies, one was chosen from a country with an economy in transition (EIT) and the other was drawn from a developing country. The EIT case study is RUSAGAS. The developing country case study is: Reduced Impact Logging (RIL) in Indonesia.

6.1 REDUCED IMPACT LOGGING (RIL) IN INDONESIA

Logging (harvesting) tropical rainforests causes damage to higher proportion of the forest relative to the amount of trees harvested, especially in Indonesia. Now, Indonesian and American partners are preparing to demonstrate that RIL is a cost-effective way of GHG emission reduction, about US$ 1.8 per Mg of CO_2.

Although Indonesian law requires loggers to minimize damage to the forests, it is unlikely that RIL techniques will soon be standard practice. In that context, the Association of Indonesian Forest Concession Holders (APHI) and the Kiani Lestari and Inhutani II Concessionaires work with American Forests of Washington, DC, USA and COPEC, a JI project developer from Los Angeles (USA) to establish an AIJ project based on RIL techniques. In 1998, the partners received both host country acceptance by the Indonesian Environment Ministry and recognition under USIJI. On behalf of the partners COPEC is presently seeking project funding.

6.1.1 Forest C sink conservation effect

The project site of 600 ha is located in eastern Kalimantan on the island of Borneo. The forests are lowland dipterocarps that have not been previously harvested, and are not densely populated by humans. The project will include developing guidelines and procedures for implementation of RIL techniques on the 600 ha. From a study in Malaysia, it is estimated that logging damage to the residual forests can be reduced by as much as 50% through pre-cutting vines, directional felling, and planned extraction of timber on properly constructed and utilized skid rails. Over the 40-year lifetime of the project, this technique could save up to 206,800 Mg of CO_2 (56,400 Mg C). Since project costs total \$380,000 this implies that the price per Mg of CO_2 avoided is \$1.8.

6.1.2 Other environmental benefits

Apart from the forest C sink conservation effect there are other benefits. This project will preserve habitat for biodiversity since it prevents degradation and conversion to other uses. By reducing the amount of forest canopy that is opened, there are fewer impacts to plant and habitat. Furthermore, RIL techniques also reduce the forest susceptibility to weed infestations (that reduce biomass recovery rates) and to destructive fires.

6.1.3 Project monitoring and reporting

The principle of monitoring and verification has been accepted by both sides but has yet to be worked out. It is expected that either a local NGO or an international forest product certification company will play a role. The aforementioned case study on Malaysia did provide a credible first estimate of the expected environmental effect but partners will conduct a more detailed baseline exercise as part of project implementation.

6.2 REDUCTION OF METHANE EMISSION FROM LEAKING GAS PIPELINES IN SOUTHERN RUSSIA (RUSAGAS)

6.2.1 Project Characteristics

This project, established in 1995, will reduce fugitive methane (CH_4) emissions, improve operational efficiency and seal the valves on the main natural gas pipelines that are contiguous to the Storozhovka and Pallasovska compressor stations in Saratov and Volgograd Oblasts. Project measures include a training program to ensure that inspection and maintenance of the facilities will be continued in the future. The project lifetime is associated with the remaining life of the compressor stations (approximately 25 years). The price per Mg of CO_2 equivalent avoided is very low, less than US$1, as methane is a very potent GHG compared to CO_2. This makes this type of project a very interesting option for potential JI investors.

The project involves several Russian and American partners and has been called the RUSAGAS-FGC (fugitive gas capture) JI project. It is being conducted by GAZPROM (Moscow office, Yugtransgas and Volgogradtransgas), Oregon State University, Sealweld Corporation, the Sustainable Development Technology Corporation (SDTC), the U.S.Environmental Protection Agency, and the Russian USA Center for Energy Efficiency. The project is still in its infancy. The project sites have been identified and a critical path schedule has been developed. Only a funding partner has yet to be identified. The project was accepted by USIJI in 1995. Both governments have reported the project to the FCCC Secretariat.

6.2.2 Project technical components

The compressor stations and contiguous main pipelines are located in Saratov and Pallasovka (Volgograd Oblast, 900-km southeast of Moscow). The Pallasovka compressor is on the main pipeline serving the Volga region and the northern part of the near Caspian Lowlands. It was built in the mid 1970s. It has 30 compressor units: 18 are gas turbine and 12 are electric motor driven. Five pipelines of 1.4 m. in diameter and about 700 valves serve the station. It's maximum capacity, it is estimated at some 90 billion cubic meters of gas pass through this station annually. The methane is currently harvested in the Orenburg Field (southwest of the Ural Mountains) and the Gazli Field in Central Asia.

Saratov with a population of 1 million is said to be the center of the Russian natural gas industry. The USSR's first long distance pipeline was built from here to Moscow in 1946. The Saratov compressor station, named Storozhovka, was built in the mid 1960s. The methane, coming from the Orenburg and Urengoi Fields, is mainly pumped into two large underground storage basins in Saratov. These are associated with the Saratov methane field, opened in 1941, and operated through 1966. The maximum estimated transmission capacity of the station is 17 billion cubic meters per year. There are 16 compressor units, of which 9 are gas turbine and 7 electric motor driven.

The valve-sealing program is composed of the following measures. The valves in and around the compressor stations will be injected with a cleaning compound to loosen old sealant and pipeline residue from the small passages inside the valve. The valves will then be injected with an appropriate lubricating/sealing compound suitable for continuous service in natural gas. Damaged and non-functioning fittings or fittings incompatible with the service equipment, supplied by the American partner Sealweld, will be replaced. Following cleaning and repair, the valves will only require a small quantity of additional lubricant/ sealant to be injected on a scheduled basis. GAZPROM will buy the monitoring equipment as part of the initial agreement and continue to buy the lubricant.

6.2.3 Baseline setting

At present, institutional barriers (governing structure, price system) cause GAZPROM to have no economic incentives for reducing the substantial leaking of methane from its pipelines. Therefore, the project partners concluded from current GAZPROM investment practice that none of the proposed fugitive gas capture measures would be implemented without external support. Consequently, they assumed that the baseline would represent a straight horizontal line at the present level of the fugitive gas emissions at the project sites. Under the valve-sealing program, the methane emissions from leaking valves will be substantially reduced or eliminated. The baseline case can thus be calculated from the initial emission of methane per leaking valve and the project case from the methane emission coming from a sealed valve.

During the project preparation stage a pre-feasibility study was conducted. It was only possible to make an approximate estimate of the magnitude of the emissions from a valve at the compressor stations since there are no statistics available. Test results revealed fugitive gas emissions ranged between 15 and 300 cubic meters per valve per hour.

TABLE 3. RUSAGAS project data for establishment of project baselines.

	Pallasovska	**Storozhovka**
number of valves	700	400
leaking valves	70	120
average CH_4 emission	250 m^3/hr	250 m^3/hr
baseline estimate (annual total)	0.15 x $10^9 m^3$	0.26 x $10^9 m^3$
percentage of maximum compressor capacity	0.2%	1.5%

During a project preparation workshop in Moscow an NGO stated that this data is unreliable as a basis for determining the baseline and thereby the size of the credit. However, direct measurements of the emissions of individual valves will be made before and after sealing, which will result in accurate figures for emission reduction. Another question was raised at the workshop. What is the appropriate length of the project lifetime? Stretching the baseline over a period of some 20 years may not be appropriate for an EIT project. Some participants felt that a static scenario, expressed in a straight horizontal line at a pre-project level, is not realistic. Instead, developments in efficiency improvement and in operational and maintenance practice when constructing a baseline should be considered. However, the disadvantage of this option is that uncertainty for the potential investor over credits increases. At the national level, the question is what amount of credit is wise for Russia to sell in the early stages of its economic transition. Presently, Russia's emissions are well below the base- year level, which means this country will easily comply with FCCC Annex I obligations.

6.2.4 Environmental benefits

The project partners assume that about 60 leaking valves at the Pallasovska compressor station may be repaired. A conservative estimate of the ER (based on random samples) would then be 0.01 billion cubic meters per year. At the Storozhovka station repair of 80 valves would result in an annual emission reduction of 0.06 billion cubic meters. Over the project lifetime these volumes of GHG emissions avoided would amount to 1.75 billion cubic meters. Assuming, furthermore, an average density of methane of about 700 g/m^3, this volume can be translated into 1.2 million Mg of CH_4 (methane). Finally, using a GWP factor of 24.5 this is equivalent to 29 million metric Mg of CO_2. According to the agreement to be signed between GAZPROM and the funding partner the latter will be assigned all credit for the reduction of GHG emissions under the project.

It is envisioned that the reduction in fugitive gas emissions will translate back to a lower depletion rate in the natural gas fields, which serve the compressor station. However, it is also possible that it will translate forward to increased industrial and residential consumption which are GHG emission producing activities. Of course, the GWP of combustion products is considerably lower than that of methane. Over the past 5 years the consumption of natural gas in Russia has decreased. Therefore, the risk of emissions leakage benefits is not expected to damage the results of the project.

6.2.5 Local project benefits

A written statement by GAZPROM indicate that the infrastructure improvements at the Storozhovka and Pallasovska compressor stations are in addition to those scheduled by GAZPROM. The monitoring equipment for the leaking and sealed valves, supplied by one of the USA partners, is common for the North American gas industry. The proposed enclosure operation is also a common industry practice in the USA. In fact, there is a chance that more accurate instruments called high flow sampler will be used in the project. The benefit for Russia is that it will have up-to-date technology and the training that goes with it at an affordable price.

6.2.6 Project monitoring

An enclosure technique will be used for the physical measurement of fugitive gas emissions to the atmosphere. The bagging procedure involves placing a plastic bag over the leaking element of a valve and securing the bag so that all fugitive emissions, which vent to the atmosphere would be captured in the bag. As the bag fills up to a known volume, the time required to fill the bag is noted and the leakage rate may be determined from these two measurements.

The operation efficiency improvement will result from the elimination of a leak in the seat (part of a machine that supports or guides another part) of a bypass valve associated with bypassing natural gas around a gas turbine. Seat leaks may be measured with an electronic flow rate indicator. Once the seat leak is quantified in terms of a flow rate the energy required to recompress the gas which has leaked from the high-pressure outlet pipe to the low-pressure inlet pipe of the turbine may be calculated.

Initially, GAZPROM, SDTC, and Sealweld Corporation will be responsible for monitoring and reporting the GHG emission reductions that occur over the three-year project period. Later on, GAZPROM will take over full responsibility for monitoring, reporting and periodically updating the emission reduction estimates, and providing annual reports on these activities.

6.2.7 Cost-effectiveness

The CO_2 equivalent associated with the capture of fugitive natural gas at the Pallasovka and Storozhovka compressor stations over the 25-year period is 29 million Mg. The total project cost is expected to be around US$ 300,000. Thus, the cost per metric ton of CO_2 equivalent for a 1996 present value funding partner, paying around US$ 300,000, are estimated at US$

0.01. The fuel efficiency that will be realized in the gas turbine and the ERs to the atmosphere associated with venting the main pipelines are not included in this estimate. Finally, the proceeds of the extra natural gas that GAZPROM can supply to its customers is not taken into account, nor are the relatively low costs that GAZPROM has to make for implementing the maintenance and reporting. Consequently, the unit cost presented may be overestimated. On the other hand, the emissions to the atmosphere for an individual valve may be overestimated. However, even if the true emission figure would appear to be an order of magnitude smaller, the unit cost would still be substantially lower than the cost per Mg of CO_2 equivalent for other mitigation options.

It will be clear from the above that RUSAGAS costs are shared between GAZPROM and the USA funding partner(s). GAZPROM will pay for the valve-sealing program with a loan from the funding partner. The latter will pay the costs involved in issuing the loan and all transaction costs. In return, the USA investor(s) will receive all GHG credits at a very low price. GAZPROM will dispose of the methane that is recovered through the valve-sealing program.

6.2.8 Project prospects

The RUSAGAS project has a large demonstration potential with regard to the state of practice in Russia. Monitoring, measuring, calculating and capturing fugitive gas emissions, as well as, the fuel efficiency improvement techniques realized in the valve-sealing program can all be demonstrated and transferred to Russian collaborators. The project is innovative in the Russian context and the concept has a potential to be applied in many of the over one thousand compressor stations and thousands of kilometers of pipeline in the former Soviet Union. The project partners have plans to reproduce the project when the work in the Saratov and Volgograd Regions is well underway.

In the summer of 1995, Russian methane sold for US$ 50 per 1,000 m^3 in the industrial and utility sectors and US$ 4 in the residential sector. However, prices are rapidly conforming to world market prices and experts expect this process to take only a few years. This development is likely to have two consequences. First, the economic incentive to conduct valve-sealing activity is expected to become obvious to GAZPROM, as the unit price of Russian methane increases. Second, as time and learning occurs, and the energy prices increase there will be more incentive for Russia to implement fugitive gas capture measures as a normal practice and with its own finances.

REFERENCES

Anonymous (1999) CDM Workshop on Baseline for CDM, NEDO, Tokyo.

Anonymous (1997) Proceedings of Regional Workshop on Activities Implemented Jointly Under the UN Framework Convention on Climate Change, Cairo, Egypt, ICED, Cairo.

Carraro, C. (Ed) (1999) International Environmental Agreements on Climate Change, Kluwer Academic Publishers, Dordrecht, in press.

Chatterjee, K. (1997) Activities Implemented Jointly to Mitigate Climate Change, Development Alternatives, Delhi.

CCAP/SEVEn (1997) Joint Implementation Projects in Central and Eastern Europe, Center for Clean Air Policy, Prague.

Dixon, R.K. (1998) The U.S. Initiative on Joint Implementation: An Asia-Pacific Perspective. Asian Perspective 22:5-19.

Dixon, R.K. (1997) The U.S. Initiative on Joint Implementation. Int. J. Environment and Pollution 8:1-18.

Ellis, J. (1999) Experience with Emission Baselines Under the AIJ Pilot Phase, Information Paper, OECD, Paris.

Foundation JIN, Joint Implementation Quarterly, Volume 1,3:7 Groningen.

Goldemberg, J. (1998) The Clean Development Mechanism: Issues and Options, UNDP, New York.

Harmelen, van A.K., S.N.M. van Rooijen, C.J. Jepma and W. van der Gaast. (1997) Joint Implementation met Midden- en Oost-Europa: Mogelijkheden en beperkingen bij de realisatie van de Nederlandse CO2- reductie doelstellingen in de periode 2000-2010, Petten/Groningen.

Jackson, T. and Begg, K. (1999) Accounting and Accreditation of Activities Implemented Jointly, European Commission, Brussels.

Jepma, C.J. and van der Gaast, W (1999) On the Compatibility of Flexible Instruments, Kluwer Academic Publishers, Dordrecht, in press.

Nordic Council of Ministers (1997) Criteria and Perspectives for Joint Implementation: Ten Nordic Projects in Eastern Europe, TemaNord, Copenhagen.

SEVEn and JIN, 1997. The Experience with Joint Implementation in Central and Eastern Europe during the AIJ Pilot Phase, Prague/Groningen.

Chapter 3

INTERPRETATION AND APPLICATION OF FCCC AIJ PILOT PROJECT DEVELOPMENT CRITERIA

A. MICHAELOWA,[1] K. BEGG,[2] S. PARKINSON,[2] R. DIXON[3]
[1]Hamburg Institute for Economic Research; [2]Centre for Environmental Strategy, University of Surrey; [3]Institute for Global Environmental Strategies

Key words: criteria, acceptance, additionality, baselines, reporting, externalities

Abstract: Many national programs in both investor and host countries have defined criteria for selection or acceptance of activities implemented jointly (AIJ) projects. In many countries the criteria far exceed the standards of the criteria defined in Decision 5 of the UN Framework Convention on Climate Change (FCCC) First Conference of the Parties (COP-1). In some counties these criteria have been uniformly and rigidly applied while in others the AIJ pilot has been a period of learning and criteria evaluation. Some criteria, some as those dealing with additionality and greenhouse gas (GHG) emissions baseline determination have been the subject of considerable analysis and debate. Project reporting has been uneven and among the Parties in spite of the establishment of a Uniform Reporting Format (URF). Lessons learned regarding the AIJ pilot criteria can be applied to the development of the Clean Development Mechanism (CDM).

1. OVERVIEW AND COMPARISON OF PROJECT APPROVAL PROCEDURES

The FCCC COP-1 Decision 5 established the AIJ pilot phase in 1995.

Decision 5 states several criteria for evaluation, selection and approval of AIJ by the FCCC Parties. These criteria have been analyzed and reviewed extensively by others (Jepma and van der Gaast 1999; Carraro 1999). This chapter will address cogent lessons learned from application of AIJ criteria by developing, developing and transition (EIT) countries during the pilot phase.

Criteria were established by the Parties to help provide a framework for the AIJ pilot. It became obvious during early stages of the FCCC negotiations that minimum criteria were needed for joint implementation (JI) activities. For example, Decision 5 states that an AIJ project has to be approved by the governments of all participating countries. This was intended to assure that participation is voluntary and that the project was in line with host priorities. Moreover, this criteria aimed to prevent the recurrence of situations encountered when a USA company publicized its forest carbon (C) offset project in Malaysia in 1993 and the Malaysian government was surprised as it had never been notified of the project (Singh 1993). This was due to the fact that the state government had approved the project but did not inform the central government.

The conditions for the AIJ pilot phase established a set of approval criteria that had to be satisfied if projects were to be viable and accepted by all Parties. The criteria were intended to provide a framework for the practical experience of identifying and implementing projects which might eventually be eligible for emission reduction (ER) credits. The AIJ pilot phase criteria are listed and discussed in the next section.

1.1 AIJ PILOT PHASE CRITERIA ESTABLISHED BY FCCC PARTIES

Decision 5 in total is presented in Chapter 2. A summary of Decision 5 is presented in this section. The primary AIJ pilot criteria based on COP-1, Decision 5:

- projects (and/or activities) must be compatible with and supportive of the relevant national environment and development priorities and strategies, as well as, contribute to cost-effective global environmental benefits and encompass all GHG sources and sinks;
- projects under the pilot require prior acceptance, approval or endorsement by the governments of the Parties participating in these activities;
- projects must bring about real, measurable and long-term environmental benefits related to the mitigation of climate change that would not have occurred in the absence of such activities (additionality);
- finance of AIJ projects should be additional to Global Environment Facility (GEF) funds and official development assistance (ODA); and,
- no Party may accrue credits to their own obligations from the FCCC with regard to GHG ERs achieved by AIJ in the pilot phase.

There are other criteria embodied in the FCCC and sometimes employed by Parties. This extended set of criteria include:

- contribute to capacity building and transfer of environmentally sound technologies;
- provide for external verification; and,
- minimise the risk of adverse effects of projects on health, the economy, or the environment.

The AIJ project acceptance procedures and criteria help tailor projects to meet the needs of both investor and host countries (Dixon 1998). Some countries have been very successful at harmonizing criteria while others have not been very successful. Some analysts have suggested that the COP-1, Decision 5 criteria fall short of ensuring equity for all parties. For example, some analysts suggest that additional criteria should more fully address and consider:

- minimizing non GHG environmental and social impacts of projects;
- ensuring capacity building and public participation in the planning process;
- ensuring that the project is in line with country development priorities; and,
- ensuring choice of technology to be transferred is appropriate and contributes to a sustainable development path

For future development of the CDM it is likely additional definition or more extensive criteria will become crucial in a successful program of cooperation between developing and developed countries (Goldemberg, 1998). Developing country host countries have called for assistance to generate their own criteria and strategies so they can identify and offer projects, which will be in line with their strategy for a sustainable path, and with development priorities. (Chatterjee, 1997). This approach could speed up the host country acceptance process in some countries and help lower transaction costs. One lesson learned from the AIJ pilot is that current project acceptance criteria are not fully adequate and substantial revision is sorely needed before implementing CDM and JI as envisioned in the Kyoto Protocol.

The AIJ project review and acceptance procedures constitute a significant hurdle for AIJ developers (Dixon, 1998). In a number of cases it has led to long delays and high transaction costs for project developers and an uneven distribution of viable projects among developing and transition country Parties (Begg and Jackson, 1999). Often conflicting views within the host country government are the cause of delays and forced project developers to continuously negotiate with different ministries and departments (Chatterjee, 1997)

The problem of crafting a set of defining criteria and measures for the implementation of viable GHG emission reduction projects without incurring a high administrative and implementation transaction cost has been a challenge for all Parties active in the AIJ pilot. The fact that there is no agreed set of criteria and different countries have different approaches reflects the tension in selecting AIJ pilot phase criteria (Jepma and van der Gaast, 1999). Project criteria define the tone of the AIJ process and the possible extent of the burden of the institutional requirements to administer it. The trade-off for each Party is between practicality and socioeconomic considerations, ensuring that the environmental risks are reasonable and that there is an equitable distribution of costs and benefits. In the following section the AIJ review and acceptance procedures employed by a range of donor countries are reviewed.

2. AIJ PROJECT ACCEPTANCE PROCEDURES

2.1 APPROVAL PROCEDURES IN INVESTOR COUNTRIES

Table 1 illustrates and summarizes the AIJ pilot project acceptance activities and procedures of selected donor countries. Different countries naturally have different institutional arrangements. Most programs have only a small number of dedicated staff drawn from relevant government agencies and a modest annual budget. A large number of countries have a review panel consisting of representatives from different ministries that evaluates and accepts project proposals. Others countries evaluate proposals on a more discretionary basis. The evaluation process may be continuous as required or at specific times. The actual project evaluation criteria, as will be discussed later in this chapter, in many cases exceed the COP-1 minimum specifications.

Some Parties have adopted criteria that encompass concerns expressed by developing countries during the FCCC negotiations. How project evaluation criteria are actually implemented in practice and the success of their implementation has been the subject of considerable analysis (Begg and Jackson, 1999; Jepma and van der Gaast 1999). A very brief summary of AIJ pilot operational project evaluation procedures, as well as, specific attributes or peculiarities of AIJ criteria are presented below for selected donor Parties. Each of the Parties discussed in the following section have published extensive information regarding their AIJ pilot and interested readers are referred to that literature.

TABLE 1. Investor country institutions and framework for AIJ acceptance (Jepma, 1997; Boucher et al. 1998; Switzerland 1998; FCCC web sites).

Investor	Projects (#)	AIJ office/ Staff[1]	Interdepartmental panel for approval (#)	Decision rounds	Criteria	Private proposals accepted
Australia	2	✔ (1)	✔ (5)	R		✔
Belgium	1		✔ (4)[2]	C		✔
Canada	0	✔ (1)	✔ (4)	C		✔
France	1			C		✔
Germany	2	✔ (1)[3]		C	✔	✔
Japan	0		✔ (3)[4]	R	✔	✔
Netherlands	8	✔ (1)[5]	✔ (3)	C	✔	✔
Norway	6	✔ (3)[6]	✔ (4)	C		✔
Sweden	50			C	✔	
Switzerland	0	✔ (1)	✔ (4)	C	✔	
USA	25	✔ >(8)	✔ (8)	R	✔	✔

1 Difficult to determine as staff often works on other issues besides AIJ
2 The central and three regional governments
3 The staff was only available from 1994-1996
4 MITI responsible for industry projects, Environment Agency for NGO projects and Ministry for Agriculture and Forests responsible for forestry and agricultural projects
5 Joint Implementation Registration Centre. Besides, several people in the Environment and Economic Ministries partly work on AIJ.
6 AIJ secretariat at the World Bank fully sponsored by Norway.

2.1.1 Australia

The government of Australia AIJ pilot closely follows the COP-1 Decision 5 criteria. An AIJ project is additional if it involves specific measures initiated as a result of AIJ. A high degree of transparency should exist. Emission reduction estimates should be regularly reassessed. Environmental and social impacts of the project should be accounted for. The Australian GHG Office administers the AIJ program.

2.1.2 Canada

The government of Canada AIJ pilot follows the COP-1 Decision 5 criteria and the domestic and international GHG emission reduction efforts are closely coordinated. Canada currently registers domestic and international voluntary initiatives in the Voluntary Challenge and Registry (VCR). The Canadian Joint Implementation Initiative (CJII) operates in conjunction with the VCR as participants describe their project to CJII and then report annually to VCR. Registration with VCR does not require compliance with the AIJ criteria established by COP-1. However CJII will assist developers to modify projects so that they do in fact meet these criteria.

2.1.3 Germany

The German AIJ program uses eight criteria (Bundesumweltministerium 1996). The first six criteria are drawn from COP-1 and the following their two specific criteria:

- focus of the AIJ pilot program is on emission avoidance; emphasis should be on promoting the use of modern technology such as heat/power cogeneration, gas and steam turbine power plants or renewable energy; build up of biomass as CO_2 sinks and emission reductions are also possible although the primary emphasis should be on energy sector measures; and,
- AIJ pilot projects have to be scientifically documented.

2.1.4 Japan

Japan has a detailed set of extra criteria that go beyond COP-1 criteria:

- predictions of with and without project emissions clearly demonstrate reduced emissions or increased absorption;
- cumulative GHG emission reductions shall not be negative;
- regular review and modification of above predictions by project implementation parties;
- the proposed project shall not cause greater increases in GHG emissions in other areas compared with the reduction of GHG emissions expected from the project; and,
- environmental, social and economic impact assessment of the proposed project shall be completed.

The Japanese program is cooperatively co-managed and implemented by the Environment Agency, the Ministry of International Trade and Industry and other agencies.

2.1.5 The Netherlands

The Dutch program has a set of eight criteria (Ministry of Housing, Spatial Planning and the Environment et al. 1998), some of which surpass the COP-1 criteria:

- projects have to be approved by all participating governments through a Letter of Intent.
- emission reduction shall be real compared to a baseline. Project proposals should include monitoring requirements; periodic reports are necessary;
- GHGs sources as well as sinks are eligible;
- projects shall be compatible with sustainable development priorities of the host country and must not introduce any conditionality;
- projects shall be screened for environmental benefits; they should result in clear local environmental benefits as well;
- projects should include a training component for local authorities and/or companies in the host country;
- funds shall be additional to development aid and GEF funds; and,
- projects should be economically and environmentally sound which would not have been set up, for whatever reason, without AIJ funding.

Moreover, Dutch AIJ activities shall include project in transition (EIT) and developing countries. Baselines are to be negotiated by the participating governments on a project basis.

2.1.6 Norway

Norway (Norway 1997) tries to maintain a balance between projects executed on a bilateral basis and a multilateral basis. The reason for this approach is to acquire experience with various ways of executing AIJ and assess their respective merits. Existing institutional capacity in host country is viewed to be an asset, but does not constitute an eligibility criterion. Private sector participation in order to leverage technology transfer and financing is given priority as an add-on to the funding provided over the Norway Climate Change Fund established separate from and in addition to the development assistance accounts. A detailed monitoring program for each project is to be established. Moreover, environmental impacts shall be analyzed.

2.1.7 Sweden

Sweden's criteria for its AIJ program (Sweden 1997) closely follow the COP-1, Decision 5 and are pragmatic:

- projects shall be implemented quickly, meaning that priority is given to small and medium-sized projects, which do not call for complicated coordination or require lengthy feasibility and design works;

- projects shall be affordable, meaning that the avoided costs of the formerly used fossil fuel pay for the new equipment and the new fuel within a reasonable time.; and,
- projects shall be reliable, meaning that the technology used should have proved to function well in earlier projects and that there will be no experimenting on the behalf of the borrower.

2.1.8 Switzerland

Switzerland (1997) states that it wants to concentrate on AIJ investment projects that:

- limit emissions caused by energy production and end-use (e.g., fuel-switching to low- or no-C fuels, renewable energies, enhanced energy-efficiency);
- can be easily monitored and verified (e.g., focus on projects that result in significant CO_2 reductions; no sink enhancement projects);
- have demonstration character in the sense that the estimated emission reductions are credible and the project has the potential to be replicated;
- projects are no regret (multiple benefits) and relatively cost-effective.

In 1998, the following criteria were added (Switzerland 1998):

- provide for sufficient training and/or other forms of capacity-building with the aim of ensuring that local capacities are adequate to properly manage, maintain and repair technology;
- Result in net C sequestration at the national level, local benefits generation (taking into account the interests of indigenous and local populations) and sustainable management of natural resources (e.g. conservation of ecosystems, biodiversity, forests and soils; substitution of fossil fuels).

Switzerland (1997) stated that during the pilot phase it will strive to find credible and user-friendly approaches to operationalizing these criteria, taking advantage of the previous and ongoing analytical work performed by other programs, institutions and individuals.

2.1.9 United States of America

The USA program is administered through the U.S. Initiative on Joint Implementation (USIJI) (Dixon, 1998). USIJI specifies in detail the information that project proposers must provide. The USIJI Evaluation Panel composed of senior officials from eight U.S. government agencies evaluates projects. The basic criteria focus on four primary principles:

- additionality in relation to GHG emissions;
- additionality in relation to funding;

- host country approval ; and,
- project monitoring and external verification.

The full set of USIJI/AIJ project acceptance criteria are as follows (US Dept. of State 1994):

A project will be accepted if it...

- is acceptable to the government of the host country;
- involves specific measures to reduce or sequester GHG emissions initiated as the result of USIJI, or in reasonable anticipation thereof;
- provides data and methodological information sufficient to establish a baseline of current and future GHG emissions:
- in the absence of the specific measures [...]; and
- as the result of the specific measures [...];
- will reduce or sequester GHG emissions [...], and if federally funded, is or will be undertaken with funds in excess of those available for such activities in fiscal year 1993;
- contains adequate provisions for tracking GHG emissions reduced or C sequestered resulting from the project on a periodic basis; for modifying such estimates and for comparing actual results with those originally projected;
- contains adequate provisions for external verification of the GHG emissions reduced or sequestered by the project;
- identifies any associated non-GHG environmental impacts/benefits;
- provides adequate assurance that GHG emissions reduced or sequestered over time will not be lost or reversed; and
- Provides for annual reports to USIJI on the emissions reduced or sequestered, and on the share of such emissions attributed to each of the participants, domestic and foreign, pursuant to the terms of voluntary agreements among project participants.

In determining whether accept projects into the USIJI portfolio, the Evaluation Panel also considers:

- potential for the project to lead to changes in GHG emissions elsewhere;
- potential positive and negative effects of the project apart from its effect on GHG emissions reduced or sequestered;
- whether the U.S. participants are emitters of GHGs within the USA and, if so, whether they are taking measures to reduce or sequester such emissions; and
- whether efforts are underway within the host country to ratify or accede to the FCCC, to develop a national inventory and/or baseline of GHG emissions by sources and removals by sinks, and whether the host country is taking measures to reduce its emissions and enhance its sinks and reservoirs of GHGs.

2.1.10 World Bank (not a FCCC Party)

While the World Bank AIJ program is financed by country contributions (mainly from Norway) and the projects thus seen as Norwegian ones, it has set an elaborate set of criteria that by far surpass most investor countries. The World Bank program has specific goals and supporting criteria (Karani 1997). Criteria beyond the COP-1 standards:

- maximise learning about JI;
- completeness and quality of the analysis and evaluation of project costs, benefits and methodologies;
- sufficient generation, documentation and dissemination of information throughout the project;
- degree of innovation in project design;
- emphasis on educational and training workshops on AIJ projects and mechanisms;
- promote the long-term objective of the FCCC;
- relative reductions in GHG emissions;
- replication value of project;
- contribution to transfer and development of GHG efficient technologies;
- contribution to expansion and creation of new markets for GHG efficient technologies;
- promotion of low- or non-C energy technologies;
- contribute to client country development;
- environmental benefits in addition to GHG emission reduction;
- contribution to institutional building;
- promotion of technological innovation, development, diffusion and research and development;
- significance of project for client country;
- explore solutions to methodological issues of AIJ;
- credibility of the additionality of GHG emissions reduction;
- number of sources and sinks to be reported;
- adequacy of arrangements for monitoring, assessing and evaluating GHG emissions reduction;
- facilitation of realistic discussions on pricing and distribution of implicit credits
- adequacy of assessment of project risks and measures to address negative outcomes;
- promote partnerships and increase private sector participation;
- leveraging of private investments;
- long-term commercial potential private sector participation in host countries;
- level of host-country NGO participation;

- creation of new partnerships across countries and sectors;
- implement a variety of AIJ projects within a given timeframe;
- maturity of project development;
- contribution to the diversity of the pilot project portfolio;
- host country project ownership;
- probability of project success;

3. HOST COUNTRY AIJ PROJECT CRITERIA AND ACCEPTANCE PROCEDURES

After the opposition of most developing countries against JI in the FCCC negotiations leading to COP-1, it was expected that many countries would remain skeptical and would not accept AIJ projects easily (Dixon et al, 1997). This expectation has proved to be only partially correct. While some host countries have invested strongly in institution building and have set up quick and effective approval procedures, others have very opaque criteria and decision structures (Chatterjee 1997). Some host countries such as China and India did not approve any projects in the early stages of the pilot (Jepma and van der Gaast, 1999).

AIJ projects have mostly developed in Latin America and Eastern Europe with a small number in Asia or in Africa (Dixon 1998; Dixon et al, 1997). The skewed regional distribution of AIJ projects in developing and EIT countries is due to two primary factors: general distrust of the AIJ mechanism and lack of institutional capacity to cope with the demands of the mechanism (Chatterjee 1997). Developing country opposition to JI arises from:

- concern that donor countries will try to buy their way out of domestic emissions reduction;
- fear of undermining ODA flows (despite the criteria in the FCCC);
- an apparent lack of equity in the process; and
- concern that safeguards will be insufficient to protect the host from, e.g., inappropriate technologies.

Lack of institutional capacity in the host country leads to an inability to obtain and/ or make use of information on AIJ and thus creates problems in implementing the mechanism. The following section reviews the current information about host country procedures. Some information is difficult to obtain but the countries reviewed, Costa Rica, Czech Republic and Poland, give a flavor of the different approaches taken by host countries.

3.1 COSTA RICA CASE STUDY

Costa Rica has by far the most developed set of criteria of any host country participating in the AIJ pilot. Fulfilment of all criteria is the stated prerequisite for official project acceptance by the Costa Rican Government (Costa Rican Office for Joint Implementation 1998). The criteria are meant to meet the following objectives:

- minimize bureaucratic red tape; as few criteria as possible, and highest possible levels of consistency with existing sets of criteria in established national programs of industrialized nations;
- meet current international standards; criteria should meet current pilot phase standards set by the FCCC;
- represent Costa Rica's particular interest (country driven); criteria should address Costa Rican development priorities, as distinct from the considerations of the investor country or of other developing nations;
- address GHG abatement benefits sharing among participants; criteria should address quantification and monetary valuation of the GHG abatement, including the sharing of the monetary surplus between the buyers and the sellers;

The following supplemental set of criteria are also applied by Costa Rica:

- legal Compatibility; is the project consistent with applicable Costa Rican laws and regulations?
- investor home acceptance; is the project acceptable to the home country government, or does the project proponent intends to apply for such acceptance?
- national sustainable development priorities; is the project compatible with and supportive of Costa Rican national environment and development priorities and strategies, including:
  - biodiversity conservation,
  - reforestation and forest conservation,
  - sustainable land use,
  - watershed protection,
  - air and water pollution reduction,
  - reduction of fossil fuel consumption,
  - increased utilization of renewable resources,
  - support for Costa Rica's efforts to fulfil its obligations under the UN's Biological Diversity Convention and Agenda 21.
- enhancement of income opportunities and quality of life for the Costa Rican civil society;
- minimized or acceptably low level of adverse consequences of the project through site selection, scale adjustment, timing, attenuation, and mitigating measures;

- local capacity building such as the transfer and adaptation of know-how and high quality technologies;
- local or community support; will the local community support and participate in and/or benefit from the project?
- offset additionality; will the project bring about real, measurable and long-term environmental benefits related to the mitigation of GHG that would not have occurred in the absence of such activities? proposal should include a defensible reference or baseline case for emissions reductions or sequestration processes in the absence of the project;
- monitoring; does the project have a monitoring plan that includes the participation of organizations capable of successfully monitoring the project? monitoring plans should include actual measurements of the project's emissions or C sequestration in order to establish a high degree of certainty that the predicted benefits were achieved by the project;
- verification; will the proposal allow for the verification of the project's progress through inspection by qualified third party verification?
- durability or quality of offset; does the project have a high likelihood that the GHG offset will be maintained over the life of the project? proposal should include a work plan for project start-up: provide the timeline for starting or completing significant phases or stages of the project, including but not limited to: feasibility studies, development and beginning of operations, and completion of advanced stages of the project;
- GHG benefits; what methodologies were used to calculate GHG emissions reductions (avoidance) and carbon sequestration (fixation), and what are the key uncertainties affecting those estimates?
- financial additionality; is the financing of the project additional to the financial obligations of FCCC Annex II Parties, as well as to the current ODA flows?
- cost estimates and financial feasibility; does the project include an accounting of all the costs of operation and economic benefits associated with the project, including organizations or entities, other than official project participants, that may contribute to the project's operation? does the project address the cost issue (in US$) of the per avoided Mg of CO_2 equivalent? does the project developers state the AIJ financial component (in US$)? does the proposal include the financial projections (cash flow, profitability, rate of return, benefit/cost relationship, etc.) with and without the AIJ additional financial contribution? does the proposal address the issue of sharing the monetary surplus related with the project GHG abatement benefits?

- institutional infrastructure and governmental Role; does the domestic Costa Rican institutional framework (legal, administrative, and technological) exist to adequately implement and administer the project?
- reliability and credibility of project participants; what is the prior experience and track record of the project partner(s) and intermediaries? is each partner's role in the project's development and implementation made explicit in the proposal?

Through a executive decree of April 1996, the President of Costa Rica and the Ministry of Environment and Energy (MINAE) increased the Costa Rican Office on Joint Implementation (OCIC) status to a technical-administrative highest deconcentration organ from the Ministry of Mines and Energy (MINAE). This permits OCIC to become the national authority to address national policies and guidelines for the implementation of AIJ projects. It is thus a one-stop-shop and operates very efficiently.

3.2 CZECH REPUBLIC CASE STUDY

The Czech Republic has the following specific criteria (Czech Republic 1997):

- evidence must be given that a significant reduction of GHG emissions (at least 10 % per annum) in comparison with the initial state (emission baseline) per unit amount of the final production shall occur through:
  - replacement or modification of existing technology or parts of it,
  - The addition to the existing technology by addition of end of pipe equipment (denitrification, waste gas incineration, trapping of volatile organize compounds (VOCs), etc.);
- as for the projects resulting in long-term CO_2 sequestration through afforestation, the project should increase the overall stability of forest ecosystems and respect the principles of biodiversity protection; and,
- foreign firms investments into subsidiaries located in the Czech Republic made solely to meet emission limits such as those outlined in the Air protection Act, shall not be considered as AIJ projects.

3.3 POLAND CASE STUDY

The Polish AIJ program has the following set of criteria (Poland 1997):

- For any AIJ project, it must be possible to estimate the expected reductions in GHG emissions and monitor ex post the actual reductions in GHG emissions. Since the distinguishing feature of AIJ projects is their contribution to GHG emission reductions, it is critically important that these benefits can be predicted and monitored. The project developer or sponsor should provide analysis of expected emission reduction and a project monitoring system to measure/calculate emission reductions after the project is implemented.

- AIJ projects should not lead to increases in other local/regional environmental quality indicators at the expense of achieving reductions in GHG emissions. Where proposed AIJ projects might lead to increases in air pollution, waste water discharges, or waste disposal, appropriate mitigation measures should be incorporated into the AIJ project.
- AIJ projects should directly or indirectly result in cost-effective realisation of environmental goals. Where AIJ projects involve the installation of new capital equipment, they should also lead to a net reduction (or at least no increase) in the facility's costs of meeting current and anticipated environmental standards (e.g. resulting from harmonisation with the EU environmental directives and/or other international treaty commitments). Thus, process changes and new technologies, which prevent pollution, are desirable.
- AIJ projects should encourage the economic use of natural resource and reuse or recycling of waste materials.
- AIJ projects should be compatible with, and promote to the greatest extent possible, utilisation of modern production processes. To compete in European and international markets, Poland needs to identify and implement the most effective technologies, while simultaneously improving or maintaining product quality.
- AIJ projects should be sensitive to and compatible with macroeconomic policies at national and local levels. Where Poland is making a financial commitment in the AIJ project, total benefits and costs as well as distribution of those benefits and costs should be taken into account. Both direct and indirect effects on employment and the viability of enterprises should be considered.
- AIJ projects should only be undertaken by Polish enterprises, which can reasonably be expected to be economically viable in the long term. In effect, to maximise the prospects for achieving and sustaining GHG reductions, a financial audit of the enterprise should be conducted. Where there are alternative sites for the AIJ project, an assessment of the relative advantages of implementation of the AIJ projects on the proposed sites should be made.

4. ADDITIONALITY, AN ELUSIVE CONCEPT

The COP-1 decision defines two kinds of additionality: environmental and financial. Environmental additionality exists if the emissions of the project are lower than emissions under a baseline scenario that calculates emissions had the project not taken place. As many project participants have an incentive to overstate emissions reductions from the project, the determination of environmental additionality has to be done carefully. Financial additionality is meant to ensure that no funds are diverted to AIJ from ODA or the funds GEF.

From the theoretical consideration of additionality, there could therefore be no beneficial environmental effect of AIJ projects unless they are additional. Emission reductions, which are secondary effects of normal investment practice, are essentially without cost. The function of the additionality criterion is therefore to ensure that real reductions are derived from the project i.e. that they would not have been carried out anyway by normal market operations. This is sometimes described as the free-rider problem.

Additionality is expected to ensure that the supply of projects was restricted and that the host did not sell the credits too cheaply (Jepma and van der Gaast 1999). When the host has a reduction target, non-additional projects mean that the credited reductions will be added to the host's emissions inventory and if the credits are not real then the host will have to take reduction action elsewhere to meet the target or be in non-compliance. In the end the environment may well pay, as many hosts cannot afford to take action elsewhere. The same argument holds for estimations of reductions, which are highly uncertain and have a high risk of over estimation. If the host accepts these as a basis for credits the same problem of non-compliance will arise unless additional action is taken. For hosts with no targets the environment will pay if the reductions are not real.

Non-additionality may therefore lead to extra costs of meeting national commitments which could in turn be passed on to consumers. Furthermore, economic sectors that provide non-additional projects may gain a competitive advantage relative to the same sector in other countries or other sectors in the same country because of the extra financial support provided. Having said that, additional projects can have trade advantages.

Financial additionality has another meaning with in a different sense to its use referring to the ODA. It is sometimes used in purely economic market terms where the project, which is additional, would be expected to cost more than the normal market choice but would deliver more reductions. In practice it was thought that this increased cost or financial additionality for the JI project would be offset by the value of the credits which would ensue from the transaction, thus making the project still attractive to investors.

The concept of additionality has been problematic to implement in practice. This is due to the counterfactual base of comparison. How can an autonomous efficiency increase be covered by the baseline methodology?

How would overall ODA and GEF funds develop in the absence of AIJ? Thus the baseline issue has become a major focus of discussion in the further development of AIJ, JI and the CDM (Begg and Jackson 1999).

The difficulties with additionality can be shown using the example of the USIJI, which tried to define both types of additionality (Dixon 1998). It allowed projects to have a positive financial return while they had to demonstrate how they were a result of USIJI and hence additional. Nevertheless, in USIJI practice the issue became blurred. For example, USAID funds were allowed to be used for project proposal preparation. CJII also takes a relatively liberal approach to the issue of additionality of project finance. The CJII Guidelines note that projects 100% financed by the Canadian International Development Agency (CIDA) would not qualify as AIJ. However, the Guidelines do not exclude partial funding or 100% CIDA funding of feasibility studies and management training, which do not relate directly to a particular JI project.

4.1 VARIOUS APPROACHES TO ADDITIONALITY DEFINITION

Different approaches to the operationalization of additionality have been proposed such as those outlined for the USIJI and for CJII above. These approaches particularly for the environmental additionality aspect are summarized and discussed in the context of examples or case studies.

The IEA /OECD (1997) have proposed a barrier removal method as a test for additionality of projects. They have identified the different types of barrier involved, when they may be encountered in the project development and the different groups of barriers involved. These can be project specific, technology specific and general locality specific. They point out that barriers usually impact by increasing project costs and or project risks hence automatically make the projects additional and mean that the projects would not have happened anyway. They suggest that a project is additional if it can be demonstrated that:

- institutional, financial, technological, or informational barriers exist which inhibit the implementation of projects and would not be removed under circumstances in which there is no incentive to reduce greenhouse gases for an investor;
- barriers do not have the same effect for the baseline project;
- the design of the projects effectively addresses these barriers; and,
- financing of the AIJ related part is additional.

The U.S. Environmental Protection Agency (EPA) has also tried to develop options for the determination of additionality (Carter, 1997). Several options were suggested:

- narrow categories of projects, which a priori must be additional;
- define additionality as overcoming project specific barriers;
- measure additionality from quantitative sector specific guidelines;
- additionality defined by a program such as USIJI;
- combination of options ;
- limit efforts to determine additionality and take measures to limit overall JI/CDM such as discounting, limiting the life time or the scope of JI/CDM

Each option can be considered relative to a set of criteria related to development costs/immediacy, transaction costs, participant certainty, accuracy, comprehensiveness, and robustness. The first option demands identification of appropriate project types. The second option refers to the IEA approach. Using sector specific baselines suggests that a project is deemed additional if it is not already planned in the baseline in a particular country. There are many problems with this particular definition of additionality (Begg and Jackson 1999).

The next option of a general set of guidelines refers to the USIJI program. The two issues there are program additionality and emissions additionality. With program additionality, it means that the project should be specifically developed because of the USIJI. Emissions additionality deals specifically with the basis for the emissions reduction compared to the baseline or what would have happened in the absence of the project. This considers barriers, requirements in the host for emissions reduction and the difference between the project and prevailing technologies in the host.

Another option, the combination of options from the set above, may offer the best alternative. The final option, limit efforts to define additionality, is a more pragmatic approach. Additionality definition is limited on the grounds that there will always be uncertainty associated with it. This limitation can be achieved through discounting emission reduction credits at some standard rate, limiting the crediting lifetime of the project, or by limiting the amount of a donor's obligations which could be met with flexible mechanisms such as the CDM (now expressed as supplementarity). These ideas were attributed to Fritsche (1994) and others at international negotiations.

Another option would be to discuss additionality on two levels: country and company (business or firm) level. Due to indirect effects, these approaches will differ. A project that is clearly additional from for a company may not be additional for a country as a whole. Under fossil fuel subsidies, for example, a wind power plant might be clearly additional due to higher costs compared with the subsidized fossil fuel. If the subsidy was phased out, it could become non-additional. Thus non-additionality on a country level will enhance the supply of projects that are additional on a company level.

Additionality on the company level could in theory be measured according to the following criteria (see Figure 1). They assume that the discount rate and the degree of risk are known, which allows a calculation of risk-neutral costs[1] (for a nice discussion of the effect of different discount rates see Varming et al. 1998):

- accept all projects that reduce or sequester emissions (Rentz 1998);
- prove that the internal rate of return (IRR) of the project is lower than that of a commercial alternative;
- prove positive incremental cost of the emission-reduction related part of the project similarly to procedures used by the GEF; loosening of these procedures in 1998 show that they have been extremely difficult to apply; and,
- prove positive costs of the full project (e.g. by investment modelling) (Bedi 1994)

Determining additionality on a company (business or firm) level may be difficult. A narrow additionality definition might lead to the choice of marginal technologies that are not always appropriate and depend on having an undistorted market, which often is not the case in host countries. Experience from existing AIJ projects reveal that autonomous technology shifts depend on hard-to-observe parameters. On the other hand, country-level additionality might be easier to assess. Such an assessment would also have the advantage that there are no perverse incentives to prolong inefficient policies. The best approach would be to phase in strong country-related additionality rules over a certain period of time, e.g. 5 years, to allow countries to change policies. In that period, all projects that prove a GHG reduction from status quo for a period of 5 years should be accepted to account for higher transaction cost and barriers in the set-up phase of the CDM. Those would kick-start the CDM and reward those businesses that have already started plans for less emission intensive projects.

5. BASELINE DEFINITION

The issue of baseline determination is the key to a successful AIJ pilot. The COP-1, Decision 5 criteria did provide specific guidance for development of guidelines (Begg and Jackson, 1999; NEDO 1999).

5.1 PRACTICAL EXPERIENCES IN BASELINE DEVELOPMENT

Despite the Uniform Reporting Format (URF), that was formulated by the FCCC and specifies the information requirements from the AIJ project, information on baseline methodologies is far from excellent and remains to be improved before the end of the pilot phase. Many projects whose baselines are discussed below have not yet started because of lack of financing. With over 100 approved projects, one could expect a wealth of practical experiences with baseline setting. We discuss energy and forestry projects in distinct sections as they involve very different baseline issues.

While almost all AIJ investor countries have published their set of approval criteria, the only one that has also set detailed baseline criteria is the U.S. Initiative on Joint Implementation (USIJI) with 31 accepted AIJ projects. Baselines have to be consistent with (USIJI 1996).

- Prevailing standards of environmental protection in the host country;
- Existing business practices within the particular sector of industry; and,
- Trends and changes in these standards and practices.

Baselines have to include indirect effects such as activity shifting, price effects and lifecycle effects in products. They shall also provide information on other environmental effects of the project. The descriptions of the approved USIJI projects include detailed baselines. Despite the criteria, indirect effects have not been covered to any extent. Changes in the legal framework are only covered in some projects, whereas others do not take them into account.

5.2 LAND-USE CHANGE AND FORESTRY (LUCF) PROJECTS

In the Belize Rio Bravo USIJI project it is not clear whether the baseline deforestation rate is just an extrapolation of the historical rate or an estimate. In contrast, in most Costa Rica projects the deforestation rate is explained. In case there are uncertainties about existing forest stocks on deforested lands which can easily reach an order of magnitude, the BIODIVERSIFIX project chose the upper bound as baseline (USIJI 1996) while CARFIX sets the stock after deforestation to be zero (USIJI 1996). The difference is 10

Mg C/ha. If BIODIVERSIFIX had chosen the same value as CARFIX its C sequestration would have been 585,000 Mg higher. The RUSAFOR project established three different baselines but did not make them public. It chose the most probable scenario that probably entails conservative C sequestration rates.

The first European projects which could be classified as large-scale AIJ (but so far have only partially been approved). The afforestation projects are developed by the FACE Foundation in the Netherlands. These projects used a very simple baseline, i.e. the state of the region without forest cover or the existing, damaged forest. Indirect effects were not taken into account. The C sequestration was then calculated with a simulation model using data from a small number of field studies and scientific literature (FACE Foundation 1996; FACE Foundation 1995). In recent years, the FACE Foundation has considerably refined its baseline methodology (Emmer et al. 1999). It has submitted baseline studies to the Dutch JI Registration Centre that are based on detailed project-related simulation, taking into account the host country economic and political framework. For its project in the Czech Republic, financial additionality was confirmed by an external financial audit. Nevertheless, it is shown that the baseline for the Czech project is much more stringent now than before.

It is particularly interesting that most of the LUCF projects try to define baselines for afforestation and preservation of existing stocks without accounting for sequestration/emission of the soil as uncertainties about the latter remain high. There are two exceptions, the ECOLAND project in Costa Rica and RUSAFOR in Russia. In the former case not accounting for the soil C would result in lowering the emission reduction estimates by 44%.

The complexity of measuring forest C sequestration and the effect of changing land use has resulted in this plethora of alternative approaches with their attendant problems. This issue will now be the subject of an IPCC report due in late 2000.

5.3 ENERGY PROJECT BASELINE EXAMPLES

Renewable project baselines do not always include life-cycle emissions of the plant material. A very interesting aspect of the renewable energy projects in Costa Rica is that because of the commitment of the Costa Rican government to phase out fossil fuel electricity production by 2001 the baseline is zero emissions after 2001. USIJI questioned this outcome and required the baselines to take that scenario into account. Thus, some renewable energy projects may not become creditable under a full-fledged JI regime after 2000. All the renewable energy projects now approved are no-regret projects as JI benefits may accrue only for a part of the economic project lifetime. The fossil fuel baseline assumes that from 1994 to 1997 fossil fuel emissions remain constant and will be reduced by 44% in 1998, 86% in 1999, 99% in 2000 and 100% in 2001 (USIJI 1996). The figures chosen differ for 1998 and 1999, though, as some projects take 33% and 66% reduction, respectively (USIJI 1996).

A biomass-fired plant in Honduras is proposed to substitute for a fossil power facility. This project has a baseline that is difficult to determine. The project lifetime of 20 years may be too long. Moreover, it is stated that there is a severe demand surplus for electricity in Honduras. Therefore, it is unlikely that the existing fossil plant will be shut down. In the baseline, it is assumed that the biomass would not be used. The possibility that deforestation will rise to supply biomass to the plant is not fully considered.

Small-scale solar electrification in rural Honduras defines average kerosene burning in lamps as the baseline. This baseline seems to be conservative as the solar panels can also be used to recharge batteries that substitute dry cells. Interestingly, the baseline for a geothermal power station in Nicaragua is hypothetical diesel fuelled power plants with a lifetime of 35 years assuming rising power demand. Emissions from the geothermal power station are taken into account.

The RUSAGAS project that entails sealing of valves on natural gas pipelines takes current emissions as baseline and estimates a lifetime of 25 years. This baseline seems somewhat insecure, though, as the national regulatory framework is likely to change in future years. So far, the Russian natural gas firm (GAZPROM) is paid only for the quantity of gas extracted, but not for the quantity delivered. In case the latter would apply because of regulatory changes, the incentives to seal the valves would be very high for the company. That means that then the baseline might have to be set to zero.

A cogeneration project in Decin in the Czech Republic uses a projection of heat use which sees a decrease of demand by 13% in 2001 and a constant demand thereafter. The existing coal-fired plant was taken as baseline for heat production for the technical lifetime of the project. This baseline might be high since it is unlikely that this plant would continue for another 27 years. Moreover, the existing average emission factor of the Czech electricity production was taken as baseline for the electricity production of the new plant. Again, this seems to be on the high side as this emission factor will surely be reduced in the business-as-usual case because of incentives to improve environmental performance as a requirement for EU membership. A retrospect comparison of the officially reported ER estimates with a range of other baselines was completed (Begg et al, 1999) and the baselines were shown to be too high.

Many Swedish small-scale boiler conversion projects in the Baltic States take the status quo before project implementation as the baseline (CCAP/SEVEN 1996). This does not take into account subsidies and market distortions or the drive for reducing local pollution levels. A combination of phase-out of subsidies and improvements in local air pollution has led to a series of fuel switches from heavy fuel oil to electricity and then to gas in Estonia.

Begg et al. (1999) conducted an assessment of baselines of eight AIJ energy sector projects in Estonia and the Czech Republic. They have defined a number of possible baselines for each project based on consideration of the pre-project situation and the local and national conditions. For example, the Valga AIJ project entails the replacement of a district-heating boiler fuelled by heavy fuel oil (HFO) with one fired with woodchips. Three baselines are specified, one where HFO continues to be used for the project lifetime, one where the HFO is replaced by gas after 10y, and one where the HFO is replaced by woodchips after 10y. The first baseline is only considered likely if refurbishment of the plant to extend its lifetime and the addition of measures to reduce SO_2 and NO_x emissions is carried out. The third baseline, which is essentially the JI project itself delayed by 10 years, is considered reasonable because of the low investment cost of the plant, the lower price of woodchips and pressure due to local air pollution regulation. Not surprisingly, the estimate of emissions reduction using the first baseline is less than 50% of that using the third baseline. Similar estimates are also found for the other projects. Taking into account measurement uncertainty, they have calculated that the uncertainty in the emissions reduction of the five heat supply projects studied is about ±80% and is higher where there are uncertainties concerning the interactions with a system, i.e. electricity supply or heat demand.

Figure 2 shows a range of AIJ projects with the range of emission reductions calculated by examining the effect of equally likely baselines. Also shown are the FCCC officially reported values as emission reductions from the project for which the donors may expect credits. It can be seen that there is a wide range of values and that it is possible for gaming to occur to maximise the credits. A standardised procedure would help to avoid this problem.

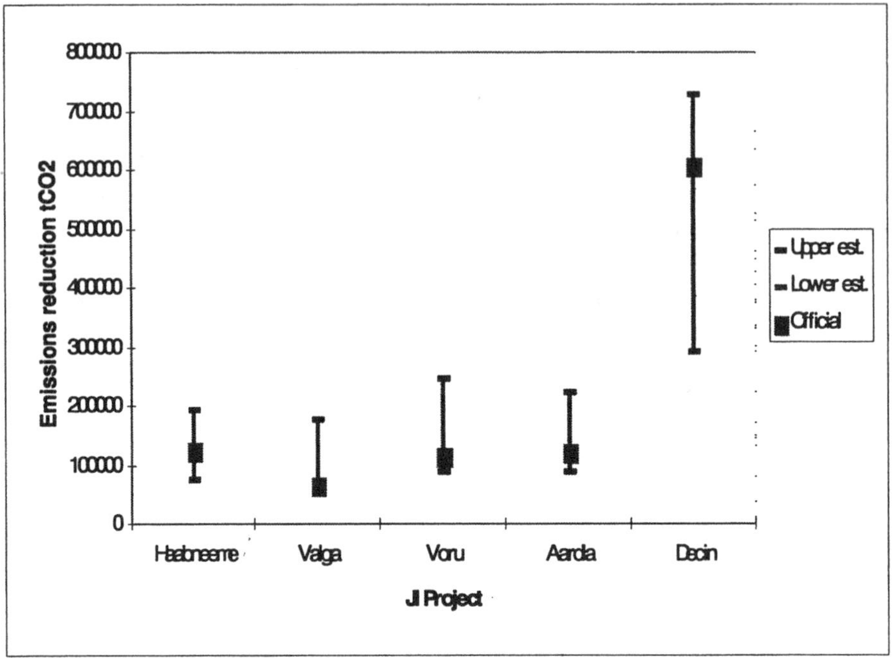

Figure 2. Comparison between uncertainty in emissions reduction due to baseline and UN FCCC reported values for 5 AIJ projects (Begg et al., 1998).

5.4 SIMULATION OF BASELINES USING EXISTING NON-AIJ PROJECTS

The Nordic Council of Ministers has simulated five existing projects in Eastern Europe as JI projects. While focusing on the costs and differences between project design and performance, they also discussed baselines (Nordic Council of Ministers 1996). Sectoral official development plans are considered as appropriate baseline for power plant projects whereas in the heat sector some improvements of the efficiency are predicted. In the case of the Czech Republic, a draft law on heat metering and thermostats is regarded to lead to business-as-usual improvements. In other locations, the availability of substitute fuels defines the baseline.

In several cases, continued operation of the existing plants was considered unlikely. Therefore, the projects would have been undertaken autonomously after some years, which means that by that time baseline emissions would equal the project emissions. In the case of a geothermal heating plant in Poland, there were five realistic baselines and the choice depended on the probability of certain frame conditions. The baseline emissions varied by a factor of five.

5.5 THEORETICAL BASELINE SIMULATION

The Wuppertal Institute simulated four projects: a coal-fired power plant in China, a solar-thermal hybrid power plant in Morocco, a DSM project in Poland and a cement plant in the Czech Republic (Luhmann et al. 1997). Generally, baselines for power plants would be derived from the official expansion plans of the respective countries. In the China case, Luhmann et al. defined the baseline as the prevailing domestic coal fired power plant with a Westinghouse licence. That plant has an efficiency of 40.3% while the latest German technology including denox would reach 43%. Therefore, the reduction would be rather low. Substituting outdated domestic power stations with an efficiency of less than 25% was not considered as acceptable JI as there is a demand surplus, which would prevent shutting down the old plant. Project lifetime should not surpass 15 years.

For A solar-hybrid plant, three baselines were developed on the base of a coal-fired, an oil-fired and a gas-fired plant. The latter was excluded because of very high costs whereas the oil-fired plant would not correspond to the expansion plans of the Moroccan government.

The DSM case was compared to an existing coal-fired plant and a new, more efficient one. Indirect effects of DSM (e.g. rebound of energy consumption or enhanced technology diffusion) were excluded. In the cement case, the baseline depends on the market situation as the emissions of an existing plant varies according to its mode of operation. In case of the substitution of a plant, the average annual emissions of the existing plant during a period of several years are taken as the baseline. An additional plant shall be evaluated in a feasibility study.

5.6 SUMMARY OF EXPERIENCES WITH BASELINE DEVELOPMENT

The AIJ pilot phase suffered from the absence of uniform baseline methodologies but provided a forum for rich experimentation and analysis of this topic. Unfortunately, many projects developed ad-hoc baselines. Even in the case of USIJI, which had clear criteria for baselines, experiences were uneven. Moreover, baseline establishment was less than transparent, as many reports did not explain how the baseline was developed. The apparent shortcomings of ad-hoc project baseline-setting led many observers to call for standardized baseline methodologies which avoid confusion and would reduce transaction costs.

TABLE 2. Summary of key authors, baseline types, method and degree of specificity of AIJ project baselines.

Authors	Type	Level	Method
1. Performed by individual investors where necessary	Project specific	*Project*	Individual project characteristics and all relevant information included to construct specific project and baseline emissions path
2. Luhmann et al. (1997)	Technology matrix	Project (specific to technology/ sector/ host country)	*Standard technologies with associated pre-selected standard baselines ; periodically updated*
3. Michaelowa (1998)	Project type matrix	project (specific to sector/ host country)	Bundles several technologies as one project type
4. CCAP (1998)	Benchmarks	Project or Sector or Country	Quantitative measures of performance
a)	Forward looking (Static or dynamic)	As above	*Based on planned future pattern of emissions for lifetime of plant*
b)	Historic (Static or dynamic)	*As above*	Based on historic performance of substituted plant
5. Begg et al (1999)	Package approach including standardised *Baseline approach where appropriate*	Project or Sector	Given the counterfactual nature of baselines this is a package of measures to limit uncertainty and to limit gaming. It includes a range of standardised baselines using mainly dynamic benchmarks depending on whether substituted plant is known as well as other measures. Based on uncertainty analysis of typical projects
6. Heister (1999)	Investment analysis approach	*Project*	*Based on financial additionality of investment. Simulates financial/ behaviour aspects*
7. Wietschel et al (1998)	System model	*Sector*	Projects future emissions of sector based on assumptions regarding demand. No regrets measures are part of the baseline
8. Puhl et al (1999)	Top down baseline	*Country or Sector*	Aggregated approach either using absolute emissions or tonnes C/ GDP. National or sectoral planning models

Neither the baselines of current projects nor the simulation studies included no-regrets opportunities (i.e. emission reduction opportunities which are profitable either for a company or for a country as a whole). A good example is the ILUMEX project in Mexico, which would lead to a macroeconomic gain of 0.05 US$ Mg CO_2 by distributing subsidized compact fluorescent lamps. The gains accrue to the participating households. Furthermore, the project accelerates the market development of the lamps. From a microeconomic perspective of the Mexican utility the project has net costs of 30 US$ Mg CO_2 (Anderson 1995).

Table 2 is a brief summary of the different approaches to baseline construction with some of the terminology being used.

6. EMISSIONS MONITORING, VERIFICATION AND REPORTING

Three activities which are often confused or thrown together are monitoring (i.e. measuring emissions or C sequestration), verification (checking of the data output of the monitoring by a third party) and reporting (publication of monitoring and verification results and other features of the AIJ project).

Until now, most countries separate these activities. The Netherlands, however, have set up a Joint Implementation Registration Centre (JIRC) that controls all three. A project has to be registered with JIRC by providing information that the criteria have been fulfilled and that it has been approved by investor and host country (basic reporting). Then a baseline has to be defined using common rules and be approved by the Environment Ministry. Afterwards, an annual monitoring study has to be submitted which is verified by JIRC. JIRC then certifies the reduction (JIRC 1996).

6.1 MONITORING

The use of feasibility data in calculations of emissions reduction from JI projects has been shown to produce over-optimistic results (Begg et al 1999). Their analysis illustrates this problem. Projects tended to under-perform relative to their predicted emissions reduction for a variety of reasons, including lower than expected demand, problems with the design of the plant, operator error, fire and flood.

To calculate emissions reductions, the output from the plant e.g. in kWh has to be monitored. This monitoring should be done at the plant and should be a normal part of its operation. Unfortunately, monitoring methodology so far has been neglected in the AIJ pilot phase.

6.2 VERIFICATION

Verification is a crucial component of AIJ but so far has not made headlines. As most projects have only started recently, verification has not been done. It also seems that project participants are reluctant to pay for the costs of verification, as they do not get any value in return. Once again, Costa Rica has been a pioneer in verification when it commissioned a thorough verification of its Protected Area Project (PAP) by the certification firm SGS in 1996. The verification was published in early 1998 (SGS 1998) and came to the conclusion that 39% of the sequestration should be kept in a buffer to cover risks and uncertainties.

6.3 REPORTING

Whereas COP-1 Decision 5 did not contain fixed rules on reporting of AIJ projects, common reporting rules were decided by SBSTA in March 1996 and guidelines for Uniform Reporting Format (URF) were issued. The URF is quite detailed but lacks crucial information such as baseline parameters and methodology, as well as, project lifetimes and costs. Both investor and host countries report on their AIJ projects annually.

The UNFCCC secretariat compiled two synthesis reports of the AIJ pilot phase in 1997 and 1998 (UNFCCC 1997, 1998). The URF reports are also available on the UNFCCC website http://www.unfccc.de/fccc/ccinfo/aijcont. Some countries have established transparent guidelines for reporting projects (Canada JI Office, 1996).

From the beginning project reporting offered some surprises. For example, a project to install small-scale renewable systems in Indonesia was fully reported by both Germany and Japan in 1996 due to the fact that a German and Japanese utility were involved. Fortunately, this case of double reporting remained the only one. After some time, this project was removed from the list of German AIJ.

Despite the URF the quality of reports remains uneven. Many countries do not submit annual reports. It has not been possible, for example, to assess costs or baseline methodologies from many of the reports. Moreover, the reports often do not state or deliberately obfuscate whether a project is actually implemented or just a paper project. Thus project reporting is a weak spot of the AIJ regime and should be much improved for the CDM or other Kyoto Protocol flexibility mechanisms.

7. EXTERNALITIES: THE ROLE OF NON-GHG ENVIRONMENTAL BENEFITS

External effects can be both positive and negative and it is this concern about the overall equity of who is bearing the burdens and who is receiving the benefits of projects can be a concern. The overall impact of most projects (depending on size), when considered singly, would be expected to be small in respect of the overall environment and social impacts prevailing in the locality, region or country resulting from other activities. Their significance resides in any cumulative effect that would result from a large increase in the number and size of such projects that would be necessary to implement really substantial GHG emission reductions by JI/CDM. Therefore, it is important and it should be possible at an early stage in the process to ensure that the benefits are maximised and that adverse effects are minimised or ameliorated in some way. That these trade-offs are transparent is an important aspect of the process.

7.1 EXAMPLES OF EXTERNALITIES

There can be both positive and negative externalities associated with AIJ projects. However positive externalities are the only incentives for host countries to engage in AIJ. The following externalities give a general indication of some of the impacts that are important:

- formation of human capital;
- transfer of technology adapted to host country capacity;
- capital transfer;
- foreign currency transfer;
- improvement of income distribution;
- local environmental and social benefits and costs such as the reduction of local; pollutants and job creation or loss of arable land or displacement of people
- protection of biodiversity; and,
- contribution to sustainable development.

The most critical negative externality of AIJ could be that it reduces incentives for innovation. For a detailed discussion of this aspect see Michaelowa and Schmidt (1997). Many negative externalities are linked to poor management of the project and can be affected by an unstable political situation. It is very difficult to quantify these externalities. Most of them are interlinked and operate on different time scales. Feedbacks depend on the local situation. While it is obvious that AIJ will lead to capital and foreign currency transfer the net effects on jobs are unclear. The transfer of modern technology could well lead to a loss of jobs, at least locally and in the short and medium term. Formation of human capital is a long-term effect and dependent on the social and political framework. Improvement of distribution also depends on the local political and social situation.

Preliminary analysis reveals that ER benefits through reduction of local pollutants, especially SO_2, are comparable to the value of C credits under a high C tax of 20-200 US$ Mg C. As the critical loads of local pollutants have not yet been reached in many developing countries, the benefit stemming from carbon emission reduction would be lower compared to industrialized countries. Nevertheless, it seems that reduction of local pollutants will be a relevant externality particularly for densely populated countries in transition and newly industrializing countries, for example in Asia. Projects that offer such benefits will be preferred. In fact, the first AIJ project in Central Europe, the local authorities exactly for that reason promoted the Decin project in the Czech republic.

Biodiversity will only be protected if the social and political framework is conducive to forest protection and prevents relocation of damaging activities. Thus, only countries with a strong administrative capacity are able to take advantage of biodiversity-related AIJ. Costa Rica is an example for such a trend as it heavily focuses on extension of national parks through AIJ funds. It is likely that the capital and technology transfer will be decisive for those host countries where official development aid is declining. Countries with high private capital flows will try to use JI funds to maximize positive environmental and social externalities.

Begg et al. (1999) examined local environmental impacts from a range of energy plant AIJ-type projects in Estonia and the Czech Republic. This involved site visits and a retrospective assessment of a range of impacts related to the fuel delivery, plant operation and waste disposal. Though it is difficult to make a comprehensive retrospective assessment due to data problems, there has been no significant change in many impact categories between the reference and the project. These areas include water, land use, soil, visual, noise, forestry, energy consumption and socio-economic aspects though the latter required much more information before final conclusions could be drawn.

The primary benefits from the conversions include SO_2 and NO_x reductions of up to an order of magnitude for conversions to biomass or gas from coal or heavy fuel oil. Particulate reductions are clear for conversions from coal. This means that there will be significant air quality and health effect improvements from a reduction in air pollutants. The extent of this reduction varies with the type of conversion. The main costs of conversions do not apply to all project types.

There are changes in CO emission levels that are already low and these increase for oil to biomass conversions. There is no change for coal conversions to biomass but conversions to gas or wind decrease CO emissions. The overall levels are nevertheless low. Waste production also increases for oil or gas conversions to biomass. However for coal to biomass there is a reduction benefit. Conversions to gas or windpower have no waste.

Transport of fuel can contribute to air pollution and traffic noise and disruption. Transport distance and frequency of trips is site dependent but appears to be greatest for the conversions to biomass. This factor could easily be improved by increasing the capacity of the fuel transport so that the number of trips was decreased. There are no fuel transport problems with wind and gas fired plant.

Begg et al (1999) suggests a method for implementing an assessment of environmental and social effects at the approval stage of the project. It involves a staged process for screening for project impacts and then an assessment process not necessarily at the level of an environmental impact assessment (EIA). Participatory processes with the local community are crucial in the implementation of such assessment procedures. Such assessments are required at the project and if possible strategic level. Feedback into the design of the project to minimise detriments and maximise benefits is necessary followed by auditing of the impacts.

8. ASSURANCE OF LONG-TERM EMISSION REDUCTIONS

Ensuring that projects are long-term projects supplying emission reductions on a continuous basis will require that the original design and implementation measures have fully engaged the local community affected by the project and addressed their concerns. It is most relevant for the CDM in developing countries but also true of JI projects in EITs.

The long-term nature of ER or C sequestration of AIJ projects is often unclear. Generally speaking, ER projects in the energy supply sector do not have the risk of reversal of the reduction. Only the lifetime of the project can be shortened due to equipment failure but that does not impact on past reductions. Sequestration projects in the LUCF sector, on the other hand, entail major risks of loss of sequestered C through forest destruction or harvest after the end of the project. Project proponents have acknowledged these risks. The FACE Foundation issued contracts running for 99 years and penalizes any C loss during this period. Other forest projects have lifetimes of 30 to 40 years.

REFERENCES

Anderson, R. (1995) *Joint Implementation of climate change measures*, The World Bank, Washington, DC.

Andrasko, K., Carter, L. and van der Gaast, W. (1996) *Technical Issues in JI Project*, UNEP AIJ Conference, UNEP, Geneva.

Anonymous (1996) Cleaner buses and energy efficiency, *Joint Implementation Quarterly* 2: 11-12

Bedi, C. (1994) No-regrets under Joint Implementation?, in: Ghosh, P. and Puri, J. (Eds). *Joint Implementation of climate change commitments*, New Delhi.

Begg, K., Jackson T. and Parkinson S. (1999) *Accounting and Accreditation of Activities Implemented Jointly*, European Union, Brussels.

Begg, K., Parkinson, S., Jackson, T., Morthorst, P.-E. and Bailey, P. (1999) *Overall Issues for Accounting for the Emissions Reductions of JI Projects*, CDM Workshop on Baseline for CDM, NEDO, Tokyo.

Billharz, S. (1997) Lessons in AIJ in the Czech Republic, in: UNEP, BMU (eds.), *Proceedings of the International AIJ workshop*, Leipzig.

Boucher, A., Moore, B. and Legg, J. (1998) AIJ-Canandian experience, in: Riemer, P., Smith, A. and Thambimuthu, K. (Eds) *Greenhouse gas mitigation. Technologies for Activities Implemented Jointly*, Elsevier, Amsterdam.

Bundesministerium für Umwelt, Naturschutz und Reaktorsicherheit (1996) *Gemeinsam umgesetzte Aktivitäten zur globalen Klimavorsorge*, Bonn.

Canadian Joint Implementation Office (1996) *Canadian Joint Implementation initiative on climate change*, Ottawa.

Carter, L. (1997) Modalities for the operationalization of additionality, in: UNEP, BMU (Eds) *Proceedings of the International AIJ workshop*, Leipzig.

Center for Clean Air Policy; SEVEn (Eds) (1996) Joint Implementation projects in central & Eastern Europe, Prague.

Chatterjee, K. and Fecher, R. (1997) India´s expectations, opportunities and strategies, in: Chatterjee, K. (Ed), *Activities Implemented Jointly to mitigate climate change – developing country perspectives*, New Delhi.

Chomitz, K. (1999): Baselines for greenhouse gas reductions: problems, precedents, solutions, in: *CDM Workshop on Baseline for CDM*, NEDO, Tokyo.

Costa Rican Office on Joint Implementation (1998) *Activities Implemented Jointly National Programme (Costa Rica)*, Report to the UNFCCC Secretariat, San José.

Czech Republic (1997) *Second National Communication to the UNFCCC*, Prague

Dixon, R.K. (1998) The U.S. Initiative on Joint Implementation: An Asia-Pacific Perspective. *Asian Perspectives* 22:5-19.

Dixon, R.K. (1997) The U.S. Initiative on Joint Implementation. *Int. J. Environment and Pollution* 8:1-18.

Ekins, P. (1996) How large a carbon tax is justified by the secondary benefits of CO_2 abatement. *Resource and Energy Economics* 18: 161-187

Emmer, I., Verweij, H. and de Ligt, K. (1999) *Reforestation and flexible instruments – from project implementation to certification*, Arnhem.

FACE Foundation (1996) Annual *report 1995*, Arnhem,

FACE Foundation (1995) *Face foundation in practice*, Arnhem

Fritsche, U. (1994) The Problems of Monitoring and Verification of Joint Implementation. In: *Joint Implementation from a European NGO Perspective*, Climate Network Europe, Brussels.

Galon-Kozakiewicz, J. (1997) Lessons learned in AIJ projects in Poland, in: Proceedings *of the International AIJ workshop*, Leipzig.

Goldemberg, J. (1998) The *Clean Development Mechanism: Issues and Options*,
UNDP, New York.

Gorbitz, A. (1997) Cost Rica's Activities Implemented Jointly Programme, in: Chatterjee, K. (Ed*) Activities Implemented Jointly to mitigate climate change – developing country perspectives*, New Delhi.

International Energy Agency *(*1997) *Activities Implemented Jointly — Partnerships for Climate and Development*, Paris.

Jepma, C. (1997) Progress with AIJ during the pilot phase, in: Chatterjee, K. (Ed.) Activities Implemented Jointly to mitigate climate change – developing country perspectives, New Delhi.

Jepma, C.J. and van der Gaast, W. (1999) On *the Compatibility of Flexible Instruments*, Kluwer Academic Publishers, Dordrecht.

Joint Implementation Registration Centre (1996) *Registration and certification procedure*, The Hague.

Karani, P. (1997) Pilot project selection criteria for the World Bank, in: Chatterjee, K. (Ed) *Activities Implemented Jointly to mitigate climate change – developing country perspectives*, New Delhi.

Luhmann, H.-J., Ott, H., Beuermann, C., Fischedick, M., Hennicke, P., Bakker, L. (1997) *Joint Implementation - Projektsimulation und Organisation. Operationalisierung eines neuen Instruments der internationalen Klimapolitik*, Erich-Schmidt-Verlag, Berlin.

Michaelowa, A. (1998) Joint Implementation - the baseline issue. *Global Environmental Change* 8:81-92

Michaelowa, A., Schmidt, H. (1997) A dynamic crediting regime for Joint Implementation to foster innovation in the long term. *Mitigation and Adaptation Strategies for Global Change* 2: 45-56

Ministry of Housing, Spatial Planning and the Environment, Ministry of Economic Affairs, Ministry of Foreign Affairs (1998) *The Netherlands'Programme on Activities Implemented Jointly*, The Hague.

Nordic Council of Ministers (ed.) (1996) *Joint Implementation of commitments to mitigate climate change*, Copenhagen.

Norway (1997) *National programme on Activities Implemented Jointly under the pilot phase*, Oslo.

Poland (1997) *National programme on Activities Implemented Jointly under the pilot phase*, Warsaw.

Rentz, H. (1998): Joint Implementation and the question of additionality — a proposal for a pragmatic approach to identify possible Joint Implementation projects. *Energy Policy* 4: 275-279

SGS (1998*) Certification of the Protected Area Project in Costa Rica commissioned by OCIC*, Oxford.

Singh, G. (1993) The Sabah slip-in. Eco special report, *Geneva.*

Sweden (1997) *National programme on Activities Implemented Jointly under the pilot phase*, Stockholm.

Switzerland (1998) *Swiss AIJ pilot program*, Berne.

Switzerland (1997) *Swiss AIJ pilot program*, Berne.

UN Framework Convention on Climate Change (1998) *Activities Implemented Jointly under the Pilot Phase. Second* synthesis *report* FCCC/CP/1998/2, Buenos Aires.

UN Framework Convention on Climate Change (1997) *Activities Implemented Jointly under the Pilot Phase. Synthesis report* FCCC/SBSTA/1997/12/Add.1, Bonn.

United States Department of State (1994) Announcement of Groundrules for US Initiative on Joint Implementation, Washington, DC.

USEPA (1998) *Activities Implemented Jointly: Third Report to the Secretariat of the United Nations Framework Convention on Climate Change, Accomplishments and descriptions of projects accepted under the U.S. Initiative on Joint Implementation,* Washington, DC.

U.S. Initiative for Joint Implementation (1998) *USIJI Project Fact Sheet,* September 1998, Washington, DC.

U.S. Initiative for Joint Implementation (1996) *Activities Implemented Jointly: First report to the secretariat of the United Nations Framework Convention on Climate Change, submitted by the Government of the United States* July 1996, Washington, DC.

Chapter 4

THE AIJ PROJECT DEVELOPMENT COMMUNITY

A. MICHAELOWA[1], R. DIXON[2] AND L. ABRON[3]
[1]*Hamburg Center for Economic Research* [2]*Institute for Global Environmental Strategies*
[3]*PEER Consultants*

Key words: incentives, investors, non-government organizations, project developers incentives, private sector, regulation, subsidies, taxes

Abstract: A community of activities implemented jointly (AIJ) participants from governments, non-government organizations and the private sector was established during the 1990s in response to the UN Framework Convention on Climate Change (FCCC) and the Kyoto Protocol. Both policy factors (taxes, subsidies, regulation) and non-policy (expansion of trade and commerce, learning by doing, voluntary actions) are incentives for the establishment and growing maturity of this community. Perverse disincentives, inadequate financial resources, and a general lack of international experience with AIJ project development have slowed the growth of this community in some countries. Incentives and disincentives for AIJ project development in host (non-Annex I) countries and donor (Annex I) countries are sometimes comparable but often disparate. The human and institutional resources within the AIJ project development community provide a solid foundation for implementation of the Clean Development Mechanism (CDM), joint implementation (JI) and emissions trading.

1. THE AIJ COMMUNITY

With the advent of the AIJ pilot in 1995 the community of government officials, private sector entrepreneurs, non-government organizations (NGOs), investors, project brokers, project developers and project monitors in developed, transition and developing countries focusing on JI was small. Most actors were unfamiliar with the successful development of JI projects.

After almost five years of project development and implementation, and the associated institutional and human capacity building a robust and growing community has developed to support all aspects of the AIJ pilot and the newly borne Kyoto Protocol flexibility mechanisms. The AIJ community project development community varies in depth and breadth among developed and developing countries. In some countries the private sector has played a prominent role in the development of JI projects. In contrast, national governments and non-government organizations have been leaders in project development in other countries. The incentives for governments, the private sector and other project participants are not intuitively obvious and subtle factors often motivate participation and leadership (Dixon 1997). The objectives of this chapter include:

- review the role of government, non-government organizations and the private sector in the AIJ pilot;
- consider the incentives and incentive structure for successful participation in the AIJ pilot; and,
- I dentify some of the disincentives for participation in the AIJ pilot.

1.1 THE ROLE OF GOVERNMENTS IN THE AIJ COMMUNITY

The role of national governments and the private sector in the development and monitoring of joint implementation projects varies widely among the 70 countries participating in the FCCC AIJ pilot. At one extreme, the national government of an investor or recipient country is actively involved in the formulation, finance and implementation of a project portfolio. Countries which have adopted this approach include Costa Rica, the Netherlands and Japan (Dixon 1998). Even India, at one time very critical of the AIJ pilot, is adopting a more active role in project development. At the other extreme, national governments have played a limited role, formulating rules of procedure, establishing basic institutions, and monitoring project implementation. The USA and Russia have followed the latter laissez faire model (Abron and Guy 1998; Dixon, 1998). Most governments have struck a balance between these two extremes, supporting pilot projects by providing limited financial resources and facilitating JI projects. Based on the development of the Kyoto Protocol, many national governments have become active in the regulation, oversight and development of the AIJ pilot.

The Government of Japan has been very active in the development and implementation of a national program in support of the AIJ pilot, especially in the Asia-Pacific region (Watanabe 1997). A federal interagency group, including the Ministry of Foreign Affairs, Ministry of International Trade and Industry (MITI), and the Environment Agency (EA) formulated rules of procedure and established an office to evaluate and accept eligible JI projects. Bilateral working relationships were established with developing countries in the region, including China and India, to help facilitate the formulation, establishment and monitoring of AIJ projects. Recently, MITI, working through the New Energy and Industrial Technology Development Organization (NEDO) established a substantial fund (approximately $35 million) to evaluate the feasibility of AIJ project development. The Government of Japan, especially the EA, has provided training, technical assistance and finance for institutional and human capacity building in support of the AIJ pilot. Several Japanese government agencies have also provide technical and financial leadership in development of methodologies for development of baselines, project acceptance criteria, and monitoring, verification and reporting procedures (Anonymous 1999).

In the USA, the private sector and non-government organizations have played a leading role in the development of AIJ projects. The 30+ projects accepted by the U.S. Initiative on Joint Implementation (USIJI) have attracted over $\$500\text{x}10^6$ in private sector capital (U.S. IJI 1998). These projects, if fully implemented, will mitigate over $200\text{x}10^6$ Mg of CO_2. None of these projects have received financial support from the USA government. Most projects in the USIJI portfolio have 6-8 project partners drawn from the USA private sector and non-government organization community. Despite this early success of USIJI, some of the projects accepted into this portfolio lack full financial support. In contrast to the USIJI model of private sector finance, the Government of the Netherlands currently provides financial support for the development of AIJ projects, reducing risk and providing a profit motive in the near-term (Abron and Guy 1999).

1.2 EMERGING ROLE OF THE PRIVATE SECTOR AND NON GOVERNMENT ORGANIZATIONS IN THE AIJ COMMUNITY

The private sector, especially small- and medium-sized businesses, has been very active and successful in the development, implementation and monitoring of projects in the AIJ pilot. In contrast, many large multinational energy efficiency and renewable energy firms with a seemingly large potential for investment in AIJ projects have not yet participated because of the perceived perverse incentive structure. A number of small businesses have mushroomed since 1995 to provide technical, financial and logistical services in support of the AIJ pilot. For example in the USA, Trexler and Associates (TAA) has been very active in the development, brokering and monitoring of AIJ projects on behalf of various USA investors. Land-use and renewable energy sector projects have been developed by TAA in four developing countries. EcoSecurities, based in the U.K., the U.S. and Australia, provides services in project design, monitoring and verification. Most recently, EcoSecurities helped establish the $17 million forest sector C sequestration fund for Australian Plantation Timber. In South Africa, PEER Africa (Pty), Ltd., the International Institute for Energy Conservation (IIEC) and the University of Cape Town formed a team to establish the first AIJ project in the residential home sector. This AIJ project has been formally reviewed and accepted by the Government of South Africa and USIJI but suffers from the absence of adequate finance.

New or expanded organizations were established and designed by the private sector firms or trade associations to participate in the AIJ pilot. For example, the USA Edison Electric Institute established the International Utility Efficiency Partnerships (IUEP) to develop and implement joint implementation projects on behalf of investor owned electric utilities. The IUEP has been active in the brokering, finance and implementation of projects in Central and South America. Similarly, the International Federation of Automobile Clubs, based in Paris, France, established a fund and task force to stimulate development AIJ project activities in developing countries. The FACE (Forests Absorbing Carbon dioxide Emissions) foundation, with support form the Dutch Electricity Generating Board, was established in the Netherlands to assist with the development of forest sector C conservation and sequestration projects.

Non government organizations (NGOs) have played a prominent role in the development of the FCCC AIJ pilot, especially during initial phases. The World Business Council for Sustainable Development (WBCSD), The Nature Conservancy (TNC), The Environmental Defense Fund (EDF), The Center for Clean Air Policy (CCAP) and other organizations have all been active in policy analysis, as well as, the formulation, development, implementation, monitoring and evaluation of energy sector and land-use sector JI projects. The leadership of some NGOs in the FCCC AIJ pilot is especially remarkable given the early criticism of the pilot by some environmental organizations. During the first two years of the AIJ pilot, the WBCSD sponsored a large number of project development workshops on give continents bringing together governments and the private sector. TNC and CCAP have been active in project development and acting as catalysts for project finance and technology transfer. For example, the TNC in partnership with American Electric Power (AEP) helped establish one of the largest forest parks in the world in the country of Bolivia. The 800×10^3ha forest park will conserve existing C reservoirs and stimulate future CO_2 sequestration. In the energy efficiency field, IIEC is active in AIJ project development in the Philippines, South Africa and other countries but the lack of private sector involvement has limited the capacity of AIJ pilot projects to achieve their potential.

1.3 WHAT NON-POLICY FACTORS MOTIVATE THE AIJ PROJECT DEVELOPMENT COMMUNITY?

Members of the Annex I (donor) and non-Annex I (host) AIJ community is motivated to participate by complex, disparate, and often non-financial factors. Primary incentives for investor and host country participation were reviewed earlier and will not be repeated in this chapter. Many private sector firms and non-government organizations cite five prominent factors for participating in the AIJ pilot:

- improved public relations for their organizations by attacking a global environmental problem;
- learning by doing, participation in the AIJ pilot is the best way to understand and master the mechanics of this new environmental protection instrument;
- stakeholders or stockholders of firms or organizations want to expand their portfolio of environmentally friendly activities;
- desire to influence the UN FCCC negotiations by participating in the AIJ pilot; and,
- AIJ pilot is an opportunity to expand the trade and commerce of climate-friendly technologies (Dixon, 1998).

Near-term financial profit was not necessarily a major motivating factor for early private sector participation in the FCCC AIJ pilot. Many firms took a long-term view and are preparing and positioning themselves to participate in the implementation of the Kyoto Protocol flexibility mechanisms if the Protocol enters in to force.

Developing country project developers sometimes have a different perspective towards the FCCC AIJ pilot. Sustainable development, poverty eradication, and other socioeconomic factors are often more prominent in national planning and action agendas compared with local or global environmental protection. For example, at a 1999 Ministerial meeting of the Southern Africa Development and Cooperation (SADC) on climate-friendly technologies, poverty eradication emerged still ranked as a top-priority. Sustainable economic development was the primary motivation and interest in the AIJ pilot and the CDM. In a plenary statement, the SADC Energy Ministers were willing to support GHG emission reduction activities that lead to improved housing, reliable supply of potable water, and employment opportunities. The Ministers as tools to stimulate AIJ project development in the developing countries of Africa identified credits, subsidies, special funds, and demonstration projects.

2. POLICY INCENTIVES FOR SUCCESSFUL AIJ PROJECTS

From the incentive point of view, the FCCC AIJ pilot was created with a fatal flaw: it did not allow crediting of the emission reduction abroad to the national target of the investing country (Jepma and van der Gaast 1999). Without GHG emission reduction crediting, no country had a direct incentive to invest in AIJ projects. Nevertheless, governments would have been able to set incentives at a national level and some have done so. One reason could be that a successful AIJ pilot with many viable projects could facilitate a decision by the FCCC Parties to allow crediting after 1999. Only a portion of the incentives discussed below has been implemented to any extent in the AIJ pilot. The discussion of their merits and flaws might be helpful to design future incentives for the CDM.

2.1 CLIMATE POLICY INSTRUMENTS AND INCENTIVES FOR AIJ IN THE PRIVATE SECTOR

How can the private sector be induced to invest in AIJ? A necessary condition is the existence of climate policy instruments in their home country. In the case of GHG emissions, these instruments can take the form of taxes, subsidies, tradable emission permits or new or expanded regulations. The development of AIJ projects can be encouraged by implement of these policy instruments. The incentives are allowed under the FCCC, as there are no restrictions on national policies as long as there is no crediting of emission reduction to the national target. Any crediting referred to below only applies to a respective policy instrument and not to the national target. For example, trade policy problems could arise in the case of discriminatory subsidies, but environmental subsidies are allowed under the GATT (Article 8, (3) of the Agreement on Subsidies and Countervailing Measures).

2.1.1 GHG Emissions taxes

Crediting GHG emission reductions (ERs) from AIJ projects may increase the efficiency of a national emission tax. It is obvious that the implementation of a full-scale AIJ program lowers the tax rate necessary to reach a given target. Tax credits should be proportional to the emission reductions achieved by an AIJ project. These emission reductions would be calculated from a GHG baseline, multiplied by the relevant rates of taxation, and credited to the firm's tax bill. The tax credit could be granted annually for the entire duration or as a one-off allowance on conclusion of the project. The latter approach would have the advantage of reducing evaluation costs since only one evaluation would be required. For accounting reasons however, project participants might probably prefer an annual tax credit. In the case of reduction projects, tax credits should be granted for the commercial life of a manufacturing facility or equipment produced in the course of the AIJ project. A suitable approximate value is the legally permitted depreciation period. Similar procedures would have to be applied in the case of forest sector C sequestration projects, based on the duration of the project.

From a private sector perspective, demand for AIJ projects is based on the following calculation, which is initiated by the introduction of an emission tax on the firm in proportion to its GHG emissions and so providing an incentive for ERs. The firm compares the calculated costs of emission reduction within the enterprise with the calculated tax savings. If the latter are greater than the reduction costs, a profit-maximizing firm will carry out GHG ER activities. The question will be whether the ER should be undertaken within the firm itself or externally. Whether or not demand for AIJ project continues will thus depend on a comparison of reduction costs abroad plus transaction costs with the firm's own internal reduction costs. The firm can decide to reduce GHG emissions abroad only, or to carry out part of its ER activities abroad in addition to its own internal ERs. Even if the firm's internal reduction costs are calculated to be higher than the associated tax credits, implementation of AIJ projects to achieve ERs is still possible. The development of AIJ projects is beneficial if the tax credits granted to the domestic firm for ERs abroad are greater than the domestic firm's reduction costs plus transaction costs abroad. This is remarkable in as much as an emission taxes alone would fail to achieve any ERs at all. In other words, even a relatively modest emission tax could lead to AIJ project development. The emission reduction would then take place exclusively outside the home country.

A certificate issued by state-approved inspectors for GHG ERs achieved in a project over a certain period of time could be recognized by the tax office, which then grants the corresponding tax credit. If, for political reasons, the short-term introduction of a national tax is not possible, it is conceivable that tax credits can be accumulated now to be redeemed at a later stage when national tax legislation comes into force (BMW1992). This approach would only be efficient and widely used, if the future rates of taxation were specified as accurately as possible.

2.1.2 Subsidies

If an emission tax were not introduced, it would in principle also be possible to subsidize GHG ER projects at home and abroad in general and to raise the necessary financial resources by raising indirect or direct taxes. The subsidy would be proportional to the ER achieved through an AIJ project. Nevertheless, from a fiscal point of view, subsidies are likely to raise local or national budget deficits. It is probable that in comparison to a tax credit there will be more AIJ projects and a higher level of subsidies as firms, which do not pay taxes, are eligible for the subsidy will. In comparison to a tax, the subsidy per unit of ER would have to be lower than the prevailing tax rate to achieve the same amount of reduction. It would also be possible to combine tax credits with subsidies. Firms, whose GHG emission reduction via AIJ projects is higher than their domestic emission, would pay no emission tax and receive a subsidy for the additional ER. In any case, subsidies are often more transparent than the other incentives. Within developing and transition countries, subsidies are often preferred means to stimulate sustainable economic development (Abron and Guy 1998).

2.1.3 Other taxes

If no emission tax is introduced, granting credits on direct taxes could encourage GHG emission reductions at home and abroad. Given the failure to introduce a GHG tax, such an arrangement can nonetheless be acceptable as a second best solution if the level of concessions is calculated according to recognized climate policy criteria. Accordingly, an implicit rate of subsidy per unit of GHG offset could be chosen. However, these tax concessions would result in a loss of tax revenue, as every firm would have the right to credit reductions achieved through ER projects to their income tax debt. Firms carrying out AIJ projects will not be identical with those in the case of emission tax concessions without subsidies. A particular firm's tax burden, which marks the upper limit for reductions via AIJ projects, is not determined by the company's domestic emissions. Companies with high emission levels but low profits or losses would have no incentive to participate in AIJ activities, whereas the incentives for highly profitable firms with low emission levels would be very high.

2.1.4 Depreciation

In principle, concessions for AIJ projects could only be made via depreciation of goods if the AIJ project involved a capital investment abroad. However, since investments of this kind are depreciated according to the host country's tax legislation, there can be no domestic concession. It would be conceivable at most that domestic investments of the same value as the AIJ expenditures be depreciated at a faster rate. A particular AIJ project's investment volume or the object of investment involved must determine the form of depreciation chosen here. The diminishing-balance method of depreciation could be chosen in order to reduce the tax burden during the first years of an ongoing AIJ project. Other possibilities include raising depreciation rates or shortening the depreciation period. This form of accelerated depreciation enables complete depreciation before the commercial life of the investment comes to an end. Furthermore, with regard to alleviating market entry for small and medium-sized companies, increased support should be given for extraordinary depreciation or additional tax depreciation (e.g. by granting a certain rate of depreciation in anticipation of an investment). Direct tax reductions have a greater measurable incentive effect than depreciation relief. Tax credit, on the other hand, brings the firm a profit if the cost of the AIJ project is lower than the sum of tax relief. Tax credit or subsidies are sometimes preferable to relaxed depreciation regulations.

2.1.5 Tradable emission rights

Instead of taxation, a system of tradable emission rights could be introduced on a national level. Regardless of the initial allocation of emission permits, AIJ activities can be accounted for in this system. Proof of a GHG emission reduction achieved by means of an AIJ project would lead to an additional allocation of emission permits equivalent to the volume of the proven emission reduction. The incentive for a firm lies in being able to either use or sell the additional emission rights. Naturally, the market price of emission rights will be the lower, the higher the reduction achieved via AIJ.

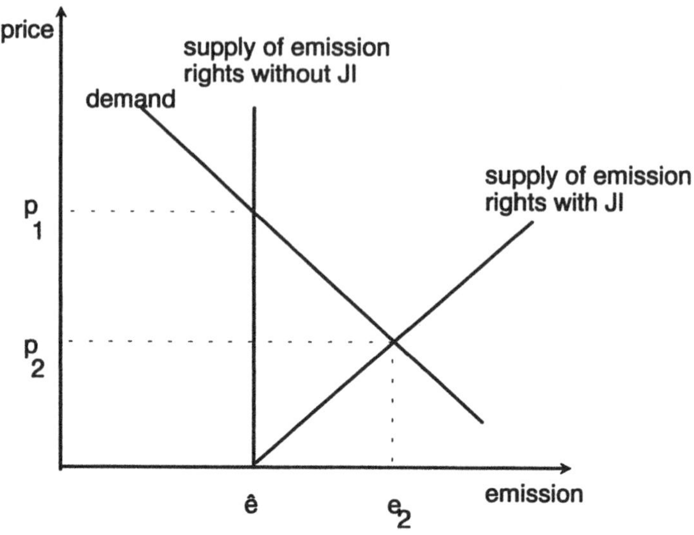

Figure 1. Potential reduction in the price of emission rights through AIJ activities. Without AIJ, the quantity of available emission rights is fixed at ê and the market price p_1 will prevail. The possibility of AIJ raises the quantity of available emission rights the more the higher the market price. As the quantity e_2 will be always higher than ê, p_2 will be lower than p_1.

2.1.6 Regulation

Despite much inefficiency, regulation is still the prevailing instrument of environmental policy as it can be administered easily. AIJ can be combined with two principal forms of regulatory measures, standards and externality adders:

- In conjunction with the GHG emission reductions achieved overseas, the authorities temporarily forego, either partially or entirely, the implementation of efficiency standards or emission limits, which normally apply to a particular plant. A trade-off between GHG and other harmful substances is prohibited. An outdated manufacturing facility with a low level of efficiency would not have to be shut down as long as the additional emissions caused by the outdated facility do not exceed the emission reductions achieved by means of the AIJ project; and,
- The proof of emission reductions achieved in an AIJ project neutralizes externality adders specified by a regulatory agency. Externality adders are applied only when new power stations are approved and thus influence the evaluation of different alternatives. The regulatory or supervisory authority approves only the most economical of the alternative power stations (New York State Energy Office 1994).

Externality adders of this sort, however, achieve only a long-term change in the structure of energy generation since they have no effect on the operations of existing manufacturing facilities, as they are never actually paid. Furthermore, in contrast to other instruments they only allow a one-time decision with regard to ERs. More favorable options for emission reductions, which arise at a later stage, are not then utilized. Externality adders could, of course, be applied to power stations' ongoing operations as well as to new investments. The supervisory authority could, for example, approve electricity prices only if they contain the externality adder. Implementation of AIJ activities could be as follows: an energy utility which intends to emit GHGs from a new power station applies for exemption from the externality adder for the volume of reduced emissions achieved by the AIJ project. If the AIJ project emission reduction is more economical than the alternative chosen when taking the externality adder into account, then a profit is made from the AIJ project.

Regulation other than externality adders is unlikely to give long term incentives for AIJ project development. If regulations are modified in step with technological change uncertainty will be created since it is impossible to tell how the previous regulatory concessions granted as a result of an AIJ project will continue in the future. Moreover, standards are not applied for all emitting appliances. As the efficiency of standards is lower than that of market-oriented or fiscal instruments, they should not be used as incentive for AIJ project development.

2.1.7 Voluntary investments and commitments

National voluntary programs and commitments have been popular first steps for many Annex I and non-Annex I countries to initiate AIJ project development. Voluntary incentives are often dismissed as weak relative to tax reform, subsidies or other regulatory measures. However, voluntary programs have been partially successful in some countries. For example:

- A successful AIJ project has a high public relations value, especially in countries with highly environment-conscious consumers. This value is likely to diminish, though, as AIJ activities become a normal business activity. Moreover, investments abroad are not as visible to the general public as conspicuous environmental projects at home; and,
- Through AIJ projects and activities firms can enter fast-growing markets and establish a their businesses.

Voluntary commitments of private companies have gained importance in a number of countries over the past few years. As a rule, voluntary commitments are established in contracts drawn up between state authorities and firms or trade associations. In comparison to regulations or taxes, voluntary commitments provide firms with a greater degree of flexibility. Firms, usually represented by a trade association, commit themselves, ideally, to achieving a reduction target. In return, the government abstains from introducing regulations or emission taxes. However, the contract does not establish any sanctions to be applied if the target is not reached. If for political reasons taxes or regulations cannot be implemented, there will be little incentive for firms to carry out any measures to increase energy efficiency, which go beyond autonomous efficiency improvements. In principle, however, the introduction of voluntary commitments is possible in connection with AIJ projects (TransAlta Corporation 1994). It could take the following structure: A branch of industry, represented by a trade association, is prepared to implement an increase in its energy efficiency, which has to be converted into a quantitative emission target. This target can be achieved by measures at domestically or abroad. The AIJ projects require no more than the approval of the foreign government and must be evaluated and verified. The can firm places relevant data and application materials for the evaluation at the disposal of its trade association with a minimum of risk.

2.2 INCENTIVES FOR HOSTS

The host country (non-Annex I) essentially is interested in the project's additional benefits such as capital, cost-free transfer of expertise, access to new technologies, local environmental improvements and other positive effects brought about by the GHG ER (Jepma and van der Gaast 1999). In countries where emissions have not previously been taxed and are not subject to regulation, as is the case in many developing countries, technologies in general and environmental technologies in particular, are not very advanced. Moreover, the technology gap between domestic companies and potential AIJ partners from industrialised countries is particularly wide. The additional benefits of AIJ project development for firms in developing countries will therefore be correspondingly great if the project entails technology transfer. However, firms offering AIJ project development opportunities will not only enjoy the positive externalities involved, but will also be faced with transaction costs. In view of the general requirements made of AIJ projects with regard to their evaluation and verification, transaction costs can be expected due to new inspection and accounting procedures. Furthermore, a firm that develops an AIJ project also enters into obligations and additional risks for years to come. Ideological aspects can also play a significant role, as the domestic political situation can be very hostile to AIJ project. The decision for or against offering a project will ultimately fall as a result of a commercial cost-benefit analysis. It can be assumed that the additional benefit of technology transfer will frequently exceed the associated transaction costs.

3. INCENTIVES FOR PROJECT DEVELOPERS IN THE PILOT PHASE

TABLE 1. Existing national incentives for AIJ project development in 1997 (X) and changes in 1998.

Country	Tax	Subsidy	Regulation	Voluntary agreement	Subsidy for project preparation
Australia	-	*	-	X	*
Canada	-	-	-	X	*
Germany	-	-	-	X	*
Japan	-	*	-	X	X
Netherlands	X	X	-	X	X
Norway	-	X	-	-	-
Sweden	-	X	-	-	-
Switzerland	-	X	-	-	-
USA	-	-	X (state level)	X	X

While many countries set up national AIJ programs, and many projects were proposed, only few have actually been implemented. This is due to a variety of reasons. Only a few countries provided real financial incentives (Table 1). Thus, the private sector restricted their activities to projects that had a high media interest, opened markets or were profitable on their own.

Before COP-3 there was no fully-fledged incentive system in place (Table 1). The most differentiated system was to be found in the Netherlands. While the USA, Australia, Canada, Japan, Germany, Sweden and Switzerland participate in the pilot phase, they apply no direct incentives besides covering some transaction costs. The USA case is especially interesting, as all large AIJ projects have been fostered by strong incentives on the state level, while the federal level has remained inactive. As the AIJ pilot program matures the incentive landscape is slowing beginning to change.

3.1 INCENTIVE STRUCTURES

Table 2. Externality adders in the USA energy production and distribution system ($ Mg) (New York State Energy Office, 1994).

State	Authority	CO_2	CH_4	N_2O	NO_x	Prices
California	Public Utility Commission	9	0	0	7,495-31,448	1992
	Energy Commission	7.6	0	0	730-16,076	
Massachusetts		22	220	3,960	6,500	1989
Nevada		22	220	4,140	6,800	1990
New York		8.6	0	n.a.	6,524	1992
Oregon	Bonneville Power	6	0	0	69-884	n.a.
Wisconsin		17	165	2,977	0	1992

The Netherlands grant about $43x10^6 in direct subsidies for the period 1996-1999, of which $25x10^6 are allocated to projects in developing countries and $18 x10^6 to countries in transition (Vos 1997). The subsidies have recently been further enhanced and may reach up to $400x10^6 by the year 2002. Private sector firms are also allowed to flexibly depreciate equipment used for AIJ projects. Income from AIJ projects is tax-exempt in the Netherlands.

Norway is currently subsidising four AIJ projects with $7.1x10^6 out of its climate fund and co-finances the World Bank AIJ program. The governments of Japan, Switzerland and Sweden also offer subsidies to evaluate and establish AIJ projects (SWAPP 1997). None of these subsidies are directed towards the private sector, and there is no tendering.

Many national AIJ pilot programs cover transaction costs such as spreading information, project development, negotiation etc. This is the prevailing kind of incentive at the moment. For example, Japan finances feasibility studies and project preparation with up to $200,000 grants or contracts for each project. USIJI offers a grant competition (up to $25-50,000) for each project proposal (DOE 1997). Transaction cost subsidies often cannot be quantified but in the case of a large AIJ Office of Secretariat, such as USIJI, this technical support can be substantial. Donor (Annex I) does not only grant such indirect subsidies but also by host countries. For example, the Costa Rican AIJ Office has a staff of seven professionals (LeBlanc 1997).

It is likely that the Scandinavian states, which apply C taxes to the private sector, will allow AIJ-related tax offsets once they have fully defined their participation in the pilot phase. Norway, where 60% of all CO_2 emissions is subject to C taxation ranging from $16 to 51 Mg C, should be the first to introduce tax incentives, especially as it has been a major driving force in the AIJ debate. Nevertheless, tax concessions are not yet feasible as the Norway

Ministry of Finance is strongly opposing them (Lunde 1997). The Swedish and Finnish private sector could have incentives around \$36 Mg C which would imply a huge demand for AIJ. So far, no plans for tax concessions have surfaced. In the Netherlands, income tax exemption for income from AIJ projects has been granted. This will be a very powerful incentive for no-regret projects. Astonishingly, crediting against the energy tax is not envisaged.

In several USA states, externality adders are a powerful incentive for electric utilities to invest in C offsets including the AIJ pilot. Over 13 states have established explicit values for the evaluation of externalities, some of which vary quite considerably (Table 2). In the majority of USA states, flat-rate externality adders expressed in percentage terms or in \$/kWh are used when approving new power stations. Five states have quantified externalities for individual substances, although not all GHGs are covered.

Externality adders have led to the biggest private AIJ projects to date with an aggregate investment of more than $\$6x10^6$. The first utility, which invested in AIJ, did it because it feared an externality adder long before the term AIJ was coined. Then, the AES Corporation invested $\$2\ x10^6$ in afforestation in Guatemala to offset its emissions of a 183 MW coal-fired power plant in Connecticut, USA. Later, AES invested the same amount in Paraguay. Tenaska, another USA firm, invested $\$1x10^6$ in the Ecoland project, located in Costa Rica (Trexler and Associates 1994, LeBlanc 1997).

The first GHG offset requirement was introduced in New Zealand in 1995 where emission from a 400 MW gas fired plant shall be compensated through (domestic) afforestation (Brasell 1996). In the USA, the state of Oregon had a fierce competition for three GHG offset portfolios in 1996. The winning proposal included an AIJ project component, a rural photovoltaic project in Sri Lanka that was accepted by USIJI (Trexler and Associates 1996). The USA State of Oregon recently put GHG offset legislation into place. In a similar policy development, the Ohio Public Utilities Commission allowed Cincinnati Gas & Electric to defer costs of their commitment to the USIJI Rio Bravo project (PUCO News 1997).

The current deregulation of the global and USA energy and electricity markets will reduce regulation of the type described in previous sections. As deregulation is also completed in other parts of the world, the possibility to use its potential for AIJ project development inventives will change. However, other incentives associated with the restructuring of global energy markets may stimulate AIJ project development.

The Netherlands registers GHG emission reduction from AIJ projects and certifies their validity (Vos 1997). If international C crediting is allowed in the future, these certified offsets could become more valuable. Thus far, no international certification system has emerged but some are being considered. When voluntary agreements with the private sector are renegotiated in 2000, AIJ projects could be included. Costa Rica has already started to certify emission reductions from three large, sector-wide AIJ projects to attract different investors, who could be interested in creditable tradable offsets (CTOs) (LeBlanc 1997). This concept of certification permits different investors to diversify or reduce their risk and increase the opportunity for a financial return. In Germany, the Federation of German Industry includes AIJ project activities in its voluntary agreements but has not yet clarified GHG emission reduction certification and baseline issues. The operating principle also applies to voluntary private sector commitments in Australia, Canada, the Netherlands and the USA.

In addition to weak incentives and some disincentives, transaction costs were very high in the beginning of the AIJ pilot. Moreover, many attractive host countries did not approve projects because of their objections to the AIJ pilot or lack of human, material and financial resources. With the advent of the Kyoto Protocol and the potential of the CDM, a growing number of host (non-Annex I) countries are beginning to participate in the AIJ pilot.

REFERENCES

Anonymous (1999) CDM *Workshop on Baseline for CDM,* NEDO, Tokyo.

Brasell, R. (1996) New Zealand's net carbon dioxide emission stabilisation target. *Agenda* 3: 329-340.

Bundesministerium für Wirtschaft, Bundesministerium für Finanzen. (1992) *Modellüberlegungen zur Ausgestaltung von Kompensationen im Zusammenhang mit einer CO_2-/Energiebesteuerung*, Bonn.

Dixon, R.K. (1998) The U.S. Initiative on Joint Implementation: An Asia-Pacific perspective. *Asian Perspective* 22:5-19.

Dixon, R.K. (1997) The U.S. Initiative on Joint Implementation. *International Journal of Environment and Pollution* 8:1-18.

LeBlanc, A. (1997) *An emerging host country AIJ regime: the case of Costa Rica*, New York.

Lunde, L. (1997) *AIJ: a case study of Norway*, ECON Report 16/97, Oslo.

New York State Energy Office (1994) *Draft New York State Energy Plan*, New York.

PUCO News (1997) *PUCO approves CG&E financing of global environmental project*, Columbus.

Swiss AIJ Pilot Program (1997) *Program overview*, Berne.

Trexler and Associates (1996) *Carbon offsets as environmental mitigation: a first regulatory case study*, Portland.

TransAlta Corporation (1994) *Pursuing greenhouse gas offsets*, mimeo, Calgary.

Trexler and Associates (1994) *Carbon offset project profiles*, Portland.

United States Department of Energy (1997) *DOE project solicitation for USIJI projects*, Washington, DC.

Vos, H. (1997) *Business, as usual?* Report for ECON, Amsterdam

Watanabe, M. (1998) Global warming mitigation through international cooperation. *Pacific Business and Industries* 40:28-44.

Chapter 5

AIJ PILOT PHASE CREDITING AND CREDIT SHARING

A. MICHAELOWA[1], K. BEGG[2], M. DUTSCHKE[1], N. MATSUO[3] AND S. PARKINSON[2]
[1]Hamburg Institute for Economic Research; [2]Centre for Environmental Strategy, University of Surrey; [3]Institute for Global Environmental Strategies

Key words: certified emission reductions discounting, emission reduction units, externalities, host countries, investors, quotas

Abstract: The activities implemented jointly (AIJ) pilot phase does not include provisions for crediting of greenhouse gas (GHG) emission reductions. The absence of crediting is a major deficiency of the AIJ pilot and has contributed to the relatively small number of projects developed. In contrast, the provisions of the Kyoto Protocol permit crediting for joint implementation (JI) or Clean Development Mechanism (CDM) projects. The issue of which entities will receive emission reduction (ER) credits is a matter of considerable analysis and negotiation. Countries that could potentially participate in the CDM may only be interested in credits if they can sell or bank them for future use. Credit sharing may raise the cost per unit of GHG reduced for investors and act like a tax on project development. Credits developed via joint implementation projects may also be traded or banked based on the mutual needs of the investor or host country entities.

1. CREDITS AND CREDITING IN THE UN FCCC CLIMATE NEGOTIATIONS

1.1 FROM JI WITH CREDIT TO AIJ WITHOUT

The original concept of JI that was launched by Norway and Germany in 1991 envisaged full crediting of emission reduction abroad (Carraro 1999;

Hanisch 1991). At that point in time this concept did not encounter much political resistance. At 1992 U.N. Conference on Environment and Development (UNCED) JI was included in the UN Framework Convention Climate Change (FCCC) as Article 4.2a but not completely defined. There were no rules or administrative precedent for crediting of GHG or carbon (C) ERs. In 1993, discord regarding the meaning and application of JI with crediting arose at the 8th session of the Intergovernmental Negotiating Committee (INC), as the developing countries rejected the concept and some OECD countries expressed reservation (INC 1993). Many countries, (e.g. France, Netherlands and Germany) called for ceilings in the share of GHG emission reduction targets to be achieved via JI, i.e. minimum quotas for domestic ER. At the UN FCCC First Conference of the Parties (COP-1) in 1995 the rules and modalities of the activities implemented joint (AIJ) pilot were promulgated. Despite opposition from the developing world, COP-1, Decision 5 established a pilot phase for AIJ projects (UNFCCC 1995). This outcome was facilitated by assistance from some Latin American countries, notably Costa Rica, that had already started experimenting with such projects. The GHG emission reductions achieved by AIJ projects cannot be credited towards the national target of the investing country. The crediting issue will be reconsidered at COP-5 and COP-6 using experience from the AIJ pilot (UNFCCC 1995).

1.2 CREDITING UNDER THE KYOTO PROTOCOL FLEXIBLE INSTRUMENTS

While the AIJ pilot phase started slowly because of weak financial and crediting incentives, the USA and other OECD made clear that they would accept legally binding emission targets only if flexibility would be allowed through emissions trading and JI (Jepma and van der Gaast 1999). While some developing countries rejected both notion of emission reduction flexibility, the Kyoto Protocol developed at COP-3 retained the option of JI with crediting involving developing countries. This outcome was partly due to insistence of Costa Rica that managed to convince skeptical members of the group of 77 (G-77) to work with the OECD countries and table a proposal for multilateral JI. To appease the opponents of JI once again the term was changed to Clean Development Mechanism (CDM) as a former, but structurally fully different proposal by Brazil had been called Clean Development Fund (CDF). Because of the lack of time in the final days of the Kyoto Protocol negotiations many crucial points were left undecided. These issues were discussed without agreement at COP-4 in Buenos Aires, Argentina, in November 1998. COP-4 did produce a work plan that prioritizes outstanding CDM issues and sets a goal that decisions shall be finalized by 2000.

1.2.1 Joint Implementation

Article 6 of the Kyoto Protocol allows Annex I countries to acquire emission reduction units (ERUs) through investment in projects in other Annex I countries. ERUs created in that way are to be considered equal to ERUs from emission trade under Article 17. ERUs are not created if annual reporting requirements have not been met or the reports do not comply with the binding rules Article 6. If a review team has doubts about additionality of emission permits they shall be tradeable but are frozen until the doubts are resolved (Article 6).

1.2.2 Clean Development Mechanism(CDM)

Under Article 12, which defines the CDM, investing countries may get credit for certified emission reductions (CERs) from CDM projects provided benefits accrue to the host country (Article 12). Credits accrue from 2000 in contrast to ERUs from JI between Annex I countries and emissions trading that accrue only in the commitment period 2008-2012. Crediting shall be only allowed until a certain percentage of the emission target is reached those remains to be defined. It is unclear whether crediting up to this quota is full or only partial. Besides countries, private sector firms are also allowed to invest and execute projects.

The CDM shall cover its administrative budget through project revenues. Moreover, a part of these revenues shall be used to assist developing country Parties that are particularly vulnerable to the adverse effects of climate change to meet the costs of adaptation. This provision was included in the Kyoto Protocol at the request of the Association of Small Island States (AOSIS). Some observers suggest this concept amounts to an adaptation tax on CDM projects.

CDM rules described above can be interpreted to allow the following creation, allocation and distribution of credits, assuming credits are tradable:

- allocating credits only after discounting ;
- allocating credits only after a share has been deducted and sold to finance adaptation projects and administrative expenses (Article. 12.8);
- allocated credits accrue to the investing country in full: if the interpretation of benefits accruing to the host country (Article. 12.3a) means investment capital and project externalities only;
- allocated credits are shared between investing and host country: benefits mean a share of the credits;

- an interesting, but daunting option would be to set the share of the investing country at zero if the host country finances the project on its own; and,
- allocated credits accrue to the investing country only until a quota is reached.

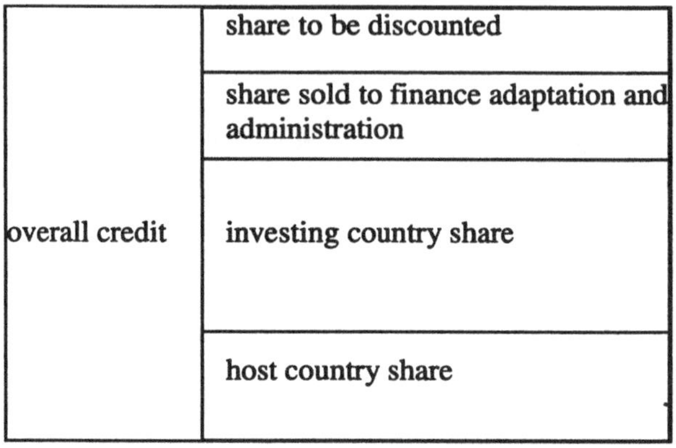

overall credit	share to be discounted
	share sold to finance adaptation and administration
	investing country share
	host country share

Figure 1. Diagram of possible options for CDM credit allocation and distribution system.

2. POSSIBLE DISTRIBUTIONS OF CREDITS

To date, the issue of credit distribution has not been discussed intensively compared to the issues of GHG baseline determination and verification of JI ERs. Nevertheless, crediting has been considered from the beginning of the negotiations. Several academic reports dealing with the issue of JI and crediting have been published (Jackson and Begg 1999; Jepma and van der Gaast 1999; Vellinga et al 1992) discussed the issue of credits in quite high detail. It called for a discounting of credits because of baseline uncertainties. Moreover, it discussed sharing credits and concluded that only countries with targets would be interested in a share assuming non-tradability of credits.

In the discussion of INC 8 design of crediting was discussed to some extent but the main issue was whether to credit at all. Some non-government organizations (NGOs) proposed that companies might only receive credits after subtracting debits from additional emitting activities abroad (NRDC 1993). Australia called for free agreement on credit sharing between investor and host country (INC 1993).

2.1 CREDITS ACCRUING TO THE INVESTING COUNTRY

JI was generally understood to involve the following type of transaction: a government or a private entity of a country with an emissions target finances a project in another, the host country. Credits from the ER accrue only to the investor. The positive externalities from the project are deemed sufficient as incentive for the host country. If credits are fully tradable domestically and internationally, they should accrue to the entity investing, even if it is a private company. If there is no domestic trading system, the credits should accrue to the government which, in turn should compensate the investor through emission tax reductions or reductions in regulatory requirements (Michaelowa 1996).

2.2 CREDITS ACCRUSING TO THE HOST COUNTRY

The host country will be interested in credits when one or more of the following scenarios apply:

- it is subject to an emissions target (as in the case of JI);
- it does not have an emissions target now but wants to bank credits for future commitments; and,
- credits can be traded on a market.

Allocating all credits to the host country would make no sense for a rational JI/CDM investor. Of course, a host country then could finance the project on its own and sell credits earned. No rule of Article 6 or 12 would prevent this. Costa Rica has already pioneered this kind of trade by financing umbrella forestry and energy projects through a fuel tax and trying to sell certified tradable offsets (CTOs).

General participation of CDM host countries in creating and trading credits would certainly lower the price of credits and alarm Parties in some quarters, especially if credits could be traded from 2000 onwards. As host countries have no GHG emission reduction targets they have an incentive to maximize credit sales. In this scenario the baseline issue becomes crucial and disincentives for developing countries to promote high GHG emissions deserves attention. This scenario is amplified by a perverse effect of the additionality rule: emission reduction measures are cheapest where there is a lack of a national sustainability policy (Michaelowa and Dutschke 1997). Application of the CDM would have to be conservative regarding baseline verification and other criteria under this perverse scenario.

These issues could be fully addressed by setting an incentive for developing countries to adopt limitation targets and participate in emissions trading and JI under Articles 17 and 6 (Carraro 1999). Such an incentive could be to prohibit the trading of host country credits now but to allow them to bank credits against future targets. One could also envisage a quota for credit trades for each country and banking for additional credits created. The national emission targets should be derived from national baselines developed using common rules and procedures. Any improvement in environmental legislation will then be beneficial for future compliance. As in the case of the investing country, credits could either accrue to the entity involved in the project or the government (Jackson and Begg 1999). The former would be only relevant if credits could be traded freely. The decision on that issue could have important credit distribution consequences.

2.3 CREDIT SHARING ARRANGEMENTS

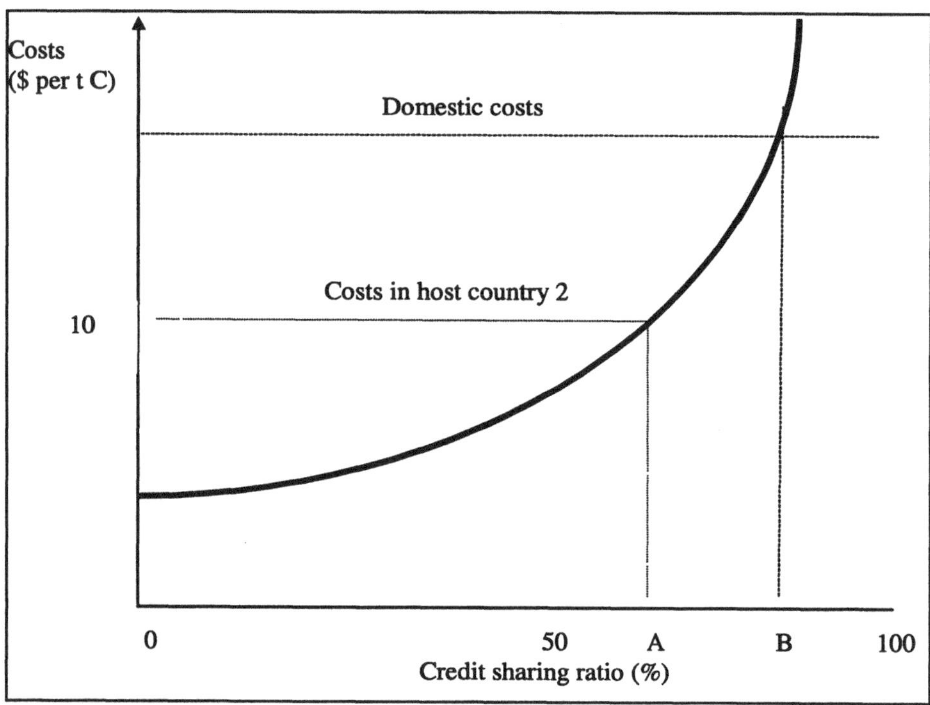

Figure 2. Possible influence of credit sharing ratio on country or firm decisions to invest in a JI or CDM project.

Article 12 of the Kyoto Protocol could be understood to provide a basis for credit sharing and it seems that some OECD countries understood it in this sense during the negotiations (Schipulle 1998). Obviously, the investor is interested in minimizing his cost per unit of credit. Any form of credit sharing raises these costs. Thus, the investor will look for other hosts where net costs after credit sharing are lower. Investors will no longer be interested in any CDM or JI project if credit sharing leads to a cost per credit that is higher than costs of domestic reduction (Figure 2).

For example, assume the project costs are $10 Mg C in host country 1, $3 Mg in host country 2 and $50 in the investor country. The bold line shows costs per credit depending on the credit-sharing ratio. If the credit-sharing ratio surpasses A, the investor will choose host country 2 and if it surpasses B the investor will choose domestic options. If one aggregates the decisions of individual investors, credit sharing will be similar to a tax on JI projects.

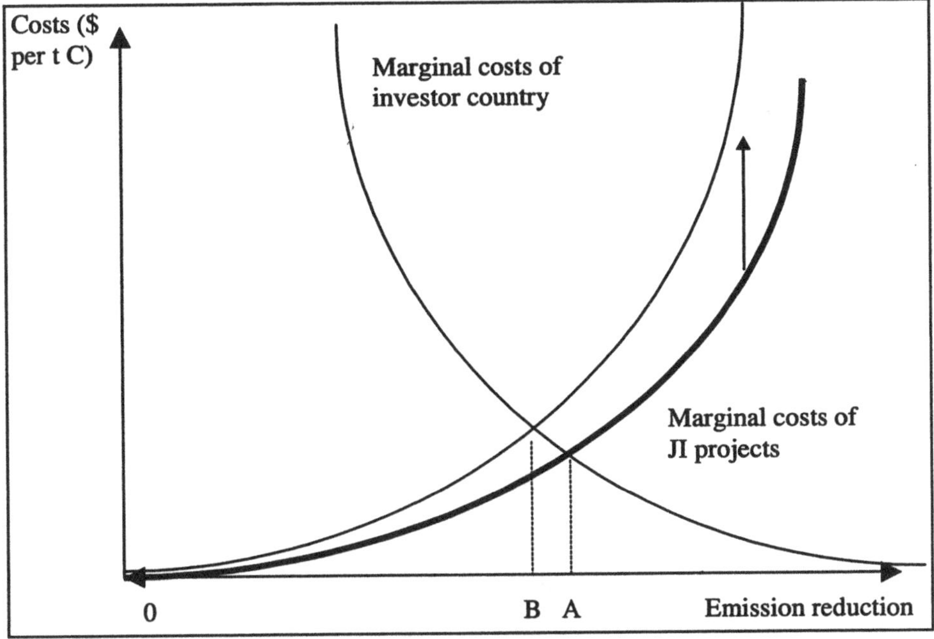

Figure 3. Possible effect of credit sharing on regional or global overall for JI projects.

Under the scenario presented in Figure 3, credit sharing raises the marginal costs of JI projects and thus reduces the amount of implemented JI projects from A to B. Thus, the credit sharing ratio is a policy variable for the host country that needs to be chosen carefully. Competition with other host countries with lower credit sharing ratios would arise. The gains of selling or banking credits have to be balanced against the loss in demand for JI projects. This applies even if the COP prescribes a uniform credit-sharing ratio as it could be circumvented by side payments. Another argument against a uniform credit sharing ratio in the case of credit banking only would be that developing countries that are unlikely to adopt targets would not be able to make any use of their banked credits.

3. INFLUENCE OF EXTERNALITIES AND UNCERTAINTIES ON CREDITING

Crediting of CDM projects should take external effects into consideration (Jepma and van der Gaast 1999). The focus should be on significant positive externalities which are unconnected with climate system protection. It is very difficult to quantify these externalities. Most of them are interlinked and operate on different time scales. Feedback depends on the local situation.

While it is obvious that JI will lead to capital and foreign currency transfers the net effect on jobs is unclear (Dixon et al, 1997). The transfer of modern technology could well lead to a loss of jobs, at least locally and in the short and medium term. Formation of human capital is a long-term effect and dependent on the social and political framework.

Reduction of local pollutants will be a relevant externality particularly for densely populated, rapidly industrializing countries of Asia (Dixon 1998). Biological diversity will only be protected if the social and political framework is conducive to forest protection and prevents relocation of damaging activities. Thus, only countries with a strong administrative infrastructure and human capacity will be able to take advantage JI projects that protect biodiversity. Costa Rica is an example for such a trend as it focuses on expansion of national parks through JI funds (Michaelowa and Dutschke 1997).

It is likely that the capital and technology transfer will be decisive for those host countries where official development assistance (ODA) is declining. Countries with high private capital flows will try to use JI funds to maximize positive environmental and social externalities (Dixon, 1998).

The most critical negative externality of JI could be that it reduces incentives for innovation (Michaelowa and Schmidt 1997). Other negative externalities could include displacement of people and loss of arable land in the case of large-scale hydro and afforestation projects. Many negative externalities are linked to poor natural resource management and an unstable political situation.

3.1 GENERAL DISCOUNTING OF CREDITS

Often, general discounting of credits by a certain percentage is proposed to cover uncertainties of baseline determination, enhanced project risks etc.

(Parkinson et al. 1998). Tattenbach (1997) asserts a 50% discounting of the credits to achieve maximum global GHG emission reductions. Discounting has the same effects on project demand as credit sharing but does not lead to enhanced revenues for the host country (Figure 4).

General discounting does not take into account differences between uncertainties and risks of projects and countries. Therefore, it is an inefficient mechanism and should not be introduced. A compulsory insurance of project risks would be a more efficient mechanism to lower risks. An approach trying to set project category specific discount factors to cover issues that cannot be solved through insurance is outlined below.

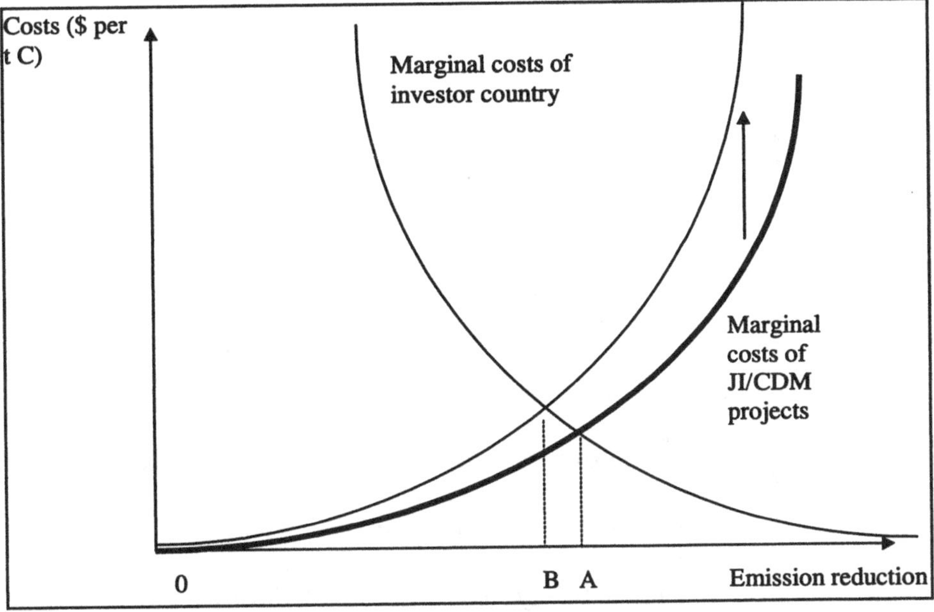

Figure 4. Possible effects of general discounting on decisions to invest in JI projects.

Discounting raises the marginal costs of JI projects and thus reduces the amount of implemented JI/CDM projects from A to B.

3.2 DISCOUNTING OF CREDITS OVER TIME

Concerning innovation on the one hand there has to be incentives for induced innovation to reach long-term efficiency gains (Dixon 1998). In contrast, short-term efficiency gains through JI have to be allowed. A strategic climate protection policy could entail a gliding reduction of exploitable short-term efficiency gains while raising an emission tax in the long run. This could be achieved through a gliding reduction of crediting of

JI (Figure 5). In the same period, either domestic carbon taxes are raised with a steadily rising tax rate in the industrialized countries or a system of tradable permits with a steadily sinking supply is introduced. This policy has the following advantages:

- investors receive long-term planning data; and,
- investors can get full crediting for JI reduction in the beginning, which allows them to invest into long-term emission reduction strategies.

The incentive to reduce domestic emission grows steadily as crediting falls while the emission tax rises/ the permit supply falls.

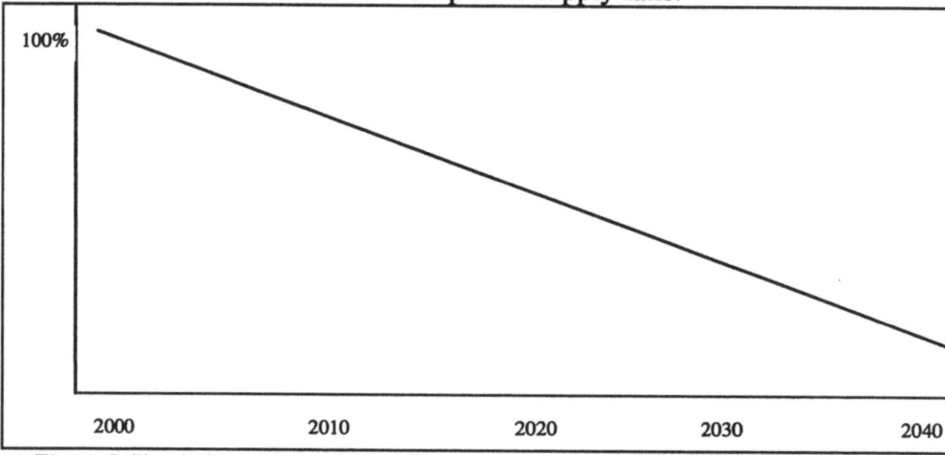

Figure 5. Simulation of decreasing crediting ratio in AIJ and JI project implementation.

An indirect discounting over time is due to the date of accrual of credits (Varming et al. 1998). If a project is credited for its whole lifetime in advance, of course, the implicit value of the credits is much higher than if it is only credited after monitoring and final verification. In the former case, credits could be sold immediately and the proceeds invested at the market interest rate. Such up-front crediting would not be advisable from a monitoring and enforcement point of view. An intermediate solution would be annual issue of credits on the base of annual verification.

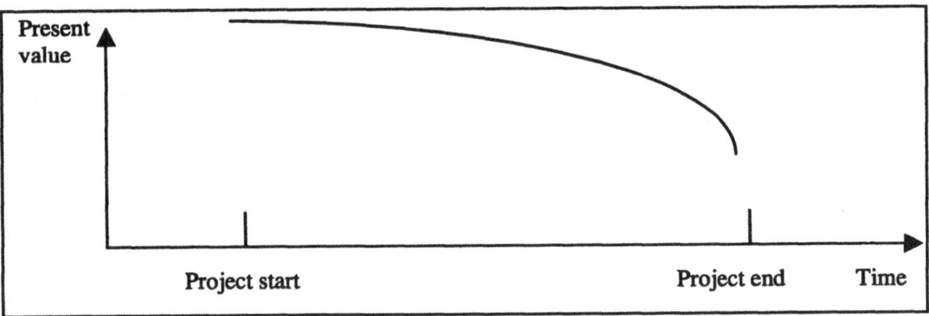

Figure 6. Possible effects of timing of AIJ/JI project credit accrual.

3.3 DIFFERENTIATING PROJECT CATEGORIES

It is probable that many AIJ or JI projects will have a mixture of positive and negative externalities. The question how to weight them will be crucial for the success of these projects. An exact quantification is impossible and the situation is different for each project. Because of high transaction costs, it is not advisable to calculate externalities for each project. Nevertheless, certain project types are more likely to entail positive externalities than other ones. For example, fossil power plants will create less employment than demand side management (DSM) programs. Application of renewable energy will result in no emission of local pollutants compared to fossil fuel substitution. Forest protection projects and afforestation may not advance technology transfer in some countries. The former are likely to entail biodiversity protection while the latter are not. Large-scale projects may be more likely to disrupt local life and displace people than small-scale ones. The following preliminary general conclusions can be used to categorize projects and differentiate crediting, provided insurance is compulsory:

- DSM and production of renewable energy and forest protection can be credited fully;
- large-scale projects such as new fossil power plants are only credited partially; and,
- afforestation should be credited at a low rate as it rarely entails technology transfer and leads to land use constraints.

3.4 EARLY PROJECT CREDITING

While the CDM crediting shall already start from 2000 onwards, JI ERUs only accrue from 2008. This has led to the demand that the start of JI crediting should be brought in line with the CDM start date. Moreover, several countries plan to introduce legislation that rewards private emission reduction activities before 2008 by carrying forward a part of the emissions budget allocated through the Kyoto targets. In the USA, proposed legislation was brought before the Congress in late 1998 to provide such early action crediting (US Congress 1998). While this legislation only provides to credit projects abroad that will result in an addition to the USA quantified emission limitation for the first compliance period, only CDM and not pre-2008 JI, it provides credit for pilot projects under the AIJ pilot phase. Similarly, at COP-4, Switzerland circulated a proposal calling for pre-2008 JI crediting, as this would reduce the amount of hot air in the countries in transition.

3.5 QUOTAS FOR DOMESTIC EMISSION REDUCTION (ER)

Many representatives of NGOs, governments and academia fear that the use of CDM and JI leads to a neglect of domestic abatement. They want to limit the possibility of a country to reach its domestic target by the use of these instruments. Thus, in the European Union (EU) context, Austria, Denmark, France and Germany argued for a buyers quota of 50% (EU 1998).

Nevertheless, the goals of quota proponents differ widely. Originally, the demand for quotas came from developing countries which felt industrialized nations might act inappropriately by buying their way out of their obligations under an eventual climate protection treaty. Furthermore, they feared that unrestricted JI could lead to long-term disadvantages as the low-cost abatement options would be used up and only high-cost options remained to fulfil future emission targets. Moreover innovation would be reduced as costs of emission reduction are lowered and thus research activities are reduced. JI and CDM would thus only achieve static efficiency gains but be dynamically inefficient. In this sense, quotas are discussed to guarantee a minimum amount of domestic reduction and innovation.

3.5.1 Buyers quotas

TABLE 1. Different definitions of a 50% quota in % of emission in the base year for selected Annex B countries.

Country	Rule a)	Rule b)	Rule c)
Australia	0	2.81	2.6
EU	4	2.39	2.6
Iceland	0	2.86	2.6
Japan, Canada	3	2.44	2.6
New Zealand	0	2.6	2.6
Norway	0	2.63	2.6
U.S.	3.5	2.42	2.6

A buyer's quota leads to a rise in average abatement cost and a reduction of demand for JI and CDM. There are different possibilities to implement a quota and the proposals discussed so far have been very unclear as to their exact implementation, especially for countries with stabilisation or growth targets. The quota could be defined as follows:

- it is proportional to the ER target and excludes countries with stabilisation or growth target (Rule a);
- it amounts to x% of the respective emission budget (Rule b);
- it amounts to x% of the average target of Annex B countries (5.2%) (Rule c); and,
- every imported emission permit will be discounted by x%; using the example of a 50% quota Table 1 shows the quotas allocated to the different countries.

It would have to be decided whether these quotas apply to the gross or net acquisition of permits as it is likely that countries will be both buyers and sellers if they have allocated permits to private companies. If governments are the only players in the permit market, the government could eliminate this problem by only buying permits until the quota is filled. Another decision would be whether the quota applies to all flexible instruments together or whether specific quotas apply for each instrument.

3.5.2 Allocation of quotas to private entities

If a quota is introduced, its allocation to private entities has to be as efficient as possible. Several procedures concerning the activities of private entities could be chosen:

- first come, first serve; (Companies can buy permits until the quota is filled. Afterwards, permits can still be bought but only be used in the next commitment period. CDM projects would be advantaged as credits already accrue from 2000. Concerning the project-related mechanisms the following problems could arise: Credits from a JI/CDM project accrue only after the quota is filled (Dudek and Wiener 1996). All credits from JI/CDM projects of one investing country could loose their value if the quota had already been filled through emissions trading. This procedure would disadvantage long-term CDM and JI projects and projects with long gestation periods);

- discounting proportional to the demand surplus (Companies can buy permits until the end of the commitment period. Then the government calculates the aggregate amount of acquired permits. If it surpasses the quota, every permit is discounted according to the demand surplus, e.g. if the quota is 1 Mt of carbon and the permits bought amount to 2 Mt, each permit is discounted by 50%. This method would lead to a high risk concerning the real price of the acquired permits as it is only known after the commitment period and depends on decisions of other companies. Thus risk-averse companies will not invest in permits. Allowing banking part of permits for the next commitment period instead of discounting could reduce the uncertainty. These banked permits would get preference in filling the next quota. Projects with long duration would thus be penalized less.); and,
- discretionary allocation of the quota according to criteria such as positive externalities, degree of innovation of the projects, diversification of sources of permits etc. Transaction costs, in transparency and uncertainties would be high.

All these allocation modes would disadvantage the project-related CDM and AIJ mechanisms. Setting a soft quota, which slowly discounts the C credits achieved beyond a certain percentage of the domestic target (Figure 7) could attenuate them. Any credit beyond another, higher percentage will still be accounted for at a minimum rate. Obviously, this mechanism would not fix the quota at an exact value. But domestic reduction would be promoted while the global reduction would be enhanced.

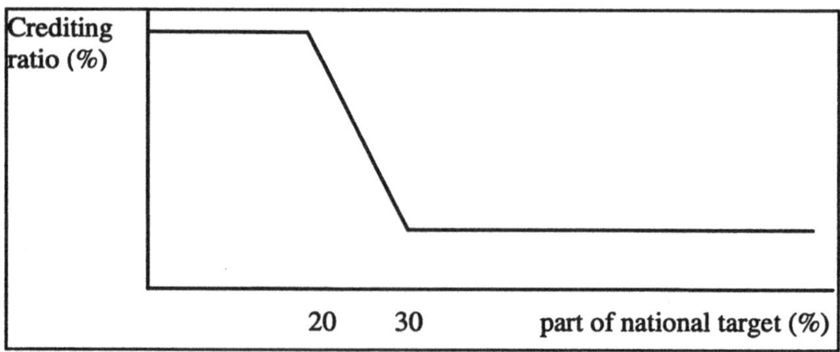

Figure 7. Impact of soft quota on C credits.

3.5.3 Sellers quotas

Seller's quotas are easier to implement than buyer's quotas. The domestic emission budget would be differentiated in a part that is internationally tradable and another that is inconvertible. This would lead to the creation of parallel markets and lower prices of the latter category of permits. Obviously, this would discourage domestic innovation.

The EU advocated limiting the sale of permits to the amount of domestic abatement. This would lead to the need of prior certification of domestic reduction and corresponding delay of trading. Private participation thus would only be possible in the case of JI projects. Moreover, creation of a bubble would reduce the stringency of sellers quotas, as the quota would not cover the internal trade.

3.5.4 Quotas as incentive for innovation

The only economically convincing argument against GHG emissions trading is that it could reduce innovation as costs of ER are lowered and thus research activities are reduced. Thus quotas are discussed to guarantee a minimum amount of domestic reduction and innovation. If additional short-term costs were lower than long-term gains of induced additional innovation, the quota would be beneficial.

Generally, it is doubtful whether a quota will lead to enhanced innovation as it has no dynamic component. Moreover, developing countries seem to have reduced their pressure for setting quotas related to the CDM, as they fear that Annex-1-JI and trading might reduce transfers. One German delegate to the Kyoto Conference, Hans Schipulle (Schipulle 1998) was surprised that G 77 countries did not support the EU in its effort to set a clear percentage for quotas in Article 12.

REFERENCES

Carraro, C. (Ed.) (1999) International Environmental Agreements on Climate Change, Kluwer Academic Publishers, Dordrecht, in press.

Dixon, R.K. (1998) The U.S. Initiative on Joint Implementation: An Asia-Pacific Perspective. *Asian Perspective* 22:5-19.

Dixon, R.K. (1997) The U.S. Initiative on Joint Implementation. *Int. J. Environment and Pollution* 8:1-18.

Dudek, D. and Wiener, J. (1996) Joint Implementation, Transaction Costs, and Climate Change, OECD (ed.), Paris, 48-49

EU Council (1998) Community strategy on climate change - council conclusions, 7125/98, Brussels.

Hanisch, T. (1991) *Joint implementation of commitments to curb climate change*, CICERO Policy Note 1991:2, Oslo.

Jackson, T. and Begg, K. (Eds.) (1999) *Accounting and Accreditation of Activities Implemented Jointly*, European Commission, Brussels.

Jepma, C.J. and van der Gaast, W. (Eds) (1999) *On the Compatibility of Flexible* Instruments, *Kluwer Academic Publishers*, Dordrecht, in press.

Intergovernmental Negotiating Committee for a Framework Convention on Climate Change (1993) *Matters relating to commitments - criteria for Joint Implementation, Comments from member states on criteria for Joint Implementation*, A/AC.237/Misc.33 and Add. Geneva.

Michaelowa, A. (1997) Considering externalities in crediting of Joint Implementation, in: Janssen, J. (ed.): *Joint Implementation -protecting the climate, maximizing joint benefits*, IWO Discussion Paper No. 49, St. Gallen.

Michaelowa, A. (1996) Incentive aspects of Joint Implementation of greenhouse gas reduction. *Mitigation and Adaptation Strategies for Global Change* 1: 95-108

Michaelowa, A. and Dutschke, M. (1997) *Joint Implementation as development policy - the case of Costa Rica*, HWWA Discussion Paper No. 49, Hamburg

Michaelowa, A. and Schmidt, H. (1997) A dynamic crediting regime for Joint Implementation to foster innovation in the long term. *Mitigation and Adaptation Strategies for Global Change* 2: 45-56

Natural Resources Defense Council (1993) NRDC on Joint Implementation, in: *Joint Implementation - an Eco special report*, Geneva.

Schipulle, H. (1998) FCCC: *Kyoto Protocol - Clean Development Mechanism. Note to explain the outcome of the Kyoto negotiations*, mimeo, Bonn.

Tattenbach, F. (1997) *AIJ: a double or nothing approach to lowering GHG emissions*, paper presented to the International Conference on Technologies for Activities Implemented Jointly, Vancouver.

U.N. Framework Convention on Climate Change (1997) *Kyoto Protocol to the United Nations Framework Convention on Climate Change*, FCCC/CP/L.7/Add.1, Kyoto.

U.N. Framework Convention on Climate Change (1995) *Report of the Conference of the Parties on its first session, held at Berlin from 28 March to 7 April 1995, Part two: Action taken by the Conference of the Parties at its first session*, FCCC/CP/1995/7/Add.1, Geneva.

U.S. Congress (1998) *Credit for Voluntary Early Action Act* (introduced in the Senate), 105th Congress, S. 2617, Washington, DC.

Varming, S., Larsen, P., Christensen, B. (1998) Possibilities of AIJ: two Polish cases, in: Riemer, Pierce; Smith, A. and Thambimuthu, K. (Eds) *Greenhouse gas mitigation. Technologies for Activities Implemented Jointly*, Kluwer, Amsterdam.

Vellinga, P., Hanisch, T., Pachauri, R., Schmitt, D. (1992) *Criteria and guidelines for Joint Implementation of commitments*, New York

Chapter 6

SCOPE AND DIMENSION OF CURRENT ENERGY SECTOR AIJ PILOT PROJECTS

C. JEPMA AND W. VAN DER GAAST
Foundation Joint Implementation Network

Key words: Activities Implemented Jointly, barriers, Clean Development Mechanism, energy, energy efficiency, fuel switching, joint implementation, incentives, renewable energy

Abstract: This chapter presents an overview and analysis of energy sector projects of the Activities Implemented Jointly (AIJ) pilot. The scope, dimensions and distribution of all energy sector projects are presented analyzed. Costs and benefits of selected projects in Eastern Europe, Latin America and Asia are examined. Barriers and incentives to energy sector project investments in the AIJ pilot are considered. General lessons learned from the AIJ pilot are discussed in the context of the Clean Development Mechanism (CDM).

1. AN OVERVIEW OF ENERGY SECTOR AIJ PILOT PROJECTS

By mid-1999, the AIJ pilot resulted in the development of approximately 130 pilot projects which had been accepted, approved and endorsed by the designated national authorities for AIJ of the host and investing countries (JIN, 1999a). Of these projects 92 had been officially reported to the UN Framework Convention on Climate Change (FCCC) Secretariat according to the Uniform Reporting Framework (URF). The full list of AIJ projects can be viewed by examining the UNFCCC Internet site at www.unfccc.de/ccinfo/aijproj.htm (Jepma and van der Gaast, 1999).

This chapter covers a relatively broad category of AIJ projects and complements other technical reviews on this topic (Jackson and Begg 1999; Jepma and van der Gaast, 1999). However, this chapter explicitly focuses on the energy sector in AIJ pilot, while other chapters in this book discuss other categories, such as land-use change and forestry (LUCF) AIJ projects. Energy sector AIJ pilot projects are listed by type and region or country in Table 1. This data suggests that a considerable part of the current AIJ projects are actually implemented in the energy sector.

Secondly, Table 1 suggests that the majority of the energy sector AIJ projects are implemented in Central and Eastern Europe (74 out of 101). Most of these projects are based on investments in the area of energy efficiency improvement or on investments that aim to switch from the use of carbon (C) intensive to relatively less C intensive fuels. Some of the projects in the latter category combine fuel switching (coal to natural gas) with energy efficiency improvement. The approach of bundling several small project components (*e.g.* energy efficiency and fuel switching) is sometimes used in Central and Eastern Europe to enlarge the scale of the investment in order to attract large investors that each component could not do on its own (Jackson and Begg 1999).

The number of renewable energy projects in the current list of energy sector AIJ projects is relatively small. Probably, the reason for this pattern is that during the pilot phase for AIJ it has turned out to be relatively easy for mainly the Western European countries to implement projects in the field of energy efficiency and fuel switch in co-operation with Central and Eastern Europe (Dixon, 1998; Dixon 1997). Similarly, it was relatively easy for USA firms to implement LUCF AIJ projects in cooperation with Central American countries because of the already existing cooperation contacts between these entities in the framework of other environment programs. The Netherlands, for example, had already started a program for cooperation with Eastern Europe (the PSO program) in the early 1990s. This program consisted to a large extent of elements dealing with energy efficiency improvement. Prior to the start of the AIJ pilot phase in 1995, the Netherlands initiated a JI simulation study on the basis of a number of the projects carried out under the program. Thus, exisiting Dutch projects could be analysed as if they had been set up as JI projects.

An example of existing cooperation between countries that was continued during the AIJ pilot phase is the Swedish government program for an Environmentally Adapted Energy System (EAES) in the Baltic region and Eastern Europe. Prior to the start of the AIJ pilot phase Sweden had initiated a program aimed at an improvement of energy systems through energy efficiency measures and increased use of renewable energy sources for energy production. Part of this program focused on domestic investment and another part dealt with assisting the neighbouring countries in the Baltic region and Eastern Europe to follow a path of sustainable economic growth. This initiative was to a large extent motivated by Sweden's objective to contribute to climate change mitigation. However, from a regional perspective it was also in Sweden's interest to reduce the emissions of other non-GHG pollutants that would create environmental problems for neighboring Scandinavian countries. The cooperation under the EAES programme led to a significant number of energy efficiency improvement and fuel switch projects in the Baltic States. The participating countries have reported a large number of these projects to the FCCC Secretariat as AIJ investments.

TABLE 1. A summary of energy sector AIJ projects reported to the FCCC Secretariat.

Type of energy sector AIJ	**Region/country**	**Number of projects**
Fuel-switch + co-generation****	Central Europe*	3
Wind energy	Central America	3
	Baltic states**	1
Solar energy	Central America + South America	2
	Asia	2
Energy efficiency	Central Europe*	5
	Africa	2
	Eastern Europe***	7
	Russian Federation	5
	Baltic states**	20
	Asia	7
Demand-side management (DSM)	Mexico	1
	India	1
	Central America	1
Geothermal	Central America	1
Biomass	Central America	3
Hydroelectricity	Central America	1
	Asia	1
Fugitive gas capture + gas transport improvement	Russian Federation	2
Renewable energy systems	Asia	1
	Mexico	1
Fuel switching****	Eastern Europe***	1
	Baltic states**	30

* Czech Republic, Hungary, and Poland.
** Estonia, Latvia, and Lithuania
*** Bulgaria, Romania, Slovak Republic, Ukraine
**** Some of the fuel switch projects listed here (Decin, Czech Republic; Sventupe and Ziegzdriai, Lithuania) consist of both a boiler conversion and an energy efficiency component.

The absence of the possibility to credit the GHG emission reductions achieved through AIJ pilot projects to Annex I Parties' commitments under the FCCC may possibly explain the relatively small number of new projects, such as in the renewable energy field (Jepma and van der Gaast, 1999). FCCC Parties, and entities acting under their auspices, are reluctant to implement projects with new, renewable energy, technologies in countries with economies in transition (EIT) and developing countries without acquiring GHG emission reduction credits for it. As a result, investing Parties have mainly opted for a wait and see approach as far as renewables are concerned and focused on energy efficiency and fuel switching projects for which they could use existing cooperation frameworks.

This chapter considers energy projects from a relatively broad perspective of the current AIJ pilot. Section 2 of this chapter briefly illustrates what the costs per Mg CO_2 equivalent of the AIJ projects. Section 3 identifies a number of barriers to energy sector investments that currently exist in potential AIJ host countries. Section 4 considers to what extent the AIJ pilot projects have contributed so far to assisting host countries to lower or even remove those barriers. Finally, section 5 lists a number of general lessons learned from the current AIJ energy sector projects. Throughout the chapter the analysis is illustrated with concrete examples of AIJ projects. Furthermore, in the sections references are made to the future implementation of JI and CDM projects to be developed in the same regions where the current AIJ projects take place.

2. COSTS OF ENERGY SECTOR AIJ PROJECTS

The primary rationale for JI projects and the AIJ pilot is efficiency and cost-effectiveness. By investing in an emission reduction project in and in cooperation with a country where the marginal abatement costs are lower than in the investing country itself, the investing country can achieve emission reductions at lower costs than through domestic action. To what extent has the AIJ pilot phase provided evidence of the cost-effectiveness potential that many observers expect to exist for cooperative international GHG abatement projects? This section provides an overview of the costs of the current energy sector AIJ projects (Table 2).

Finally, it should be noted that the cost information provided by the project developers to the FCCC Secretariat provides useful data for some evaluation purposes but the utility of this data has limitations (Jackson and Begg 1999). It is not altogether clear whether the AIJ project abatement costs are truly representative for the marginal abatement costs in the energy sector in the host countries. As was explained above several energy sector AIJ projects in, for instance, Central and Eastern Europe have, mainly because of the absence of crediting, been implemented on the basis of already existing cooperation program. As a result for a considerable part of the projects the main reason for AIJ approval is not the investments' cost-effectiveness, but more the opportunity the projects provide to gain experience with technical AIJ issues such as baseline determination, monitoring and verification, and certification procedures.

TABLE 2. Cost of reducing emissions of CO_2- equivalents in current energy sector AIJ projects as reported to the FCCC Secretariat.

Type of energy sector AIJ	**Region/country**	**Costs (as reported to the UNFCCC Secretariat; US\$ Mg CO_2 eq.)**
Fuel-switch + co-generation****	Central Europe*	
Wind energy	Central America	
	Baltic states**	
Solar energy	Central America + South America	
	Asia	
Energy efficiency	Central Europe*	
	Africa	
	Eastern Europe***	
	Russian Federation	
	Baltic states**	
	Asia	
Demand-side management (DSM)	Mexico	
	India	
	Central America	
Geothermal	Central America	
Biomass	Central America	
Hydroelectricity	Central America	
	Asia	
Fugitive gas capture + gas transport improvement	Russian Federation	
Renewable energy systems	Asia	
	Mexico	
Fuel switch****	Eastern Europe***	
	Baltic states**	

* Czech Republic, Hungary, and Poland.

** Estonia, Latvia, and Lithuania

*** Bulgaria, Romania, Slovak Republic, Ukraine

**** Some of the fuel switch projects listed here (Decin, Czech Republic; Sventupe and Ziegzdriai, Lithuania) consist of both a boiler conversion and an energy efficiency component

3. BARRIERS TO ENERGY SECTOR CO_2 EMISSION REDUCTION PROJECTS

At the first FCCC Conference of the Parties the AIJ pilot was established and the rules of the game were developed and approved (Jepma and van der Gaast 1999). COP-1 Decision 5 states the reduction of GHG emissions of AIJ projects should be...real, measurable, and verifiable. Although the technical and policy dimensions of AIJ project measurability and verifiability (as well as of future JI and CDM projects) are considered to be important as well for successful projects, the word real in this provision can be considered as the biggest topic of discussion during the AIJ pilot phase. This topic has been review and discussed by other authors (Carraro 1999).

The question of whether an AIJ project results in real emission reductions relates to the issue of whether the project is additional to the situation that would have occurred under business as usual circumstances at the project site (Dixon, 1998). The additionality requirement included in Decision 5 implies that project developers have to prove that without the AIJ (or in future JI and CDM) project the investment would not have taken place. This is not an easy issue to deal with. In general, an AIJ project aims at replacing a situation with relatively high C emissions with a situation with relatively lower emissions of C. For the determination of this additionality a reference scenario (the baseline) of the emissions of the old situation has to be formulated. The emission reductions achieved through the project are calculated by comparing the emissions in the new situation with the emissions estimated under the baseline. The complication is that the reference scenario describes the situation that will, because of the AIJ project, never exist. In other words, the scenario is counterfactual and therefore cannot have a future verification point.

For emission reductions via AIJ projects to be real, it is therefore important to clearly identify those factors that prevent GHG emission reduction investments from being implemented anyway. In this section an overview is given of a number of barriers that have been identified during the AIJ pilot phase as obstacles for commercially feasible energy sector investments in Central and Eastern Europe. The overview is to a large extent based on information collected by the Prague office of the US Center for Clean Air Policy (CCAP), the Czech Energy Efficiency Centre (SEVEn), and the Dutch JI Network (CCAP/SEVEn, 1996; SEVEn/JIN, 1997). This information consists of financial and technical project information based on 30 AIJ projects carried out in Central and Eastern Europe and of an assessment of barriers for project implementation in the region. The information was collected via the project reports provided by the project developers as well as via personal communication with both experts at the project level and the government level. The situation in Central and Eastern Europe may to some extent be considered as representative for the energy sector in developing countries. Where necessary a reference will be made to specific factors characteristic for the developing world.

3.1 ENERGY EFFICIENCY IN A PROCESS OF ECONOMIC TRANSITION

During socialism and under the corresponding system of central planning, energy policy in Central and Eastern Europe gave little priority to energy efficiency. On the contrary, the use of energy was highly subsidized which resulted in low energy prices for the end users. During the transition period governments of Central and Eastern European countries have started, mainly for economic, welfare and health reasons, to reformulate their energy policies to make the use of energy more efficient. Also for companies, who are currently faced with hard budget constraints instead of the former soft constraints to budgets, energy efficiency has become an increasingly important part of business. However, changing the legislation to achieve all this is easier said than done and often requires long processes of government decision making and institutional changes in the countries' energy sector.

Although as a result of the process of transition from central planning to market liberalisation energy prices in Central and Eastern Europe have risen during the 1990s to levels that are closer to world market levels, too steep rises are difficult to achieve. In Central and Eastern European countries, as well as for many developing countries, energy costs make up a large part of households' income, (e.g. 15 to 20%). A quick removal of energy subsidies and a strong increase in energy prices would create social problems for households, especially those in the low-income segment, as well as serious financial problems for the business community. It can therefore very feasibly take place in combination with energy efficiency improvement measures, which could offset the lower subsidies.

3.2 SHORT PAYBACK PERIODS

Several barriers that exist in the region's market for energy efficiency investments have hampered energy efficiency improvements in Central and Eastern Europe. First, in the financial markets in the region the interest rates for loans are often high, due to the relatively strong rate of inflation and the scarcity of capital. As a result, the market can take up only investments with relatively short payback periods. As many energy sector projects in the region (as well as in developing countries) have relatively long payback periods, the investments through the market are often discouraged. This in particular holds for investments aiming at energy efficiency improvement. For example, an electric utility in the region who is faced with an increase in energy demand is under current financial market circumstances more likely to invest in a new power plant to meet the extra energy demand than to establish a program to increase energy efficiency. The reason for this could just be that the financial returns of the first investment option are, at least in the short run, larger than in the second option.

3.3 LACK OF BUSINESS AND RISK MANAGEMENT SKILLS

Second, the lack of knowledge of and experience with business and risk management that exists in several potential AIJ host countries can be a serious barrier to the preparation and implementation of business plans and to a proper management of the risks involved with the investments. This, in general, has hampered project developers to find access to international financial markets. As a result, project developers or plant owners in Central and Eastern Europe often lack the capital to fully finance investments via the (international) financial market.

3.4 SMALL SCALE PROJECTS

Third, the scale of energy efficiency projects in the region is often small (Jackson and Begg 1999). Not only may these projects lack the capacity to attract the capital from the financial markets, they typically are also too small to become part of investment portfolios of multilateral development banks, such as the European Bank for Reconstruction and Development, or, for the other world regions, the World Bank, the Asian Development Bank (ADB) and the African Development Bank. Multilateral institutions often set standard minimum lending requirements for the projects they support and energy sector investments suffer.

In the March 1999 issue of the *Joint Implementation Quarterly* (JIN, 1999b) an ADB official, Dr Prodipto Ghosh, explained the Bank's experience with clustering projects that are in itself too small to raise funding themselves. Under the auspices of the ADB several of such projects have been bundled together and implemented as a group by specialised intermediaries in a program rather than project approach. In an example explained by Dr Ghosh the ADB provided a US$100 million loan to a program in India, which aimed at disseminating renewable energy technologies. This program consisted of numerous small subprojects, each investment ranging from a few thousands to several hundred thousands USA dollars. The program was managed by the Indian Renewable Energy Development Agency (IREDA), a specialized local intermediary.

3.5 LACK OF CREDIT MUNICIPALITY CREDIT WORTHINESS

Finally, a fourth example of a financial barrier to market-based investments in the energy sector in Central and Eastern Europe could be the lack of credit worthiness of several municipalities or other entities that own power plants or other energy sector elements. The financial resources of municipalities in potential AIJ host countries are often limited and too small to provide satisfactory financial guarantees. In some countries, (eg, the Russian Federation) energy suppliers may after a first glance of their balance sheet seem solvent, but may turn out to be insolvent when the expected amount of non-paid bills is taken into account.

3.6 DELAY OF INVESTMENTS DUE TO BARRIERS

An attempt to answer the question to what extent the above barriers prevent energy sector projects from being implemented via the market was made by the Nordic Council of Ministers (1997). A study was established based on ten energy sector projects carried out by the Nordic countries in and in cooperation with the countries in the Baltic region. In its conclusions the study distinguishes between projects aiming at investments carried out in power plants (energy supply) and investments in energy efficiency improvement in city or district heating systems (energy demand). According to the study, under business-as-usual circumstances the host countries would probably have carried out energy demand side management (DSM) projects at an earlier stage than investments in power plants. First, energy supply side investments are often larger than investment at the energy demand side and thus require more capital. Second, the study suggests that in the Baltic region and other regions in Central and Eastern Europe the pressure of energy consumers to improve energy systems at the demand side is higher than that of owners of power plants to increase energy efficiency at the supply side. One of the reasons given in the study is that households increasingly wish to enjoy better and more comfortable living circumstances in apartment blocks to be achieved by better isolation systems and comfortable heating systems.

Another conclusion of the study by the Nordic council is that several of the projects carried out by the Nordic countries in the Baltic region and Central Europe were additional at the time that they were implemented, but that they would likely have been implemented anyway by the host countries themselves at a later date. For example, in 1996 the Czech Republic adopted a law with respect to energy efficiency and energy production which set norms for the use of fuel prescribing a maximum C content for coal. As part of the law, meters for heating systems were to be installed in public buildings and apartment blocks. Furthermore, it was concluded that in some potential AIJ host countries in Central and Eastern Europe alternative fuels (eg, biomass in the Baltic States) are relatively available. So, fuel conversion projects (from coal to biomass) initiated by the Nordic countries in the region probably only have speeded up investments that the countries would have made themselves later on.

4. AIJ AND THEIR CONTRIBUTION TO REMOVING BARRIERS IN THE ENERGY SECTOR

Although the number of AIJ projects has remained limited, an impression can be obtained of the extent to which energy sector AIJ projects have been able to contribute to removing some of the (energy and financial) market barriers in potential AIJ host countries. On the basis of case studies, this section describes what contribution AIJ pilot projects have made to support the energy policies of the host countries where they were implemented. Furthermore, this section aims to explain what concrete solutions AIJ projects have provided to the problems related to the barriers described in section 3.

4.1 SHORTENING THE PAYBACK PERIOD

In section 3 it was explained that an obstacle for energy sector investments in many potential AIJ host countries could be that the payback period for such investment is often too long. This is where AIJ (as well as JI and CDM) can play a role. An energy sector project the required investment capital of which can only for 85 to 90% be obtained through commercial financial markets, may become commercially feasible if the GHG emission reductions achieved through the investment is valued. An example of an AIJ project that was made possible in this way is the Decin-Bynov project in the Czech Republic. This project was initiated in April 1993 when the City of Decin began with the development of an investment to reduce the emissions of pollutants from the Bynov district heating system. Due to the heavy air pollution caused by the coal fired plant the Decin region suffered from health problems and damage to forests.

The project that was developed consisted of a fuel switching investment via which the fossil fuel plant would be replaced with a natural gas boiler. In addition, the project developers aimed at improving the efficiency of energy production and consumption in Decin via the installation of modern metering equipment and techniques to make use of the potential of the project for co-generation. The project's feasibility study was to a large extent carried out by Danish consultants who also took part in the handling of technical and engineering aspects of the project. The costs involved with this part of the project were covered from the Danish side. The total cost of the Decin-Bynov project amounted to US$ 9,058,000 of which 74.5% has been covered by the Czech Environment Fund. The Danish contribution amounts to 11% of the project cost and 7% has been collected through commercial investment.

Although the Decin-Bynov project had many helpful partners about 7.5% of the investment costs to be financed through alternative means. The idea to add this funding to the project's capital as a JI component was brought up on the basis of an initiative by three USA utility companies: Wisconsin Electric Power Company, NIPSCO Power Company, Inc., and the Edison Electric Institute. Several USA utility companies have been in consultation with the USA government regarding reductions or offsets of their GHG emissions. One of the opportunities investigated in this respect was the setting up of C offset projects abroad. In the late 1980s and early 1990s this approach resulted in several LUCF projects in Central America and Latin America funded by USA firms. Subsequently, firms began focusing on GHG abatement projects in Central and Eastern European countries, such as the Czech Republic, Poland and the Republics of the former Soviet Union. The participation of the three companies in the Decin project can be considered as an example of this development.

The three companies each contributed US$ 200,000 to the investment capital of the project in Decin for which they acquired 133,827 Mg of the estimated total of 608.952 Mg CO_2 emission reduction. Thus, by explicitly valuing the CO_2 emission reductions achieved primarily as a second benefit of the project, the investment became financially feasible and could start in 1995. In other words, adding an AIJ component (the CO_2 emission reduction) to the project made the investment's payback period shorter. It is envisaged that similar constructs and partnerships will be developed for future JI and CDM projects as well.

Another example of how the energy sector project internal rate of return (IRR) was increased (or the payback period shortened) via AIJ is provided by the Swedish EAES in the Baltic region and Eastern Europe. The projects under the EAES program mainly deal with energy efficiency improvement and fuel switching projects in block and district heating systems. The projects are financed through loans to the plant owners provided by the Swedish side. These loans often have a payback period of 10 years (usually with a grant period of two years) and cover the investment costs (the Swedish government covers the technical assistance for the projects). The loans are provided under the same conditions as those of World Bank loans. The interest rate is linked with the Swedish STIBOR rate (Stockholm Interbank Offered Rates) and is adjusted every six months during a loan's payback time. Thus, the host countries participating in the program are enabled to borrow capital for energy sector projects at an interest rate, which is far below the domestic interest rate and under conditions, which are more favorable than those they would meet when borrowing in the international financial markets.

4.2 INFRASTRUCTURE IMPROVEMENTS VIA DSM AIJ PROJECTS

The AIJ pilot phase project portfolio contains a small number of DSM projects. In section 3 it is explained, via an example, that potential host countries when confronted with higher energy demand sometimes opt for enlarging the energy production capacity of power plants instead of improving energy efficiency. The first option is from a cost-effectiveness perspective often more attractive than the second one, which usually requires long procedures for adjustments of the infrastructure of energy transport and consumption, for example via DSM. An example of a DSM AIJ project is provided by a low cost housing project in South Africa. Under an AIJ project agreed upon in July 1998, low-cost energy-efficient homes are developed in South Africa in order to address urgent social and economic needs, while simultaneously reducing GHG emissions.

South Africa is in the process of reconstructing and developing its society and economy, which has been formulated and elaborated on in the Reconstruction and Development Program (RDP). Among the socio-economic goals included in RDP are meeting basic human needs, developing the country's human resources, building the economy, and democratising the state and society. The most high profile of the RDP implementation programs are surely the mass housing and electrification schemes. In 1993, the housing backlog was estimated at 3 million units. The demand for housing has been estimated as growing at a rate of 200,000 units annually. Many of the targeted one million low-income households (the Department of Housing in South Africa has a goal to provide 1,000,000 housing opportunities to households in the end of 1999) rely on inefficient stoves that burn high GHG emitting fuels, such as coal, wood and kerosene, for cooking and heating. Currently, only 55% of homes in South Africa have access to electricity. Approximately 27% of rural households have access to the electric power grid (as compared with 79% of urban households).

As part of an AIJ project the construction of so-called ECO™ houses has been initiated. These houses provide energy efficient and comfortable living circumstances. During the implementation of the ECO™ project several initial barriers had to be overcome. For example, it was difficult for the project developers to obtain credit to finance the building of the homes. The South African banking community has been hesitant to fund development in historically disadvantaged communities. The Department of Housing is working with financing institutions to develop innovative lending solutions for the low-income sector. Another barrier is the apparent lack of available expertise in energy efficiency and passive solar design in housing in South Africa. The project developers learned that in South Africa energy efficiency and low-cost housing are generally considered to be mutually exclusive. As a result a historically disadvantaged community such as the one in which the project is implemented would normally be one of the last to be provided with innovative technology to overcome problems like high use of energy, pollution and discomfort.

4.3 TRANSFER OF TECHNOLOGY AND ESTABLISHING A TRAINING PROGRAM

Obviously, as part of the energy sector AIJ projects a transfer of new environmentally-sound technology has taken place to the AIJ host countries, which would not have been able to install it themselves or only with a delay. The project in Decin described above led to the installation of a state-of-the-art boiler in a plant providing energy to a district heating system, and to the installation of modern metering equipment. The project in South Africa leads to the construction of eco-houses in rural areas, a technique, which probably would not have been feasible for application in the short or medium term. This project also helps develop the human capital that is required to manage and maintain the technologies established under the AIJ program. This investment in human capital enables host countries to implement energy sector projects themselves in the future and helps removing the barrier explained in section 3. Many potential AIJ host countries currently lack the management and technical skills to implement energy efficiency investments themselves (Dixon 1998).

An example of what a training component may look like is provided by the energy homes project in South Africa described above. The developers of this project have aimed at involving the community in the project management, in the decision making process and in the negotiations about investment conditions. As such the project contribute to a so-called Community Development Framework, which is used as a basis for providing the leaders of the community where the construction of the houses takes place with the opportunity of further developing their leadership and negotiation skills. This is done through interactions with a wide range of stakeholders in the implementation of the project. The community has already started to market their newly acquired skills and knowledge with other communities.

The project is also providing paid employment (and on-the-job training) for over 120 residents of the community. After the project, the skills acquired are likely to be useful in other jobs. Moreover, a local industrial business zone is planned close to the village, which is expected to provide a source of employment to the community. The business zone is part of a program initiated by the South African Department of Trade and Industry to develop business zones in parallel with housing projects. The programme aims, among others, to stimulate business growth and employment opportunities near communities gaining access to new housing. The community where the eco-housing project takes place has become part of the program.

5. CONCLUSION: LESSONS LEARNED FROM ENERGY SECTOR AIJ PROJECTS

The AIJ pilot phase has resulted in a considerable number of projects carried out in the energy sector in countries in Central and Eastern Europe and in several developing countries. This chapter has attempted to answer the question to what extent these projects have contributed to removing current barriers for energy efficiency measures the host countries would wish to implement themselves. This question is important for AIJ, JI and CDM because it strongly relates to the issue of additionality, which on its turn relates to the question of whether, or when, the AIJ host country would have implemented the project by itself anyway (Goldemberg, 1998).

This chapter has identified several barriers that currently exist in many potential host countries for AIJ, JI and CDM. On the basis of the experience with a number AIJ pilot projects and programs it has been analyzed what contribution AIJ has made so far to deal with these obstacles. A first barrier identified is the payback period of energy sector investments in Central and Eastern European and developing countries. Due to the lack of capital and the often relatively high interest rates in these countries, borrowing capital for investments is relatively difficult and expensive. As a result, the payback period of an investment must be short in order to make the project commercially feasible, a situation which does not very often occur in several potential AIJ host countries.

The AIJ phase has provided some useful examples of how the internal rate of return of a project can be improved to a level that makes the investment feasible. Fuel switching and co-generation projects in the Czech Republic reveal that an investment that is largely feasible, via a combination of commercial financing, central government support in the host country and capital provided by the municipality itself, can be realized if the GHG emission reductions of the investment are valued as well. Foreign investors that are for one reason or another interested in relatively cheap GHG emission reductions can acquire these by providing the additional funding to the investment's capital. Another option that has been provided by AIJ is that investor countries provide loans, in exchange for GHG emission reductions, to potential host countries under conditions that are more favorable than those provided by (international and domestic) commercial financial markets. The entities in the host countries, which under current circumstances of economic transition face difficulties with borrowing capital in the commercial market, have with the help of the AIJ cooperation become eligible for international financing opportunities.

Another way in which AIJ has contributed to host countries' attempts to reduce (or eventually remove) the barriers for energy sector investments is the transfer of state-of-the-art technology that projects provide to host countries. The AIJ investment improves the stock of energy sector capital goods and improves the efficiency of the energy production and consumption. The technology transfer also has the potential to support host countries' domestic priorities in terms of sustainable development or economic growth. In addition, a project's hardware investment has in some AIJ cases been accompanied with measures to improve the infrastructure of the project's surroundings. To this end, some projects include training components to ensure that sufficient capacity building takes place in order to make the projects manageable for local entities.

Finally, although not carried out as an AIJ project, a useful experience was gained with some small-scale projects carried out under the auspices of the ADB. As is explained earlier in this chapter, the scale of many potential projects is too small to mobilize the interest of international corporations and multilateral institutions. Such entities usually set minimum lending requirements the magnitude of which is often too large to make energy efficiency projects in Central and Eastern Europe and in developing countries eligible for financing under their programs. On the basis of the experience with projects in Southeast Asia the ADB has bundled small-scale projects in a certain energy sector category together to one big project so that it would become eligible for funding programs. The bundled projects are carried out by specialized implementation agencies.

REFERENCES

CCAP and SEVEn (1996) *Joint Implementation Projects in Central and Eastern Europe*, Description of ongoing and new projects prepared by Center for Clean Air Policy in co-operation with SEVEn, Prague.

Carraro, C. (Ed) (1999) *International Environmental Agreements on Climate Change*, Kluwer Academic Publishers, Dordrecht, in press.

Dixon, R.K. (1998) The U.S. Initiative on Joint Implementation: An Asia-Pacific Perspective. *Asian Perspective* 22:5-19.

Dixon, R.K. (1997) The U.S. Initiative on Joint Implementation. *Intern. Journal of Environment and Pollution* 8:1-17.

Goldemberg, J. (Ed) (1998) *The Clean Development Mechanism: Issues and Options*, UNDP, New York.

Jackson, T. and Begg, K. (Eds) (1999) *Accounting and Accreditation of Activities Implemented Jointly*, European Commission, Brussels.

Jepma, C.J. and van der Gaast, W. (1999) *On the Compatibility of Flexible Instruments*, Kluwer Academic Publishers, Dordrecht, in press.

JIN (1999a) Planned and ongoing AIJ pilot projects. *Joint Implementation Quarterly*, 5:14.

JIN (1999b) The Asian Development Bank and Capacity Building for the CDM. *Joint Implementation Quarterly*, 5:3-4.

SEVEn and JIN (1997) *The Experience with Joint Implementation in Central and Eastern Europe During the AIJ Pilot Phase*, Prague.

Chapter 7

FORESTRY AND LAND-USE CHANGE IN THE AIJ PILOT PHASE:

The Evolution of Issues and Methods to Address Them

M TREXLER, L KOSLOFF, R GIBBONS
Trexler and Associates, Inc.

Key words: activities implemented jointly, land-use change and forestry, Clean Development Mechanism, carbon offsets, flexibility mechanism

Abstract: The Parties to the UN Framework Convention on Climate Change (FCCC) established the activities implemented jointly (AIJ) pilot phase at the first Conference of the Parties in 1995. As we move toward implementation of the Kyoto Protocol and operationalization of the project-based mitigation initiatives included in Articles 6 and 12 of the Protocol, it is important to understand the experience to date with land-use change and forestry (LUCF) mitigation efforts. Experience with LUCF projects even predates the AIJ pilot phase; the first LUCF project was initiated in 1988. Because LUCF has been a contentious subject during the pilot phase and in setting the stage for implementation of the Kyoto Protocol, understanding the lessons to be gained from this experience is particularly important. This report is not intended to be a definitive view of the field, since more and more effort is being dedicated to the study of LUCF mitigation options. What is clear from the AIJ experience, however, is that most issues raised in the context of LUCF projects are generic to project-level mitigation, rather than being specific to LUCF. As such, the AIJ lessons learned from projects carried out in other sectors should be seen as relevant to policy development around LUCF.

1. INTRODUCTION

Early analyses heralded the land-use change and forestry (LUCF) sector as a major part of the solution to the climate change problem (Dyson, 1977; Sedjo, 1989a; Sedjo, 1989b). Today, however, the role of forestry and other land-use and biologically based greenhouse gas (GHG) mitigation efforts in a post-AIJ world remains less than clear. This is true at the national accounting level (Articles 3.3 and 3.4 of the Kyoto Protocol), at the project level in Annex I countries (Article 6 of the Protocol), and at the project level in developing countries (Article 12 of the Protocol) (UN FCCC, 1997). Indeed, LUCF issues have been among the most contentious issues of the AIJ pilot phase, during the drafting of the Kyoto Protocol, and during discussions of how the provisions of the Protocol will be implemented.

Proponents of forestry offset projects argue that land use-based mitigation measures in both tropical and temperate zones have an important potential role to play in future climate change mitigation efforts. Beyond their value for climate change mitigation purposes, proponents argue that biotic carbon offsets can effectively advance other policy objectives at little or no additional cost, including biodiversity conservation, rural economic development, and watershed management. However, many international environmental groups and developing countries continue to express skepticism about allowing carbon dioxide (CO_2) emitters to avoid absolute CO_2 emissions reductions by relying on C offsets generally and LUCF projects specifically (Barnett, 1992; Friends of the Earth, 1997; World Wildlife Fund, 1995; Mott, 1993; Sierra Club, 1995).

Yet forestry projects have been among the most popular mitigation projects pursued to date. LUCF projects have proven attractive to mitigation project investors not only because of their perceived climate change benefits and low costs, but also because of their significant environmental and social co-benefits. This attractiveness illustrates the fact that a problem with the current LUCF debate is that participants often have visions of project extremes rather than trying to work with the great bulk of projects on the middle ground. One participant to a recent expert workshop noted: “I see two sets of types of projects. We are interested in projects that are at the intersection of these two types. We do not want simple plantations, and we do not want projects that are so social in nature that the C benefit is virtual. In between are kinds of projects that can be done, can be verified, and are socially relevant projects. The problem is that people have visions of extremes, and it tends to overly influence policy discussion (Trexler and Associates, Inc., 1998b).

As the AIJ pilot phase winds down, most of the debate surrounding LUCF projects results from concerns voiced by a variety of governmental and non-governmental participants to the AIJ process. Many of the concerns associated with LUCF can be traced back to observers understanding or interpretation of activities undertaken in conjunction with the AIJ pilot phase. The objective of this paper is to evaluate the arguments for LUCF projects in the framework of the anticipated joint implementation (JI) and

Clean Development Mechanism (CDM) processes through a review of AIJ pilot phase experience under the FCCC. A key component to this analysis is a discussion of technical concerns that have been raised regarding methodologies for implementing project-level mitigation efforts in the LUCF arena, in conjunction with a case-study analysis to illustrate some of the experience with actual projects.

LUCF projects have come under fire in the context of project-level initiatives primarily due to a misconception that technical issues relevant to LUCF-sector projects are inherently different from or more complex than energy-sector project issues. The fact that most of the technical concerns raised are specific to project-level mitigation efforts, rather than to a particular sector, is often overlooked. The AIJ pilot shows that most of the issues raised (*e.g.*, additionality, leakage, monitoring and verification, environmental impacts, co-benefits) are project- rather than sector-specific (Sathaye, *et al.*, 1998; Trexler and Associates, Inc., 1998b). LUCF benefit permanence is a possible exception to this conclusion, as discussed subsequently.

The LUCF sector is receiving much more analytical and policy attention today than it has in the past. The Intergovernmental Panel on Climate Change (IPCC) upcoming Special Report on Land-Use, Land-Use Change, and Forestry will have one chapter focusing specifically on project-level interventions. And as this paper is completed, many LUCF-specific studies and journal articles have recently appeared or will soon be appearing (*e.g.*, Nabuurs, 1999; Schlamadinger, 1999; Brown et al, 1999) The politics of the LUCF debate are also in significant flux, with member states of the Amazonian Treaty Cooperation Countries recently agreeing that certain forestry interventions should indeed be included within the CDM (Cochabamba Declaration, 1999). The significant amount of work currently underway in this area, as well as the evolving politics of the LUCF sector, will add considerably to our understanding of LUCF projects as a climate change mitigation tool and will no doubt add to the lessons to be drawn from LUCF experience to date. With this rapidly evolving context in mind, this paper briefly examines the relationship between land-use change and climate change, reviews the experience with LUCF projects gained through the AIJ pilot phase, and seeks to apply this experience to key aspects of the debate as it is currently framed.

2. THE LUCF SECTOR AND CLIMATE CHANGE

2.1 LUCF AS A CONTRIBUTOR TO CLIMATE CHANGE LUCF

Since before the Industrial Revolution, temperate and tropical land-use changes have significantly contributed to rising levels of GHGs in the atmosphere (Houghton, *et al.*, 1983; Houghton, *et al.*, 1987; Dale, Houghton, and Hall, 1991; Wisniewski and Sampson, 1993; Brown, *et al.*, 1996; IPCC, 1996; Apps and Price, 1996; Dixon, *et al.*, 1994). Although the relative importance of land use-based emissions is declining as fossil fuel emissions continue to rise, absolute emissions from human activities remain high. Carbon losses from deforestation and other land-use changes continue in both temperate and tropical zones. Tropical deforestation accounted for an estimated 20 percent or more of all anthropogenic CO_2 emissions to the atmosphere in 1990 (Brown et al, 1993). Even today, 1-2 Pg C are annually emitted by anthropogenic land-use changes (IPCC, 1996). In many tropical countries, deforestation is the dominant source of CO_2 emissions from human activities (World Resources Institute, 1998). Vast stretches of tropical forest, currently storehouses for hundreds of billions of tons of C, remain threatened by deforestation or degradation.

2.2 LUCF AS CLIMATE CHANGE MITIGATION STRATEGY

Some early studies suggested that large-scale reforestation could remove very large quantities of C from the atmosphere every year, thus effectively offsetting current energy-related emissions (Dyson, 1977; Sedjo, 1989a; Sedjo, 1989b; Houghton et al, 1993). Prominent environmentalist Norman Myers stated that one of the most cost-effective and technically feasible ways to counter the greenhouse effect lies with grand-scale reforestation in the tropics as a means to sequester C from the global atmosphere provided. That strategy should be accompanied by greatly increased efforts to slow deforestation (Leggett, 1990). At the international level, the Noordwijk Declaration of December 1989, signed by 68 environmental ministers from around the world, proposed increasing global forest cover by 12 million ha annually (starting in 2000) to help slow climate change (Ministerial Conference on Atmospheric Pollution and Climatic Change, 1990).

More recent work has dramatically tempered the perceived magnitude of LUCF-based interventions for climate change mitigation purposes as implementation realities of large-scale forestry initiatives and the difficulty of accomplishing even small-scale initiatives in the past have sunk in (IPCC, 1996; Trexler and Haugen, 1994; Dixon, *et al.*, 1993). These studies have come from think tanks, environmental groups, government agencies, research institutes, and the IPCC (Wisniewski and Sampson, 1993). Growing world populations are straining many forestry and land resources and increasing pressures will be difficult to reverse. The IPCC concluded in its Second Assessment Report that 12-15 percent of cumulative anthropogenic emissions over the next 50-60 years plausibly could be offset through forestry interventions (IPCC, 1996). While far from a panacea for climate change, this and other analyses suggest that LUCF mitigation efforts can contribute to a global mitigation strategy by (Marland et al, 1997):

- protecting existing C reservoirs from losses associated with deforestation, forest and land degradation, urbanization, and other land management practices.
- expanding the C stores in forests, other biomass, soils, and wood products (including through agroforestry, reforestation, afforestation, and forest management efforts).
- using biomass to substitute for fossil-fuel use, whether directly (production of biomass energy) or indirectly (substituting wood for steel, cement, or other fossil fuel-intensive products).
- reducing emissions of other GHGs, primarily methane and nitrous oxide, from land-use and agricultural interventions ranging from fire management to more efficient use of nitrogen-based fertilizers.

Even this basic outline of LUCF options clarifies that there is no such thing as a homogeneous LUCF strategy or intervention. There are many LUCF options or technologies, each likely to perform differently against any given set of evaluative criteria. Some options involve reducing GHG emissions to the atmosphere; others involve increasing the removal of CO_2 from the atmosphere. The heterogeneity of LUCF interventions almost certainly makes it inappropriate to group all LUCF mitigation options under the same tent, either for purposes of embracing or dismissing their climate change mitigation potential. LUCF options presumably will need to be judged against a given set of evaluative criteria in determining their future role in climate change mitigation efforts. The continued absence of an agreed-upon set of specific criteria for evaluating climate change mitigation efforts at the project level, and continued confusion about how different types of projects will perform against such a list, makes learning from existing mitigation project experience in the LUCF and other sectors so important.

3. AN OVERVIEW OF FORESTRY-SECTOR AIJ PILOT PROJECTS

3.1 FORESTRY AS AN EARLY AIJ PILOT CHOICE

The very first climate change mitigation project was an LUCF project. This project involved the 1988 decision by U.S. independent power producer (IPP) AES Corp. to voluntarily fund an agroforestry project in Guatemala to offset emissions from a planned coal-fired facility in the United States (Trexler et al., 1989). Today, LUCF activities for climate change mitigation are underway around the world. Of the AIJ projects formally reported to the FCCC Secretariat, 17 are LUCF projects. Thirteen of the 26 projects approved by the U.S. Initiative on Joint Implementation (USIJI) involve forestry-based emissions reduction or sequestration projects (USIJI, 1996; USIJI, 1997).

There are several reasons for the early dominance of forestry projects in the climate change mitigation project arena (Trexler, 1997):

- early offset investors often wished to clearly differentiate their offset projects from energy-sector investments in which they might already be involved;
- forestry-based offsets were seen as a particularly cost-effective approach to CO_2 mitigation;
- forestry-based offsets, particularly in developing countries, offer many co-benefits, including public perception, environmental, and economic benefits. In a strictly voluntary mitigation regime, offset investors have found these ancillary benefits to be important.

TABLE 1: Selected examples of FCCC registered LUCF AIJ projects.

Project Type	Project Title	Parties Involved	Project Start Date	Lifetime (in years)
Afforestation	RUSAFOR: Saratov Afforestation Project	Russian Federation, USA	1993	40
Agriculture	Community Silviculture in the Sierra Norte, Oaxaca	Mexico, USA	NA	30
Agriculture	Project Salicornia: Halophyte Cultivation in Sonora	Mexico, USA	1996	10
Forest Preservation	Bilsa Biological Reserve	Ecuador, USA	NA	30

Project Type	Project Title	Parties Involved	Project Start Date	Lifetime (in years)
Forest Preservation	ECOLAND: Piedras Blancas National Park	Costa Rica, USA	1995	16
Forest Preservation	Forest Rehabilitation, Krkonose and Sumava National Parks	Czech Republic, Netherlands	1992	15
Forest Preservation	Reduced Impact Logging in East Kalimantan	Indonesia, USA	NA	40
Forest Preservation	Rio Bravo Carbon Sequestration Pilot Project	Belize, USA	1995	40
Forest Preservation	Territorial and Financial Consolidation of Costa Rican National Parks and Biological Reserves	Costa Rica, USA	1998	25
Forest Preservation	Noel Kempff Mercado Climate Action Forestry Project	Bolivia	1997	30
Forest Preservation	Reforestation Project CARFIX: Stabilize and Expansion of Forest Cover	Costa Rica, USA	1996	25
Reforestation	Commercial Reforestation in the Chiriqui Province	Panama, USA	1998	25
Reforestation	Klinki Forestry Project	Costa Rica, USA	1997	40
Reforestation	Reforestation and Forest Conservation	Costa Rica, Norway	1997	25
Reforestation	Reforestation in Vologda	Russian Federation, USA	NA	60
Reforestation	Scolel Te: Carbon Sequestration and Sustainable Forest Management in Chiapas	Mexico, USA	1997	27
Reforestation	Biodiversifix: Forest Restoration Project (WETFIX, DRYFIX)	Costa Rica, USA	NA	50

Table 1 lists LUCF projects officially pursued through the AIJ pilot phase. It includes a description of the project type, the countries involved, the project start date (if any), and the project lifetime as described by project developers. As with other categories of AIJ projects, however, not all the projects in Table 1 have been implemented. Only nine of these projects are in the implementation phase; two of these (Scolel Te and Klinki Pines) are being pursued at a very small scale.

Additional LUCF climate change mitigation projects have been pursued outside the AIJ pilot phase. Some of these projects preceded the 1995 initiation of the AIJ pilot phase; some have been purely domestic projects, internal to a particular country; some have been pursued by entities such as the Global Environment Facility (GEF) outside the AIJ pilot phase structure. At least one project has simply been unable to receive host country approval as required for AIJ projects. Although not part of the formal pilot phase, these projects contribute significantly to the experience base for drawing lessons relevant to the performance of LUCF projects for climate change mitigation purposes. While 9 AIJ-recognized projects are currently being implemented, with an asserted total benefit of 146 million Mg of CO_2, more than 20 non-AIJ LUCF projects are also being implemented, with asserted total benefits of well over 600 million Mg of CO_2 (*see* Table 2). While several of these projects were pursued early in the climate change mitigation process and might not fare well against project evaluation criteria being applied today, they have generated a great deal of learning about LUCF projects and their applicability to climate change mitigation objectives. This learning is reflected in the analysis presented in this chapter.

TABLE 2: Selected non-AIJ LUCF climate change mitigation projects.

Project Type	Project Title	Project Location
Forest Management	Reduced Impact Logging (New England Electric System Pilot-Scale Demonstration)	Sabah, Malaysia
Forest Management	Community-Based Rangeland Rehabilitation for Carbon Sequestration and Biodiversity (GEF)	Sudan
Forest Management	Participatory Management of Natural Forests and Village Reforestation to Reduce Carbon Emissions (GEF)	Benin
Forest Management	Olafo Project	Peten, Guatemala
Forest Protection	AES Mbaracayu Initiative	Paraguay
Forest Protection	AES Oxfam America	Amazon Basin
Reforestation	Oregon Forest Resource Trust	Oregon, USA
Reforestation	Peugeot Sink Creation Project	Brazil

Project Type	Project Title	Project Location
Reforestation	Eastern Washington Carbon Sequestration Project	Washington, USA
Reforestation	Western Oregon Carbon Sequestration Project	Oregon, USA
Reforestation	AES CARE Guatemala Forestry Project	Guatemala
Reforestation	Green Fleet Initiative	Australia
Reforestation	Innoprise- FACE Project (Infapro)	Sabah, Malaysia
Reforestation	Uganda National Parks (FACE)	Uganda
Reforestation/Affor estation	Urban Forest Afforestation (FACE)	Netherlands
Reforestation/Affor estation	Profafor Reforestation (FACE)	Ecuador
Urban Forestry	Salt Lake City Urban Forestry	Utah, USA
Agriculture	Agricultural Soils Enhancement Project	Saskatchewan, Canada

3.2 A SNAPSHOT OF THE LUCF UNIVERSE

The approximately 30 LUCF projects (both AIJ and non-AIJ) that have been or are being implemented for climate change mitigation over the last decade fall into several categories, including:

- forest conservation
- reforestation and afforestation
- reduced impact logging and forest management practices
- biomass energy deployment
- projects involving agricultural soils and crops

Despite the number of LUCF projects that have been pursued, the dollar figures associated with project implementation is still quite small. Estimates of dollars spent on LUCF projects for mitigation purposes range from $50 million to $130 million (Trexler and Associates 1998a; Moura-Costa, 1998), depending on which projects are counted and what implementation phase is assumed. When compared to the financial flows that could be associated with future implementation of the Kyoto Protocol, these numbers are clearly extremely small, even if significant at the individual project level. One must clearly take care, however, to not over-extrapolate from what is a relatively small practical experience base.

Although project economics have not routinely been reported during the AIJ pilot phase, many people believe that forestry-sector projects are a particularly inexpensive mitigation option. As discussed later in this chapter, however, the cost-effectiveness of LUCF projects, as with most AIJ projects, is difficult to interpret. A snapshot of activities pursued within the primary LUCF categories during the AIJ pilot phase is provided below.

3.2.1 Forest conservation

Forest conservation has been a popular AIJ project strategy. These projects are pursued primarily in developing countries where threats to forest resources are the most pressing and results are the most obvious. Well-known examples include the Mbaracayu Conservation Project in Paraguay (AES Corp.), the Rio Bravo Conservation and Forest Management Project in Belize (Wisconsin Electric Power Co., PacifiCorp, Cinergy, Detroit Edison, UtiliTree Carbon Company, and The Nature Conservancy), the Noel Kempff Mercado Climate Action Project in Bolivia (The Nature Conservancy, American Electric Power Co., PacifiCorp, and British Petroleum), and the ECOLAND Project in Costa Rica (Tenaska, Inc., a U.S. IPP). Forest conservation has appeared to be a cost-effective project choice; perhaps more important, however, have been the biodiversity and other benefits associated with many of these projects. One of the best-known projects in this group, the Noel Kempff project in Bolivia, is profiled below (Box 1).

Box 1: Forest conservation - The Noel Kempff Mercado Climate Action Project, Bolivia

The Noel Kempff Mercado Climate Action Project is being implemented by The Nature Conservancy (a U.S. NGO), Fundación Amigos de la Naturaleza (a Bolivian NGO), American Electric Power, PacifiCorp, and British Petroleum. The project developers□ stated goal has been to demonstrate an optimal balance between GHG mitigation, park expansion and protection, and sustainable micro-enterprise development. The project relies on several GHG benefit streams:

- *Preventing the loss of carbon sinks and emissions resulting from commercial logging.* The project will pay for cessation of logging activities in an endangered tropical production forest and will expand the boundaries of an existing park.

- *Preventing conversion of protected forested lands to subsistence agricultural use.* Conversion has historically occurred in this region. Project interventions are intended to protect the new protected stands as well as the forest in the rest of the park. The project will establish a mix of income-generating activities for local populations, establish a park protection endowment fund, capitalize a new genetic resources enterprise that will market orchids and other forest products, and construct an ecotourism infrastructure.

- *Mitigating GHG emissions in secondary areas beyond the expanded park by providing environmentally sustainable economic opportunities for local population groups and technical assistance to help loggers adopt sustainable forest management practices.* The last intervention, the benefits of which are being quantified by the project, is intended to reduce the potential leakage impacts of the displaced loggers' activities outside the project area.

The project will also fund supporting activities, including an intensive monitoring and verification program, project oversight by the Bolivian government, park protection, scientific and technical research, enforcement of forest management requirements, and local

Box 1: Forest conservation - The Noel Kempff Mercado Climate Action Project, Bolivia

government participation in the project.

Project developers report that $900,000 will be used to establish a comprehensive and verifiable GHG monitoring, recordkeeping, and reporting regime. The details of this regime will be set forth in a monitoring and verification protocol, which will be developed as a milestone during the project's preparatory phase. The effort includes developing and maintaining a refined baseline and project reference case. New aerial photography and satellite imagery of the project area will be conducted on a regular basis. Data collection also will focus on gathering pertinent information regarding secondary area populations, including immigration rates, land-use needs, management practices, demands for new land, and rates of expansion of commercial agriculture, as well as information on subsistence hunting, gathering, and agricultural practices (The Nature Conservancy, 1996).

Forest conservation projects are not sequestration projects. Instead, a successful forest conservation project will reduce GHG emissions in much the same way as other emissions reduction projects (Fearnside, forthcoming). Forest conservation projects have been questioned, however, on the basis of whether the threat to the area in question was simply displaced to another geographical location and based on concerns regarding the long-term success and viability of the projects.

3.2.2 Reforestation/Afforestation

The IPCC has identified a range of reforestation, land-use change, and afforestation options that can contribute to climate change mitigation efforts (IPCC, 1996). These approaches include pasture and cropland reforestation and regeneration, degraded or arid land reforestation, planting along highway rights-of-way and riparian corridors, and planting in windbreaks and other agroforestry applications. There is a tremendous ecological need for restoring tree cover to many degraded and marginal lands around the world that, although previously forested, are now marginally productive (Trexler and Haugen, 1994). Restoration of tree cover would certainly increase long-term carbon storage on those lands. Reforestation here is thus used in the context of converting an existing agricultural or other land-use to forestry use, rather than the reforestation of recently harvested stands. The Reforestation of Uganda National Parks project, pursued by the FACE Foundation, is a good example of a reforestation project in a developing country.

Reforestation has been a popular AIJ pilot phase strategy and is being pursued both in Annex I and non-Annex I countries. It has been the primary strategy used in Annex I countries, where forest conservation projects have not been actively pursued. Reforestation projects include long-term and short-term reforestation projects, and potentially commercial as well as non-commercial forestry projects. Concerns have been raised, however, about the quantifiability of the C benefits associated with such projects, benefit leakage if pursued within the context of functioning timber markets, and permanence of the carbon storage benefit (Trexler and Associates, 1998b).

Box 2: Reforestation – FACE Foundation, Uganda.

This project is part of the Dutch FACE Foundation's efforts to offset the CO_2 emissions of a hypothetical 600 MW coal-fired power plant in the Netherlands. Funding for the projects was provided by the Dutch Electricity Generating Board through a small self-imposed electricity tax. The projects all involve reforestation (FACE eliminated forest protection activities from consideration) and are taking place in both Annex I and non-Annex I countries.

This particular project provides for restoration and conservation of damaged forest lands in a national park in Uganda. It was created shortly after Ugandan President Museveni started a rural development program aimed at decreasing forest destruction by people who live in or near the forests. The program offers economic alternatives to traditional income-generating activities that rely on forest products (small-scale plantations, grazing livestock, and timber harvesting). When the President's program got off to a successful start, the government contacted FACE Foundation and asked for assistance in reforesting previously damaged lands. The President's goal is to plant 35,000 hectares of forest land in two national parks, Mount Elgon National Park (25,000 has) and Kibale National Park (56,000 ha.).

FACE agreed to fund and help design a two-year pilot reforestation program. This first phase targets 5,250 hectares. FACE has an ultimate goal of reforesting 27,000 hectares over a longer period of time.

A good deal of effort went into infrastructural development during the project□s pilot phase, which was completed at the end of 1996. In Kibale National Park, FACE has sponsored construction of a seedling nursery, offices, staff quarters, and nursery irrigation facilities. The nursery has a capacity for 300,000 seedlings and was producing to capacity (with 30 different indigenous species) by the end of 1994. A similar nursery was built in Mt. Elgon National Park. Project developers anticipate that seedling nurseries will eventually be privately run operations owned by local people (FACE Foundation, 1997).

3.2.3 Reduced Impact Logging/Forest Management

In some forest systems, research suggests that harvesting damage to the forest can be reduced by as much as 50 percent by reduced impact logging (RIL) measures. RIL includes removing vines before cutting, directional tree-felling, and better planned extraction of timber on properly constructed and utilized skid trails (Pinard and Putz, 1997). The best-known project in this category is New England Electric System's Reduced Impact Logging Project in Sabah, Malaysia (*see* project profile in Box 3). RIL components can also be found in the Rio Bravo Conservation and Forest Management

Project in Belize. Other forest management measures with potential carbon benefits include increasing stocking rates on poorly stocked timberlands; thinning overstocked stands where growth is being impeded; lengthening harvest rotations; and managing more actively to protect forest resources against fire, insects, and disease. Questions that have been raised regarding RIL projects include the ability to reliably quantify relatively modest changes in stand management strategies, the potential distinction between short-term and long-term C benefits, and the creation of potentially perverse policy incentives in encouraging poor harvesting practices as a means of attracting climate change mitigation funding.

Box 3: Reduced impact logging – Malaysia.

This project was one of the first climate change mitigation projects pursued, although it has not been successful in gaining formal AIJ status. New England Electric System entered into a 1992 agreement with Zakyat Berjaya Sdn. Bhd. (a subsidiary of Innoprise, a large forestry concession owner in Malaysia) to develop and implement RIL techniques on 1,400 hectares of commercial forest land the company held under concession. The project goal was to demonstrate that alteration of harvesting techniques in natural commercial forests can provide credible forest-based GHG offsets on an industrial scale.

The demonstration site is in an area of uninhabited virgin lowland dipterocarp rainforest. Historical harvest practices within the concession area have been extremely destructive, frequently destroying as many as 50 percent of untargeted trees and compacting up to 40 percent of the land area, leading to erosion and other environmental problems. This limited the potential for regrowth within the remaining forest stand.

RIL guidelines were developed with the goal of reducing logging damage by 50 percent through pre-cutting vines, directional felling, and planned extraction of timber on properly constructed and utilized skid trails. Vine cutting reduces felling damage because the tree crowns are not tied together; it reduces post-felling vine infestations because there are fewer fallen vine stems to re-sprout. Project guidelines requiring stream buffers and restricted harvesting on severe slopes (>35 degrees) also reduce damage and contribute to conservation of residual forest. The result is a management regime that is expected to reduce CO_2 emissions during and after the harvesting process; enhance sequestration potential in the residual forest; make logged forest less susceptible to fire; and reduce soil erosion.

3.2.4 Increasing Soil C Sequestration

Conventional agricultural practices tend to reduce organic soil carbon levels on croplands; additionally, forest-clearing usually leads to the loss of a significant fraction of the soil organic C. Recently, the use of no-till or low-till cropping systems have been gaining attention as a means to replenish soil organic content and sequester C (Lal, *et al.*, 1998). A number of projects are being proposed or developed in this area, although few have yet been implemented. A soil C enhancement project in Saskatchewan, Canada, funded by a consortium of Canadian companies, pays landowners to pursue no-till agricultural practices. Prairie-grass restoration projects are another means by which a great deal of soil C can be sequestered in the deep root systems of such grasses (Owensby, 1993). Concerns regarding soil C sequestration projects have been raised in connection with the relatively small accumulation of carbon per unit of land area, the time periods involved, and the potential loss of benefits should future agricultural economics dictate reversals of no-till land management strategies.

3.2.5 Biomass Energy

Biomass energy projects are a hybrid between the LUCF and energy sectors. Many of their GHG benefits result from fossil fuel displacement, but some proportion of a biomass energy project's GHG benefits can result from increasing the average standing biomass on a land area and from soil carbon accumulation. Biomass projects include open-loop projects involving agricultural and industrial wood residues and closed-loop projects based on annual crops or short-rotation trees (Wisniewski and Sampson, 1993; Princeton University, 1999). Although a great deal of technical work is underway on biomass energy options and their contribution to climate change mitigation, relatively little practical experience has been gained during the AIJ pilot phase (Marland and Schlamadinger, 1995; Schlamadinger and Marland, 1996; Schlamadinger and Madlener, 1998; Marland et al, 1996; Apps, *et al.*, 1999).

3.2.6 Other LUCF and Biotic Options

The great majority of LUCF projects pursued to date fall into the categories described above. Many other LUCF options are being investigated, particularly involving a variety of forest management interventions, increased use of wood products, and anti-desertification efforts (Schlamadinger and Waupotitsch, 1996; Brown et al, 1998; Trexler and Broekhoff, 1995). Although forest-fire suppression has been identified as one of the most cost-effective mitigation options available in Russia and fire suppression may be one of the most cost-effective mitigation options in Africa, it is difficult to pursue at the project-specific level. There are no AIJ projects in this category. Outside the LUCF arena, but still involving manipulation of biological systems, are proposals such as protection and restoration of coral reefs and ocean fertilization. Very little experience with such projects outside the LUCF sector yet exists.

4. REVIEWING LUCF CONCERNS AND EXPLORING THE LESSONS OF PROJECT EXPERIENCE

Project-based climate change mitigation efforts have been the subject of intense discussions in connection with the AIJ pilot phase. LUCF projects have received particular attention from critics. Notwithstanding the recognized environmental and socioeconomic benefits of many LUCF-based mitigation efforts, interest groups ranging from developing countries to environmental organizations have questioned forestry's role in helping achieve climate change mitigation goals and have questioned whether forestry activities can truly help as offsets through the JI or CDM mechanisms (Barnett, 1992; Friends of the Earth, 1997; World Wildlife Fund, 1995).

Most of the debate surrounding LUCF projects as the AIJ pilot phase winds down arises out of the following concerns (Trexler and Associates, Inc., 1998c):

- Social and economic concerns regarding conformance of LUCF-based mitigation projects to national sustainable development and general environmental objectives;
- Technical concerns regarding quantification and crediting of project-level LUCF interventions, *e.g.*, issues surrounding benefit quantifiability, leakage, and permanence. These concerns are coupled with the perception that assessment of GHG benefits for LUCF projects, and the benefits themselves, are qualitatively different than those associated with other mitigation project types; and,
- Concerns regarding the perception that a LUCF project will displace other mitigation efforts including needed energy-sector changes, and swamp a developing GHG mitigation market.

Although provided here in capsule form, these three groups of issues provide a good summary of key criticisms that have faced LUCF projects during the AIJ pilot phase. LUCF project continue to face these issues as their role under Articles 6 and 12 of the Kyoto Protocol is debated prior to COP-6 and potentially beyond. The lessons below are framed in terms of concerns that have been expressed and are based on empirical experience with LUCF projects (Faeth et al, 1994). Given the complexity of some of the issues, the still-limited amount of experience from which to draw, and the rapidly expanding amount of work going on in the LUCF arena, the authors recognize that the lessons proposed here deserve further evaluation and documentation. The projects and subjects covered has been the subject of extensive work and a complete review of each is beyond the scope of this chapter. The lessons below are proposed here in their current form in the spirit of advancing our understanding of what we do and do not know about LUCF project experience to date. These lessons learned contribute to the debate over how to guide future project-based mitigation efforts as the AIJ pilot phase ends.

4.1 LESSON #1, MOST IF NOT ALL EXISTING LUCF PROEJCTS ARE CONTRIBUTING TO ECONOMIC OR ENVIRONMENTAL OBJECTIVES IN HOST COUNTRIES

The formal AIJ LUCF projects pursued to date have received the support of host country governments. Most have received support from key environmental and development NGOs. The same conclusion applies to most projects pursued outside the formal AIJ pilot phase. A number of countries, particularly in Central and Latin America, have openly stated that the LUCF projects pursued there advance their long-term sustainable development and environmental agendas.

Most concerns regarding the possibility of LUCF projects impeding economic and environmental objectives of host countries are based on a combination of three assumptions:

- land occupied by forestry offsets would deprive countries of alternative economic development opportunities and potentially impede national sovereignty over their natural resources;
- resources going into forestry offsets would displace funding that otherwise might become available for activities more directly beneficial to economic development; and,
- projects would be pursued in ecologically and socially inappropriate ways, rather than being pursued on degraded or marginal lands where there is little question that projects would be environmentally beneficial.

It is certainly possible to envision circumstances in which these outcomes could result from the pursuit of particular projects. LUCF projects to date have not chosen this course. The hypothetical construct of massive plantations being pursued for climate change mitigation with negative environmental and socioeconomic impacts has been used to justify the argument that LUCF project mitigation role should be denied or limited. The nature of the concerns brought up are represented by the following excerpt from the literature (Barnett, 1992):

> Techniques involving fast-growing trees in species-poor plantation conditions would most likely be used as part of a global warming mitigation effort. If large-scale tree planting is not undertaken on the basis of meeting local needs, it could lead to severe impacts on local agricultural production,

foster land conflicts between local residents and outside interests, increase rural poverty and landlessness by denying local access to land, promote migration to urban centers, and possibly even lead to the compensatory conversion of remaining areas of natural forest to agricultural uses.

Despite these concerns, no LUCF projects pursued under the AIJ pilot phase has involved large-scale plantation forestry. There is little reason to anticipate that massive tropical reforestation will be the economically favored approach to climate change mitigation, given that experience with tropical reforestation projects has shown them to be risky and expensive (Evans, 1992; Trexler and Meganck, 1993; Trexler and Haugen, 1994). Other options, such as forest conservation, forest regeneration, modified forest management practices, and agroforestry, are likely to prove more cost-effective means of advancing climate change mitigation goals in most circumstances (Trexler and Haugen, 1994). The Electric Power Research Institute (EPRI), looking at Siberian forestry, came to a similar conclusion (EPRI, 1996). EPRI noted that there is a common tendency among forestry managers to associate carbon sequestration with high-density plantations. EPRI concluded, however, that low-density plantations reduce project costs and often allow hardwoods to grow naturally between plantation rows. This process enhances biodiversity as well as fire and pest resistance.

Indeed, empirical evidence from the AIJ pilot phase actively suggests that LUCF projects can contribute a great deal to economic and environmental objectives in host countries, including positive impacts on wildlife habitat, watershed quality, reduced soil erosion, and socioeconomic well-being of local communities (Trexler and Associates, Inc., 1998b). Several examples, drawn from a range of project types and locations, are provided in Table 3 to illustrate some of the environmental and social co-benefits claimed by project developers.

Slowing forest loss and land-use degradation, as well as returning land to forest cover, can advance sustainable development, energy production, and environmental goals in developing countries, all while adding to terrestrial carbon stores. Often overlooked in the technical climate change mitigation debate is the tremendous role that LUCF projects, appropriately designed and implemented, can play in biodiversity conservation, sustainable development activities, watershed protection, food production, and other areas that are high societal priorities. Indeed, forestry critics, while raising concerns, acknowledge that forestry projects can result in environmental and social benefits (Barnett, 1992).

Going beyond the empirical experience with LUCF projects is the fact that future projects pursued under Articles 6 or 12 of the Kyoto Protocol will require multiple governmental approvals. Moreover, these projects will presumably have to satisfy decision-makers that they do in fact advance environmental and economic objectives in the host country.

4.2 LESSON 2, MANY DETRIMENTAL PROJECT TYPES CAN BE PRECLUDED BY FORMULATION OF CLEAR PROJECT CRITERIA AT THE NATIONAL AND INTERNATIONAL LEVELS

One concern that has been voiced by some students of the Kyoto Protocol is that recognition of reforestation as a credit-generating strategy, coupled with the failure to recognize forest conservation as a credit-generating strategy, will encourage conversion of existing tropical forests to plantations. This scenario could be prevented by a simple decision rule restricting reforestation credits to lands not under forest cover in 1990 or some other specified baseline period. The same conclusion could presumably be applied to other project interventions where there is agreement that they should not be pursued for climate change mitigation purposes.

4.3 LESSON 3, LUCF PROJECTS HAVE NOT FUNDAMENTALLY INTERFERED WITH EMISSIONS REDUCTION OR TECHNOLOGY TRANSFER OBJECTIVES

Some observers have characterized forestry's likely role as a mitigation strategy as being to undercut or displace achievement of other objectives, *e.g.*, technology transfer. Seen in this way, LUCF projects could be portrayed as a negative from the standpoint of other climate change mitigation goals, even if the ability of individual forestry projects to offset CO_2 emissions were undisputed. This concern involves three assumptions:

- Land use-based emissions reductions are less real than other reductions. It is undisputed, however, that more than 1 GT of carbon is released to the atmosphere annually through land-use change, particularly from deforestation in the tropics. If these emissions can be permanently avoided through LUCF interventions, this type of emissions reduction is as real as other kinds of emissions reductions in the energy sector.
- Forestry mitigation opportunities will compose more than just part of a societal mitigation portfolio *and* will squeeze out other mitigation projects. While forestry has been a popular mitigation measure during the AIJ pilot phase, projects pursued since initiation of the pilot phase are extremely diverse. The great majority of projects have been in a variety

of energy sectors. Indeed, LUCF-sector projects have had increasing difficulty in competing with energy-sector projects that claim to deliver financial returns in addition to GHG benefits. This gives little reason to believe that LUCF sector projects will come to dominate mitigation efforts in the future.

- Forestry projects offer few or no technology transfer opportunities. A number of technical and resource technologies can be transferred through projects involving forest management, reduced impact logging, wood product substitution, forest conservation, and reforestation. Transfer of these technologies will carry positive co-benefits, much as is anticipated for technology transfer in other sectors.

Although it is possible to envision situations in which LUCF-sector projects are at odds with emissions reduction and other policy objectives including technology transfer, there is little evidence to suggest that these scenarios will come to pass. Moreover, it should again be relatively simple to prevent such outcomes through the development of suitable criteria for JI and CDM projects.

Table 3: LUCF projects in AIJ pilot phase and associated co-benefits.

Project Title	***Reforestation and Forest Conservation (Costa Rica/Norway)***	***ECOLAND: Piedras Blancas National Park***	***Reduced Impact Logging for Carbon Sequestration in East Kalimantan***	***RUSAFOR: Saratov Afforestation Project***
GHG Impact (CO_2 equivalent in metric tons)	*231,000*	*1,343,000*	*134,379*	*300,000*
Economic Benefits	*The four hydroelectric plants in the area depend highly upon the hydrological conditions found in the upper river basin, precisely where the forestry project will have its area of influence. The project will improve existing hydrological resources and water quality, increasing the efficiency of the hydroelectric generation. As a result, the dependence on fossil fuels for electricity generation in the national electric system will be reduced and the cost of electricity generation will decrease. Furthermore, the provided scenic beauty will promote ecological tourism attraction in the area.*	*The government of Costa Rica actively supports the ECOLAND project. Most tourists are attracted to Costa Rica by its natural areas. The project is compatible with and supportive of the country's economic development and socioeconomic and environmental priorities.*	*The anticipated standardization of certification guidelines after 2000 could have a significant impact on the Indonesian economy. This project will build local capacity to understand and implement sustainable logging systems in preparing for post-2000 certification standards. The project will also allow companies to begin meeting these standards and accessing niche certification markets in the United States and Europe.*	*Although technology transfer associated with the project has been minimal regarding forestry issues, the project has made significant progress in advancing the thinking of the Russian partners to the free enterprise system and capitalism.*
Social and Institutional Benefits	*The involvement of local small and medium- size landowners in forestry activities, through individual contracts, will improve the income and living conditions for residents in the watershed area.*	*The project developer states that there are no identifiable negative developmental impacts of the project. Landowners living on parcels purchased through the project are expected to relocate to existing farms outside the Piedras Blancas National Park or into urban areas where they will have greater access to amenities.*	*Concession areas are sparsely inhabited. There are no data to suggest or reasons to suspect that implementation of RIL guidelines will affect local or regional supply, demands, or price of timber products. Improved market access or price premiums may result if the area is certified by an accredited agency and is well managed.*	*Regardless of whether the project will have an economic impact on the surrounding area, it has become a source of pride for the nearby communities. Community commitment provides an additional level of stewardship, enhancing formal monitoring and risks reduction activities conducted as part of the project plan.*
Environmental Benefits	*– mitigation of GHGs and protection of biodiversity.* *– protection of aquifers, reduced soil erosion, improvement of water quality, and stabilization of hydrological regime.*	*– provision of critical habitat for mammals and birds that are extinct or threatened in other parts of their range.* *– maintenance of water quality and reduction of soil erosion.* *– most important concentration in Costa Rica of biodiversity not under adequate protection.*	*– enhancement of logging operation sustainability.* *– improvement of tree growth.* *– encouragement of natural regeneration.* *– maintenance of forest cover.* *– protection of soils from bulldozer damage.* *– reduction of erosion.* *– maintenance of bird populations.*	*– reduction of erosion.* *– enhancement of soil nutrient content.* *– provision of wildlife habitat.*

4.4 LESSON 4, MOST ISSUES FACED IN DESIGNING AND CREDITING LUCF MITIGATION OPTIONS ARE THE SAME AS THOSE FACING ANY CREDITING OF PROJECT LEVEL INTERVENTIONS, RATHER THAN BEING SPECIFIC TO THE LUCF SECTOR

A frequently voiced concern with LUCF projects is that they pose qualitatively different issues from the standpoint of evaluating their climate change mitigation benefits, and correspondingly, that LUCF projects need to be dealt with separately in the development of project-level standards and guidelines. Participants to an LUCF issues-identification workshop, which included experts versed in energy offsets, concluded that many LUCF projects are of comparable mitigation quality and involve a similar degree of implementation ease as most other projects including in the energy sector. They expressed some concern that the controversy over being able to compare the two types of mitigation measures may be the result of a naive community of forestry experts who openly share the strengths and weaknesses of measurement capabilities with a policy community not sufficiently prepared for interpreting this discussion. As one participant said: We [forestry experts] have done some damage in getting too involved in technical discussions. As a result, we have confused policymakers. The technical issues for forestry are no more perplexing than they are for energy offsets. Voicing support for this view, another participant stated that the central issue we need to address is not what our confidence level in our forestry measurements are, but to make it clear that forestry offsets can accomplish the same levels of accuracy as energy at equivalent levels of effort. The issue is comparability (Trexler and Associates, 1998b).

Technical concerns most commonly voiced are:

- additionality of LUCF projects;
- potential leakage of project benefits;
- quantifiability, monitorability, and verifiability of LUCF project benefits; and
- permanence of LUCF project benefits.

The following sections evaluate these technical issues in the context of project-level experience in the LUCF sector. This chapter makes no attempt to actually solve these issues, whether for LUCF or other mitigation projects, given the complexity of the issues and the amount of ongoing work in these areas. The discussion does explore how the issues have been addressed for selected LUCF projects. The issues are fundamentally different than those raised across the board in discussing quantification and crediting of project-level mitigation efforts, particularly in an un-capped crediting system such as the one associated with implementation of Article 12 of the Kyoto Protocol. Indeed, a clear conclusion of the AIJ pilot phase experience across all sectors illustrates that there are relatively few systematic differences among the issues that must be looked at in quantifying and crediting individual mitigation projects (Trexler and Kosloff, 1998). This, combined with the heterogeneity of options in both the LUCF and other mitigation sectors, means that it is probably inappropriate to make general statements about the quality of energy-sector vs. forestry-sector mitigation projects. Projects and sectors need to be evaluated against the criteria that are eventually established and judged on the basis of their ability to satisfy those criteria.

4.4.1 The financial and environmental additionality of LUCF projects

Additionality of most LUCF projects has not been a major source of debate during the AIJ pilot phase (as distinct from leakage, discussed subsequently). There has been little doubt, for example, regarding the financial additionality for the forest conservation and reforestation projects that have characterized the AIJ pilot phase. Projects have also faced relatively few challenges on the basis of environmental additionality. With forest conservation projects, for example, environmental additionality involves the question of whether the forest area was truly threatened with loss and whether the project would change that outcome. Projects to date have focused on areas where the case of environmental additionality could be made in a relatively straightforward way.

As the number of LUCF projects increases, and as other categories of LUCF project categories are increasingly pursued, the clarity of LUCF project additionality is likely to decrease. For some LUCF options mentioned earlier in this chapter, financial and environmental additionality are both likely to be an issue in the eyes of observers. Commercial plantations and some forest management interventions, for instance, may generate strong financial additionality concerns. Like most energy-sector projects, these projects usually have an economic rationale motivating investment. Even forest conservation projects, although posing little or no financial additionality questions, may be increasingly subject to environmental additionality concerns.

Table 4 briefly describes the additionality arguments made for several existing LUCF project types and geographic areas. Because there have been so few LUCF projects with any financial return, few empirical examples of proofs of financial additionality exist. Over time, however, the current project-by-project approach to baseline determination and additionality assessment for LUCF projects will almost certainly prove as unacceptable for LUCF-based mitigation efforts as it is likely to prove for energy-sector mitigation efforts. This is because of the unavoidable subjectiveness of estimating what would have happened but for any specific project (IEA, 1997). The difficulty with this approach is that best guesses can vary widely, allowing analysts to come to completely different legitimate conclusions as to both financial and emissions additionality of a project. It also creates an incentive for project developers to creatively overstate project benefits and encompasses high transaction costs.

The concept of benchmarks is being put forward as a potential alternative to project-by-project assessments. Benchmarks are currently being considered primarily in the context of energy-sector mitigation efforts, but they may also have application to LUCF projects. Based on some projection of business-as-usual performance at the sectoral or country level, benchmarks would establish the standard-to-beat for projects seeking CO_2 credits. Once established, projects meeting or going beyond the benchmarks would receive CO_2 credits for doing so. Although less site-specific than a project-level additionality review, and hence capable of missing nuances specific to an individual project, the standardized benchmarking approach would presumably be less susceptible to project-level gaming. This approach is also more conducive to guiding mitigation projects and activities in particular policy and project directions, and presumably entail lower transaction costs than true project-level analyses.

As with energy-sector interventions, LUCF benchmarks would be applied to all projects within a given sector, looking at historical and sectoral information to develop a benchmark by which project developers and evaluators could assess project additionality without having to develop a project-specific methodology and assessment (Carter, 1997). As with energy-sector benchmarking, this approach applied to the LUCF sector implicitly assumes that inaccuracy of the resulting benchmarks at the project level is statistically unimportant when averaged over many projects. Moreover this approach assumes that the systematic error associated with the sectoral approach is less than the random or systematic error associated with project-by-project additionality assessments. Consolidating baseline-setting efforts across an entire class of projects should also offer the opportunity to achieve increased credibility and fewer transaction costs.

Overall, LUCF additionality has not been as vexing an issue during the AIJ pilot phase as it has for energy-sector projects, where almost all the AIJ projects have been challenged on additionality grounds. Over time, LUCF projects are likely to be less distinct in this respect from their energy cousins. Long-term resolution of additionality issues for project-level mitigation interventions, whether involving benchmarking or some alternative, involves almost identical issues for all sectors.

TABLE 4. LUCF projects in the AIJ pilot phase and additionality characteristics.

Project Title	Additionality
Reforestation and Forest Conservation (Costa Rica/Norway)	The Norwegian Ministry of Foreign Affairs and the Norwegian Consortium, a private-sector participant made AIJ funding for implementation of this project available. The Norwegian Climate Fund is a separate budget line that has been established in addition to Norway's overseas development assistance budget, and is argued to satisfy financial additionality.
ECOLAND, Piedras Blancas National Park, Costa Rica	Although the forestland protected by the project is located within national park boundaries, it was threatened by logging and conversion to agricultural and pastoral lands by those who own the land. As a result of scarce funds, the Costa Rican government has been unable to buy out the private landowners, and they are legally permitted to utilize the land. It was estimated that without the project, 2,150 ha of forestland would have been deforested over the next 15 years. The project brought nearly 20 percent of the park is land under government protection.
Reduced Impact Logging for C Sequestration in East Kalimantan	In the absence of the project, uncontrolled and destructive logging practices were predicted to continue in the concession area that is the focus of the RIL project. Studies in neighboring Malaysia indicated that conventional logging practices break and uproot as many as 50 percent of the remaining trees and disturb soils on up to 40 percent of the land area.
Scolel Te: Carbon Sequestration and Sustainable Forest Management in Chiapas	This project assists farmers primarily in nine Mayan indigenous communities with developing small agroforestry and forestry enterprises. GHG benefits accrue from forest growth that would not have occurred in the absence of project activities.
Noel Kempff Mercado Climate Action Forestry Project	There are expectations that the park will be threatened by human- induced changes to the land and water ecosystems by expansion of neighboring communities. Recent colonization at the park's borders has resulted in continued illegal extraction of mahogany and cedar, hunting and trading of endangered species, and expansion of agricultural and cattle ranching on park lands. Project funding will counter these threats.

4.4.2 Potential Leakage of Project C Benefits

In the context of project-based mitigation interventions, CO_2 leakage refers to GHG effects of a project that occur outside the project's monitoring boundary. Leakage, to the extent it exists, results in a misstatement of a project's GHG benefits. Leakage can be difficult to identify and even more difficult to quantify. The Section 1605(b) guidelines under the U.S. National Energy Policy Act of 1992 recognize at least four types of effects that should be considered in reporting GHG emission projects (U.S. Department of Energy, 1994). These are:

- activity shifting (moving processes outside of a project or company boundary);
- outsourcing (purchasing commodities or services formerly produced);
- life-cycle emissions shifting (upstream and downstream changes in processes or materials used as a result of a specific project intervention); and
- market effects (due to post-intervention equilibration of supply and demand).

LUCF projects, particularly forest conservation projects, are often characterized as being leakage-prone. People who live near new forest preserves, for instance, are likely to still have a need for wood products (either to sell or to use themselves) and may simply turn to another forest resource if one is not available (Fearnside, 1996; IPCC, 1996). While this is sometimes termed physical leakage, others have noted that basic market economics would lead to leakage of some of a commercial reforestation project C benefits as other fiber suppliers respond to increasing fiber supplies and falling prices. Benefit leakage typically is thought of as leading to a net loss of project benefits. However, leakage can also be positive, however, leading to an underestimation of project benefits (Sathaye *et al,* 1998). This is particularly true for projects that might demonstrate a new technology or land-use, leading to its dissemination in an area.

Leakage is clearly a potential issue of LUCF projects, as it is for project-level mitigation efforts generally. For any mitigation project, questions can always be raised regarding how the boundary within which the project benefits are quantified was drawn and whether there might be impacts on feedbacks from outside the boundary area. Large-scale fuel switching from coal to gas, for example, would lower the price of coal and lead to the opening of new markets. With energy efficiency projects, the so-called snap-back effect refers to the situation where the purchase of more efficient appliances is likely to result in increased utilization, since their efficiency has effectively reduced the cost of running the appliance. Both of these are examples of market leakage.

Little quantitative analysis has been conducted on the magnitude of leakage from either energy-sector or forestry-sector projects, although some modeling has been done at the national and international levels (Perez-Garcia, 1994). There is no obvious reason, however, to assume that LUCF projects are inherently more or less likely to suffer from leakage, or that the leakage is likely to be of a greater or lesser magnitude (Trexler and Associates, 1998b). Both LUCF and other types of mitigation projects can affect the supply and demand of the commodity being traded (whether energy or timber products), leading to leakage.

At this point, insufficient analytical work has been conducted to determine that we should either ignore leakage or use its threat as a reason to exclude certain project types. Acceptable methods of dealing with leakage concerns need to be investigated for all mitigation projects, including those in the LUCF sector (Box 4). The World Resources Institute has created a leakage index, which identifies leakage potentials associated with different biotic project types and strategies for addressing leakage (Brown et al, 1997). Some work has also been done to attempt to systematize treatment of project-level leakage, proposing a nesting-based monitoring system intended to allow individual offset projects to be tracked both individually and within a country's national accounting system (Andrasko, 1997). Dudek and Weiner (1996) have suggested that the ideal mechanism for measuring the emissions abatement effects of any mitigation project would be general equilibrium analysis to reflect ramifications in the national and global economies. It may turn out that project-by-project leakage assessment suffers from the same problems as project-by-project additionality assessment and will require development of an alternative approach.

Box 4: Empirical approaches to LUCF leakage mitigation.

There is considerable debate as to the best way to deal with project-level leakage of carbon benefits. It has received a high profile in discussing forest protection projects, where some critics have asserted that leakage can approach 100 percent. Some attention has been devoted to solving the □leakage problem□ at the project level, as illustrated in the following examples.

Developers of the Noel Kempff Mercado forest conservation project in Bolivia argue that the project has controlled leakage by incorporating demand-based mitigation activities into the project. This reduces the need for local inhabitants to simply switch their unsustainable activities to areas outside of the new park□s boundary. These mitigation activities include agroforestry to provide sustainable sources of wood, employment generation activities, and equipment retirement schemes, each addressing a different potential leakage type.

The Noel Kempff Mercado project also employs innovative contractual approaches to leakage mitigation. The concessionaires being bought out by the project committed to technical assistance and training in sustainable forest management practices to be implemented on their remaining concessions (which could lead to positive leakage, since those benefits were not counted toward the project□s benefits); not to use money received from the project to purchase new concessions; and to abandon concession logging equipment onsite. These interventions were specifically pursued for leakage mitigation purposes.

Developers of the ECOLAND forest conservation project in Costa Rica argue that the potential for project-level leakage is largely precluded by the project□s Costa Rican location and that leakage considerations were a major factor behind the project□s being pursued there. With so little natural forest left outside of protected areas, project developers argue that there simply wasn□t much potential for leakage to occur. Landowners selling land to the project for transfer to the Costa Rican national park service were expected to buy established farms or move to the city, as opposed to invading other natural forest areas.
Costa Rica, in putting together its *Territorial and Financial Consolidation of Costa Rican National Parks and Biological Reserves* project, argued that leakage is controlled by the fact that the program is national. For leakage to occur, it argues, forces leading to leakage would have to cross international borders. To the extent that the source of most forest loss is local agricultural or logging pressures, there may be little basis to believe that such pressures would move to neighboring countries.

4.4.3 Quantifying, Monitoring, and Verifying C Benefits

In estimating total carbon benefits of an offset project, a baseline of emissions reduction or sequestration-related activity must be established against which to measure change. With energy projects, the baseline is the level of emissions that would have occurred in the absence of the GHG project. With forestry, the baseline is the level of C storage that would have existed in the absence of the offset. LUCF offsets have historically been questioned on the basis of having baselines that are more difficult to establish and measure than energy baselines (Richards, 1994). Similar arguments have been made with respect to monitoring and verification of these carbon benefits.

For a number of years, the challenges associated with quantifying, monitoring, and verifying LUCF carbon benefits were perceived as the primary challenge to pursuit of LUCF projects for climate change mitigation. Over time, however, a great deal of work has been done in this area (*e.g.*, Sathaye *et al.*, 1997; MacDicken, 1996; MacDicken, 1997; Vine and Sathaye, 1997; Rombold, 1996; Trexler and Associates, 1998b; World Bank, 1994; Moura-Costa, 1997), including at the project level (The Nature Conservancy, 1996; SGS Forestry, 1997). Quantification of benefits has varied widely for LUCF projects, ranging from simply extrapolating from desk studies to in-depth empirical research. This also applies to monitoring and verification of project benefits; some projects have simply assumed that monitoring and verification provisions will be retrofitted onto the project at a later date, while others have developed and begun implementing extensive monitoring and verification protocols. Some information regarding a number of projects reflected in Table 5.

A group of experts looking at quantification and monitoring issues recently concluded that the primary problem with quantification and monitoring is the absence of a standardized approach, rather than theoretical inability to arrive at a standardized approach (Trexler and Associates, 1998b). Workshop participants concluded after considerable discussion that accumulation of carbon in on-site vegetation associated with forestry offset projects can be measured with high precision, up to ±10 percent, with a confidence of 95 percent in most situations. They recognized that the difficulty of getting to this level of confidence would vary by project types and individual project contexts. They noted that this level of precision is comparable to that found in many energy-sector offset projects. The workshop concluded that ongoing monitoring of a project's CO_2 benefits does not pose significant problems for some offset types where remote sensing can reliably be used (Skole *et al.*, 1997). For other project types, including agroforestry projects, on-the-ground requirements of project monitoring are more intense.

Because so many mitigation projects have been in the LUCF sector, these projects have received the most attention. But the issues relevant to quantification and monitoring of project benefits including baseline and boundary setting, measurement, and leakage questions arise in almost every type of project-level mitigation effort. Importantly, there are enough pilot phase LUCF projects that have moved to implementation that experience is beginning to reveal answers to some of these questions (Brown, 1997). The Noel Kempff project (Box 1) is often viewed as an example of how empirical project-level experience has convincingly shown that issues such as quantification and monitoring of benefits can be successfully addressed for LUCF projects.

These conclusions are now widely accepted; many critics no longer focus on quantification and monitoring issues in discussions of LUCF projects. It is still common, however, to confuse quantification of national LUCF emissions and sequestration trends, *e.g.* under Articles 3.3 and 3.4 of the Kyoto Protocol, with quantification of project-level benefits.

TABLE 5. Empirical experience with LUCF benefit monitoring and verification.

Project Title	Monitoring and Verification
Reforestation in Vologda	Carbon dioxide benefits that will be measured are aboveground and root biomass C sequestration and soil C. The project will develop sampling procedures based on differences in vegetation, soils, and land-use history, and will measure tree heights and diameters.
Reforestation and Forest Conservation (Costa Rica/Norway)	SGS Forestry has developed a sophisticated carbon monitoring and verification system for the project that includes extensive satellite imagery and remote sensing. The project has been designed to ensure transparency and access for verification entities during implementation, including access to project data, methodology and verification of the carbon sequestration, and avoidance of emissions resulting from the project.
ECOLAND: Piedras Blancas National Park	The Costa Rican park service will monitor forest conservation activities in the park annually and ensure that the purchased areas are protected from incursion or logging. An Austrian group, Regenwald der Österreicher, recently established an eco-tourist lodge in the area. Lodge staff will provide on-the-ground monitoring. Periodic satellite and other photographic imaging will be used to monitor protection.
Reduced Impact Logging for Carbon Sequestration in East Kalimantan	Extensive work was undertaken during project start-up to establish project baselines. Scientists, foresters, and other Indonesian researchers and technicians will carry out field measurements to monitor CO_2 emission reduction and enhanced sequestration.
RUSAFOR: Saratov Afforestation Project	Project developers have established 53 sampling plots for monitoring. Measurements of relevant variables will be taken regularly, including stocking densities, tree heights, stem diameters, ratios of above ground to below ground biomass, and soil C contents.
Noel Kempff Mercado Climate Action Project	The monitoring and verification effort includes developing a refined baseline and project reference case. Aerial photography and satellite imagery of the project area will be conducted regularly, and new aerial photograph studies for biomass estimation are underway. Data collection also will focus on information regarding immigration rates, land-use needs, management practices, demands for land, and rates of agricultural expansion. Winrock International is managing the monitoring

4.4.4 Permanence of LUCF Project Benefits

As concern over the ability to quantify and monitor the C storage or sequestration benefits associated with LUCF projects has waned, the issue of benefit permanence has taken an increasingly high profile in political and technical debates over of the future role of LUCF projects in climate change mitigation efforts. It has also become increasingly evident that permanence is an issue where there is a qualitative difference between the LUCF sector and most other mitigation projects (Richards and Stokes, 1995; Trexler and Associates, 1998b). The C benefits of an energy-sector emissions reduction project may not be truly permanent (*e.g.*, the conserved fossil fuel may be released as economically recoverable fuel reserves are depleted). The period over which the benefit is certain is likely to be considerably longer than can be said for inherently fragile biological systems.

While the GHG impacts of many LUCF interventions can effectively be permanent (*e.g.*, long-term conservation of an existing forest or restoration of permanent forest cover on bare land). The C benefits in question certainly could fall prey to random or unforeseen events (*e.g.*, fire or pest infestation, or illegal clearing or harvesting). As a result, the GHG benefits could be released back to the atmosphere as CO_2. Indeed, even climate change itself could interfere with the long-term survival of tree cover being developed or forests being conserved as a climate change mitigation strategy. Just as important as potentially unforeseen outcomes, however, is that LUCF projects vary widely in the length of their intended benefits; some are indeed intended to be permanent (*e.g.*, forest conservation), while others are intended to provide what is effectively interim C storage (*e.g.*, commercial forest plantations).

Ambiguities surrounding the permanence of LUCF project benefits has led LUCF critics to argue that LUCF tons are qualitatively different from, for example, energy-sector emissions reductions, and that the two should not be treated as equal for climate change offset and mitigation purposes. Greenpeace, for example, has stated that it would support LUCF-sector projects as long as permanence is guaranteed over geological time-scales. Project developers, not surprisingly, feel unable to make such guarantees.

The general question of LUCF benefit permanence raises important technical and policy questions for project-level interventions:

- What is the value of temporary C storage to climate change mitigation objectives? Temporary storage certainly has value. This value, however, depends on the shape of the climate change damage function.
- At what point should temporary storage be considered permanent for project-level mitigation purposes? The value of storing C in a biological system, whether in vegetation or in soils almost certainly increases as the period of storage gets longer. At what point, however, should the period of storage be considered long enough to be equivalent to the avoided emission of a ton of fossil fuel CO_2? This question is crucial for designing and implementing LUCF projects.
- How do countries' future mitigation obligations factor into the importance of benefit permanence considerations? To the extent that

host countries adopt binding commitments that include LUCF emissions reductions, the duration of the benefits associated with individual projects will no longer be an issue from the standpoint of global environmental benefit (just as project-based additionality concerns are commonly characterized as disappearing in an emissions-capped system).

- Are avoided fossil-fuel emissions as permanent as they seem? Fossil fuels not used today could be used in the future, particularly if economic reserves of those fuels become depleted. As such, it is not clear that any mitigation project benefits can be considered truly permanent.

Answering the first question is a key objective of climate change scientists. Unfortunately, climate change damage functions are too poorly understood to properly assess the value of temporary storage. For example, if threshold effects are at work (meaning no significant climate change impacts will be felt until atmospheric concentrations of CO_2 reach a certain level), then the value of storage at one point in time is very different than at another point in time. If, on the other hand, the damage function is linear, the value of temporary storage may be a constant (Richards, 1994).

In the absence of a definitive scientific answer to this question, answering the other questions will require policy judgments. Only recently has serious analytical attention begun to be paid to these issues. Proposals are now beginning to appear examining the period of time over which storage should be maintained in order to be deemed equivalent to avoided fossil-fuel emissions. Analysis of how to credit LUCF projects in such a way that mandated storage periods are no longer necessary include the following concepts or proposals:

- Require that LUCF projects store C in perpetuity. Advocated by some interest groups, this approach would essentially eliminate LUCF projects from participating in climate change mitigation efforts. This approach does not address the legitimate value of temporary sequestration.
- Set a minimum time period for assessing LUCF projects. Under this approach, LUCF projects would have to guarantee that C benefits endure for a minimum period of time or be replaced with equivalent units. LUCF project benefits could then be considered equivalent to those of energy or other project types. Some observers have argued that storing C during a 50-75 year transition period would be sufficient to allow a global shift away from fossil fuel use and emissions and that storage over this period should be fully credited (Keystone Center, 1994; Center for Clean Air Policy, 1993). While this approach provides a clear definition of what qualifies as a CO_2 offset, it could eliminate shorter projects that still might offer some climate change mitigation benefit.
- Credit projects for the temporary storage provided. Under this approach, offset projects would be given annualized credit for storing C. One

commonly discussed way of implementing this approach, termed the unit-years approach, would be to award credits on an annual basis for the temporary mitigation value of keeping C out of the atmosphere (Fearnside, 1997). This approach still requires specification of a time period over which C storage becomes permanent for policy purposes. Credits awarded in any given year would in effect be a fraction of total project storage in that year. If permanence is defined as 50 years, for example, a mitigation project might receive two units of credit annually for each 100 units being stored. If the C were released before full credit was received, the project would receive no additional credit. Because prior credit had only been awarded on the basis of the annual value of storing the C, simply stopping issuance of credits is sufficient. There is no need for guarantees or for similar provisions. This approach has the advantage of simplicity, but has significant implications for the cost effectiveness and practicality of LUCF projects. Unit-years analytically is an intriguing approach that has received support from noted experts (Fearnside, 1997; Trexler and Associates, 1998b).

Given the lack of policy guidance on the permanence issue, its consideration at the project level has ranged widely. Many projects have simply ignored the issue. Others have set an arbitrary project life over which project tree cover is maintained. This area clearly will require a great deal of policy and technical discussion. While the permanence issue does to some extent set LUCF projects apart from other project sectors, recent analyses suggests that the issue can be resolved in a way that allows LUCF projects to participate in the achievement of society's climate change mitigation goals.

4.5 LESSON 5, EVEN THOUGH LUCF PROJECTS HAVE BEEN PERCEIVED AS INEXPENSIVE, IT IS DIFFICULT TO ASSESS TRUE COST EFFECTIVENESS OF LUCF PROJECTS

Given the ambiguities associated with specification of what will be a creditable offset, it is not surprising that there has been considerable uncertainty in determining costs of mitigation projects during the AIJ pilot phase (Sathaye, 1995). To some extent, all AIJ projects face this problem. The timeline characteristics of forestry projects makes it particularly difficult for LUCF projects:

- The sheer diversity of LUCF CO_2 mitigation options makes it difficult to consistently cost out the resulting CO_2 benefits. Benefits of some projects start immediately; for others, benefits are decades off into the future. Some projects have permanent benefits, while some have temporary benefits. Some are certain and some are high-risk.

- Different ways of crediting CO_2 benefits from the same offset project could significantly affect a project□s perceived cost effectiveness. Whether or not CO_2 benefit streams are discounted for the time value of C storage, for example, and at what rate, can significantly affect the perceived cost of a CO_2 offset (Moulton and Richards, 1990).
- One review of the economics of LUCF projects can be found in Sedjo et al. (1997).

4.6 LESSON 6, THERE IS LITTLE REASON TO SUSPECT THAT CHEAP LUCF PROJECTS WOULD SWAMP A NASCENT GHG MITIGATION MARKET, AND UNDERCUT THE PURSUIT OF OTHER MITIGATION OPTIONS

Cost-effectiveness of LUCF-sector projects will depend significantly on the accounting and crediting rules that are developed (Trexler and Kosloff, 1998). While some LUCF projects are likely to be at the bottom of the mitigation supply curve, others will not (Trexler and Associates, 1999). Some will be quite expensive compared to other mitigation options. Until criteria for CDM and other projects are formalized, any attempt to assess whether any mitigation sector will dominate the GHG market is an exercise in speculation. At the same time, it is possible to begin bounding some of the potentials based on alternative project criteria. Such work is only now getting underway.

5. RESOLVING THE ROLE OF LUCF PROJECTS IN ADVANCING CLIMATE CHANGE MITIGATION OBJECTIVES UNDER THE KYOTO PROTOCOL

The future treatment of LUCF projects under Articles 6 and 12 of the Kyoto Protocol remains ambiguous. It is to be hoped that judgements will be made with the best possible technical and policy input, including the lessons that can be learned from the last 10 years of project experience. In June 1998, the Parties, through the FCCC Subsidiary Body for Scientific and Technical Advice (SBSTA), requested the IPCC to prepare a special report on land-use, land-use change, and forestry (UN FCCC, 1998). The Special Report will address methodological, scientific, and technical issues associated with land-use trends and project-level LUCF interventions under the Kyoto Protocol. Although focused particularly on providing Parties with insight into interpreting and implementing the provisions of Article 3 of the Protocol,

Chapter 6 of the Special Report will specifically look at project-level LUCF interventions and grapple with many of the issues raised in this chapter. The Special Report, due to be delivered in mid-2000 in conjunction with the IPCC's treatment of LUCF issues in its ongoing Third Assessment Report, should significantly contribute to defining the role sinks will play under the Kyoto Protocol or any other international policy structure. Nevertheless, the report will not itself answer the difficult questions that must be addressed to finalize international policy for project-level interventions in the LUCF sector.

Based on the experience of the last 10 years, the AIJ pilot phase, and continued concerns of some observers regarding pursuit of LUCF-sector projects for climate change mitigation, several priorities can be identified with respect to advancing the state of the debate:

- Continue refinement of the lessons to be learned from the empirical project experience of the last 10 years, both in the LUCF and other mitigation sectors, with specific attention given to increasing recognition of the heterogeneity associated with project-level interventions across all sectors.
- Pay specific attention to concerns that LUCF projects might swamp a mitigation market, either by showing the concern to be misplaced or addressing it in development of project evaluation and crediting criteria under the Kyoto Protocol. Work in both areas is currently underway.
- Increase the general understanding of the differences and similarities of different mitigation sectors from the standpoint of project-level interventions. It may simply not be possible to make systematic statements at the sectoral level about the quality of mitigation options available from one sector or another. Given the heterogeneity of potential mitigation interventions in all sectors, characteristics of individual mitigation projects may need to be considered in assessing compliance with any evaluation and crediting system established under the Kyoto Protocol. Whether LUCF or energy-based, projects should be required to prove individual compliance with the standards that are established.
- Increase the amount of thinking going into development of standardized approaches for quantifying and evaluating project-based mitigation efforts, including in the LUCF sector. Ranging from resolving the additionality conundrum to the definition of permanence, and including issues of leakage and the fate of forest products in a carbon accounting system, there is a considerable amount of work still to do as the pilot phase comes to a close.
- For LUCF projects, clarify for policymakers that the technical and policy issues associated with implementation of Articles 3.3 and 3.4 of the

Protocol are different from those faced with implementation of Articles 6 and 12.

- For LUCF as well as other projects, develop project design and evaluation criteria that will avoid pursuit of projects with unacceptable environmental, economic, or social outcomes. It should be relatively easy, for example, to design criteria to prevent the potential practice of replacing existing forests with plantations in order to claim reforestation-based carbon credits.

One fundamental conclusion is that there is plenty of work to be done in this area and plenty of opportunities to move the debate forward in a constructive way.

6. CONCLUSIONS REGARDING LUCF LESSONS FROM THE AIJ PILOT PHASE

Technically and politically credible answers to questions surrounding LUCF-sector projects will be needed if LUCF projects are to play a significant role in climate change mitigation efforts. For several years, relatively little attention was given to these questions. This situation is now changing. One of the early workshops to specifically address project-level LUCF issues was convened in 1997 in Baltimore, Maryland. Attended by forestry and climate change experts and a range of NGOs the workshop focused on experience with and the status and future of LUCF-sector projects. (Trexler and Associates, Inc., 1998b). The workshop concluded that quite a bit has been learned from AIJ pilot phase experience regarding performance of the LUCF sector for climate change mitigation purposes. The workshop produced the Policymakers Summary of Conclusions presented in Box 6. It also generated two tables, reproduced here, that help summarize the status of LUCF knowledge. Table 6 compares the relative performance of several LUCF project types against a generic energy-sector project. The matrix could be made more complex by comparing a range of forestry-sector projects against a range of energy-sector projects. Table 7 identifies tentative priorities assigned to a range of LUCF-specific analytical and policy issues.

More recently, a number of workshops, papers, and special journal issues have tackled a range of issues associated with project-level LUCF interventions (*e.g.*, Nabuurs, 1999; Schlamadinger, 1999; Brown et al, 1999; Fearnside, 1999. Full treatment of this topic is beyond the scope of this chapter.

There is extensive experience with LUCF mitigation projects over the last 10 years and during the AIJ pilot phase. This experience cannot resolve all technical and policy issues regarding the effectiveness of project-level interventions for climate change mitigation. Indeed, important policy decisions remain to be made in a number of areas. At the same time, experience amassed in the LUCF sector indicates that project-level interventions can advance climate change mitigation objectives while simultaneously advancing sustainable development and socioeconomic objectives. Empirical AIJ evidence in the LUCF sector suggests little cause for concern that LUCF projects are generically contrary to the objectives of the FCCC and the Kyoto Protocol. Rules and standards governing project-level interventions generally cannot be designed to prevent pursuit of projects that do not satisfy agreed-upon standards. This is an important AIJ lesson as policy development under the Kyoto Protocol gains more momentum.

Box 6: Biotic offsets assessment workshop – policymakers summary.

Forestry and land use-based mitigation projects are more similar to energy-sector projects than is generally recognized. They face similar analytical challenges and can offer similar climate change mitigation benefits.

Forestry and land use-based carbon offset projects can be an effective tool that can provide an important component of a domestic or international climate change mitigation strategy.

There is no question that LUCF offset projects can help slow the rise in atmospheric concentrations of CO_2 for several decades or more. Among forestry's benefits, this will provide time to implement CO_2 mitigation policies and measures that require long lead times, including conversion to more efficient electrical generation technologies.

Employing a diversity of carbon offset projects will increase the likelihood of the desired carbon benefits, as it will reduce the influence of any single assumption (whether biophysical, socioeconomic, or policy) on the total accrued benefit.

Inclusion of biotic carbon offset projects among the strategies for addressing international concern about global climate change may increase resources available to support sustainable land-use and forestry practices, both of which are unlikely to be adequately funded in the absence of such a mechanism.

Biotic carbon offset projects, which include both forestry and land-use management options, provide an opportunity to support efforts to reduce deforestation and protect vulnerable forest ecosystems, many of which will be lost or degraded in the near to mid terms (within ~20 years) without additional support.

We have the ability to measure direct changes in carbon flows and stocks associated with individual forestry based carbon offset projects and the issues involved are clear. As such, we can quantify the benefits of many forestry projects as readily as we are those of many kinds of energy-sector offset projects.

The precision with which we can measure C accumulation in onsite vegetation associated

Box 6: Biotic offsets assessment workshop – policymakers summary.

with forestry C offset projects is very high -- up to 10 percent, with a confidence of 95 percent in most situations.

Whatever the concern associated with the level of measurement precision achievable for a given project, it is always possible to report the net C benefit based on the lower bound of the achieved confidence interval. Doing so makes the carbon benefits claimed highly credible in relation to energy projects, for which estimates may be more precise.

Some categories of biotic projects are capable of meeting a crediting regime, whatever that regime might be.

Accounting for leakage of benefits, if any, associated with biotic C offset projects is similar to that associated with many energy projects.
In developing baseline and reference cases for individual projects, the capability exists to identify and monitor reference plots, including on the basis of socioeconomic variables.

Third-party verification of accrued carbon presents similar technical issues in both forestry and energy projects. For both types of projects, verification improves the accuracy of C claims; it can enhance and verify the environmental and social benefits of biotic offset projects.

TABLE 6. Comparing forestry- and energy-sector mitigation projects on the basis of commonly used evaluative variables.

	Forest Conservation	Forest Management	Forest Regeneration	Watershed and Other Ecological Reforestation/ Afforestation	Commercial Plantations	Biomass Energy	Wood Product Substitution
Additionality	Easier	Easier	Easier	Easier	Similar	Similar	Similar
Leakage	Similar/Harder	Similar	Easier	Easier	Similar	Similar	Similar
Permanence	Harder	Harder	Harder	Harder	Harder	Similar	Similar
Quantification	Similar						
Baseline Establishment	Similar to Easier	Similar	Similar to Easier	Similar to Easier	Similar	Similar	Similar
Monitoring and Verification	Easier to Harder	Harder	Easier to Similar	Easier to Similar	Similar to Harder	Similar	Similar
General Protocol Standardization	Similar	Similar	Similar	Similar	Similar	Similar	Similar

TABLE 7. Prioritizing analytical and policy issues associated with LUCF-sector mitigation efforts.

	Forest Conservation	**Forest Management**	**Forest Regeneration**	**Watershed and Other Ecological Reforestation/ Afforestation**	**Commercial Plantations**	**Biomass Energy**	**Wood Product Substitution**
Additionality	Low Priority	Medium Priority	Low Priority	Low Priority	Medium Priority	Medium Priority	Medium Priority
Leakage	High Priority	High Priority	Low Priority	Low Priority	High Priority	Medium Priority	Medium Priority
Permanence	High Priority	High Priority	High Priority	High Priority	High Priority	Low Priority	Low Priority
Quantification	Low Priority	Medium Priority	Low Priority	Low Priority	High Priority	Low Priority	Low Priority
Baseline Establishment	Medium Priority	Medium Priority	Low Priority	Low Priority	Medium Priority	Medium Priority	Medium Priority
Monitoring and Verification	Low Priority	Low Priority	Low Priority	Low Priority	Low Priority	Low Priority	Low Priority
General Protocol Standardization	High Priority	High Priority	High Priority	High Priority	High Priority	High Priority	High Priority

REFERENCES

Andrasko, K. (1997) Forest Management for Greenhouse Gas Benefits: Resolving Monitoring Issues Across Project and National Boundaries. *Mitigation and Adaptation Strategies for Global Change* 2:117-32.

Apps, M.J., and Price, D.T. (Eds) (1996) *Forest Ecosystems, Forest Management, the Global Carbon Cycle.* NATO ASI Series I (Global Environmental Change), vol. 1, Springer-Verlag Academic Publishers, Heidelberg.

Apps, M.J., *et al.* (1999) Carbon Budget of the Canadian Forest Product Sector. *Envtl. Science and Policy* 2:25-41.

Barnett, A. (1992) *Deserts of Trees: The Environmental and Social Impacts of Large-Scale Tropical Reforestation in Response to Global Climate Change,* Friends of the Earth, London.

Brown, S., Bloomfield, J. and Ratchford, M. (Eds) (1999) *Land-Use Change and Forestry in the Kyoto Protocol. Mitigation and Adaptation Strategies for Global Change,* in press.

Brown, P., Cabarle, B., and Livernash, R. (1997) *Carbon Counts: Estimating Climate Change Mitigation in Forestry Projects,* World Resources Institute, Washington, DC.

Brown, S., *et al.* (1993) Tropical Forests: Their Past, Present, and Potential Future Role in the Terrestrial Carbon Budget. *Water, Air & Soil Pollution* 70:71-94.

Brown, S., *et al.* (1996) *Management of Forests for Mitigation of Greenhouse Gas Emissions.* In: Intergovernmental Panel on Climate Change, Climate Change 1995 Impacts, Adaptations, and Mitigation of Climate Change: Scientific-Technical Analyses, Geneva.

Brown, S., Lim, B., and Schlamadinger, B. (1998) Evaluating *Approaches for Estimating New Emissions of Carbon Dioxide from Forest Harvesting and Wood Products,* IPCC/OECD/IEA Programme on National Greenhouse Gas Inventories, Dakar.

Carter, L. (1997) Modalities for the Operationalization of Additionality, in U.N. Framework Convention on Climate Change, *Determination of Environmental Benefits: Part Two, The Actual Project Case,* Paris.

Center for Clean Air Policy (1993) *The Role of Forestry Offsets in Reducing CO_2 Levels,* Washington, DC.

Cochabamba Declaration (1999) *Meeting of Ministers of Environment and Forestry of the Amazonian Countries on the Clean Development Mechanism,* Cochabamba, Bolivia.

Dale, V., Houghton, R.A., and Hall, C.A.S. (1991) Estimating the Effects of Land-Use Change on Global Atmospheric CO_2 Concentration, In: *The Global Carbon Cycle,* Oak Ridge.

Dixon, R.K., *et al.* (1994) Carbon Pools and Flux of Global Forest Ecosystems. *Science* 263:185-90.

Dixon, R.K., *et al.* (1993) Forest Sector Carbon Offset Projects: Near-Term Opportunities to Mitigate Greenhouse Gas Emissions. *Water, Air & Soil Pollution* 70:561-77.

Dudek, D.J., and Wiener, J.B. (1996) *Joint Implementation, Transaction Costs, and Climate Change,* ENV/EPOC/GEEI (96) 1/REV1, OECD, Paris.

Dyson, F.J. (1977) Can We Control the Carbon Dioxide in the Atmosphere? *Energy* 2:287-91.

Electric Power Research Institute (1996) *CO_2 Offset Opportunities in Siberian Forests.*

Evans, J. (1992) Plantation *Forestry in the Tropics,* Oxford University Press, New York.

FACE Foundation (1997) *Forests Absorbing Carbon Dioxide Emission.* FACE Foundation Annual Report, Arnhem.

Faeth, P., Cort, C., and Livernash, R. (1994) *Evaluating the Carbon Sequestration Benefits of Forestry Projects in Developing Countries,* World Resources Institute, Washington, DC.

Fearnside, P. (1996) *Socioeconomic Factors in the Management of Tropical Forests for Carbon, Forest Ecosystems, Forest Management, and the Global Carbon Cycle,* NATO 40:349-61.

Fearnside, P. (1997) Monitoring Needs to Transform Amazonian Forest Maintenance into a Global Warming Mitigation Option. *Mitigation and Adaptation Strategies for Global Change* 2:285-302.

Friends of the Earth (1997) *Forests and Climate Change,* Washington, DC.

Houghton, R.A., *et al.* (1983) Changes in the Carbon Content of Terrestrial Biota and Soils Between 1860 and 1980: A Net Release of CO_2 to the Atmosphere. *Ecol. Monographs* 53:235-62.

Houghton, R.A., *et al.* (1987) The Flux of Carbon from Terrestrial Ecosystems to the Atmosphere in 1980 Due to Changes in Land Use: Geographic Distribution of the Global Flux. *Tellus* 39:122-39.

Houghton, R.A., Unruh, J.D. and Lefebvre, P.A. (1993) Current Land Cover in the Tropics and its Potential for Sequestering Carbon. *Global Biogeochemical Cycles* 7:305-20.

Intergovernmental Panel on Climate Change (1996) *Climate Change 1995: Impacts, Adaptations, and Mitigation of Climate Change: Scientific-Technical Analyses,* Cambridge Univ. Press, Cambridge.

International Energy Agency (1997) Key Issues on the Design of Activities Implemented Jointly; Baseline Determination and Additionality Assessment, in U.N. Framework Convention on Climate Change, *Determination of Environmental Benefits: Part Two, The Actual Project Case,* Paris.

Keystone Center (1994) *Voluntary Reporting of Greenhouse Gas Emission Reductions and Offsets by Electricity Generators: Final Consensus Report of a Keystone Policy Dialogue* Keystone.

Lal, R., *et al.* (1998) *Management of Carbon Sequestration in Soil,* CRC Press, Boca Raton.

Leggett, J., ed. (1990) *Global Warming: The Greenpeace Report,* Oxford University Press, Oxford, U.K.

MacDicken, K. (1996a) *A Guide to Monitoring Carbon Sequestration in Forestry and Agroforestry Projects.* Forest Carbon Monitoring Program Working Paper 96/04, Winrock International, Morillton.

MacDicken, K. (1996b) *The Utility of Remote Sensing Technology in Monitoring Carbon Sequestration Agroforestry Projects.* Forest Carbon Monitoring Program Working Paper 96/02, Winrock International, Morillton.

MacDicken, K. (1997) Project Specific Monitoring and Verification: State of the Art and Challenges," *Mitigation and Adaptation Strategies for Global Change* 2:191-202.

Marland, G. (1988) *The Prospect of Solving the CO_2 Problem Through Global Reforestation,* U.S. Department of Energy, Office of Energy Research, DOE/NBB-0082, Washington, DC.

Marland, G. and Schlamadinger, B. (1995) Biomass Fuels and Forest-Management Strategies: How Do We Calculate the Greenhouse-gas Emissions Benefits? *Energy* 20:1131-40.

Marland, G., Schlamadinger, B., and Canella, L. (1997) Forest Management for Mitigation of CO_2 Emissions: How Much Mitigation and Who Gets the Credits?" *Mitigation and Adaptation Strategies for Global Change* 2:303-318.

Marland, G., Schlamadinger, B., and Feldman, D. (1997) Reforestation: What Happens When the JI Project Ends? *International Conference on Technologies Implemented Jointly,* Vancouver.

Marland, G., Schlamadinger, B., and Leiby, P. (1996) Forest/Biomass-Based Mitigation Strategies: Does the Timing of Carbon Reductions Matter? *Envtl. & Resource Economics.*

Ministerial Conference on Atmospheric Pollution and Climatic Change (1990) *The Noordwijk Declaration on Atmospheric Pollution and Climatic Change*, Noordwijk.

Moura-Costa, P. (1998) Forestry-Based Greenhouse Gas Mitigation: A Short Story of Market Evolution. in: Asian Timber, 1998:44-51.

Moura-Costa, P., *et al.* (1997). *SGS Forestry: Carbon Offset Verification Services Introduction*. SGS Forestry, Oxford.

Mott, R.N. (1993) *Joint Implementation and Carbon Offsets,* World Wildlife Fund. Washington, DC.

Moulton, R.J.,, and Richards, K.R. (1990) *Costs of Sequestering Carbon Through Tree Planting and Forest Management in the United States,* U.S. Department of Agriculture, U.S. Forest Service, GTR WO-58, Washington, DC.

Nabuurs, G.J., *et al.* (1999) Resolving *Issues on Terrestrial Biospheric Sinks in the Kyoto Protocol.* Dutch National Research Programme on Global Air Pollution and Climate Change, NRP Report. Wageningen.

Owensby, C.E. (1993) Potential Impacts of Elevated CO_2 and Above- and Belowground Litter Quality of a Tallgrass Prairie, in: *Terrestrial Biospheric Carbon Fluxes: Quantification of Sinks And Sources of CO_2*. Kluwer Academic Publishers, Dordrecht.

Perez-Garcia, J. (1994) *An Analysis of Proposed Domestic Climate Warming Mitigation Program Impacts on International Forest Products Markets,* CINTRAFOR Working Paper #50, University of Washington, Seattle.

Pinard, M., and Putz, F. (1997). Monitoring Carbon Sequestration Benefits Associated with a Reduced-Impact Logging Project in Malaysia. *Mitigation and Adaptation Strategies for Global Change* 2:203-215.

Princeton University, Woodrow Wilson School of Public and International Affairs (1999) Strategies *for Promoting Innovative Energy Technologies in the Electric Power Sector, Princeton.*

Richards, K.R. (1994) *Large-Scale Carbon Sequestration: Preliminary Considerations for Instrument Choice,* Pacific Northwest Laboratory, Washington, DC.

Richards, K.R. and Stokes, C. (1995) *National, Regional, and Global Carbon Sequestration Cost Studies: A Review and Critique,* Pacific Northwest Laboratory, Washington, DC.

Rombold, J. (1996) *A Bibliography on Carbon Sequestration and Biomass Estimation,* Forest Carbon Monitoring Program, Working Paper 96/03, Winrock International, Morillton.

Sathaye, J.A. (1995) *Estimating Costs and Benefits of Forest-Sector Mitigation Options,* Lawrence Berkeley Laboratory, Berkeley.

Sathaye, J.A., *et al.* (1997) Sustainable Forest Management for Climate Change Mitigation: Monitoring and Verification of Greenhouse Gases. *Mitigation and Adaptation Strategies for Global Change* 2:101-339.

Sathaye, J.A., *et al.* (1998) Concerns *About Climate Change Mitigation Projects: Summary of Findings from Case Studies in Brazil, India, Mexico, and South Africa.* Ernest Orlando Lawrence Berkeley National Laboratory, Berkeley.

Schlamadinger, ed. (1999) Special Issue: Land-Use, Land-Use Change, and Forestry in the Kyoto Protocol. *Envtl. Sci. & Pol.* 2:89-207.

Schlamadinger, B. and Madlener, R. (Eds) (1998) Effects of the Kyoto Protocol on Forestry and Bioenergy Projects for Mitigation of Net Carbon Emissions, *in: Greenhouse Gas Balances of Bioenergy Systems,* Rotorura.

Schlamadinger, B. and Marland, G. (1996) Full Fuel Cycle Carbon Balances of Bioenergy and Forestry Options. *Energy Convers. Mgmt.* 37:813-18.

Schlamadinger, B. and Waupotitsch, M. (Eds) (1996) *Greenhouse Gas Balances of Bioenergy Systems: A Bibliography Including Greenhouse Gas Implications of Wood Products, Forestry and Land-Use Change.* Joanneum Research, Graz.

Sedjo, R.A. (1989a) Forests: A Tool to Moderate Global Warming? *Environment* 31: 14-20.

Sedjo, R.A. (1989b) Forests to Offset the Greenhouse. *J. Forestry* 12-15.

Sedjo, R.A., Sampson, R.N., and Wisniewski, J. (Eds) (1997) Economics *of Carbon Sequestration in Forestry.* Special Issue of Critical Reviews in Environmental Science and Technology, Lewis Publishers, Boca Raton.

SGS Forestry (1997) *Certification of The Protected Area Project (PAP) in Costa Rica for OCIC (the Costa Rican Office for Joint Implementation),* Carbon Offset Verification Report, Oxford.

Sierra Club (1995) *Risky Business: Why Joint Implementation is the Wrong Approach to Global Warming Policy,* Washington, DC.

Skole, D.L., *et al.* (1997) A Land Cover Change Monitoring Program: Strategy for an International Effort. *Mitigation and Adaptation Strategies for Global Change* 2:157-75.

The Nature Conservancy (1996) The *Noel Kempff Mercado Climate Action Project Bolivia: A United States Initiative on Joint Implementation Pilot Project Proposal,* Washington, DC.

Trexler and Associates (1998a) *Innovative Forest Financing Options and Issues: Forest Conservation and Management for Climate Change Mitigation,* UNDP, New York.

Trexler and Associates (1998b) The *Role of Forestry as a Climate Change Mitigation Strategy,* U.S. Environmental Protection Agency, Washington, DC.

Trexler and Associates (1998c) *Forestry as a Climate Change Mitigation Option: Reviewing the State of the Art. A Report of the Land Use and Biotic Mitigation Policy Project,* Portland.

Trexler and Associates (1999) Assessing *the State of the GHG Market: Using the Greenhouse Gas Offset Cost Assessment and Decisionmaking Model (GGOCAD*[□]*).* Research Report 99-05, Portland.

Trexler, M.C. (1997) *Considerations in Forecasting the Demand for Carbon Sequestration and Biotic Storage Technologies.* Trexler and Associates, Portland.

Trexler, M.C., and Broekhoff, D.J. (1995) *Arid Lands as Carbon Offsets - A Viable Option?* Paper presented at UNEP Workshop on Combatting Global Warming by Combatting Land Degradation, Nairobi.

Trexler, M.C., Faeth, P.E., and Kramer, J.M. (1989) *Forestry as a Response to Global Warming: An Analysis of the Guatemala Agroforestry and Carbon Sequestration Project,* World Resources Institute, Washington, DC.

Trexler, M.C. and Haugen, C. (1994) *Keeping it Green: Global Warming Mitigation Through Tropical Forestry,* World Resources Institute, Washington, DC.

Trexler, M.C. and Kosloff, L.H. (1998) The 1997 Kyoto Protocol: What Does it Mean for Project-Based Mitigation? *Mitigation and Adaptation Strategies for Global Change* 3:1-58.

Trexler, M.C. and Meganck, R. (1993) Biotic Carbon Offset Programs: Sponsors of or Impediment to Economic Development? *Climate Research* 3:129-36.

U.N. Framework Convention on Climate Change, Third Conference of the Parties (1997) *Kyoto Protocol to the United Nations Framework Convention on Climate Change,* FCCC/CP/1997/L.7/Add.1.1997, Geneva.

U.N. Framework Convention on Climate Change, Subsidiary Body for Scientific and Technological Advice (1998). *Methodological Issues: Issues related to land-use change and forestry,* FCCC/SBSTA/1998/6, Geneva.

U.S. Department of Energy (1994) *General Guidelines for the Voluntary Reporting of Greenhouse Gases Under Section 1605(b) of the Energy Policy Act of 1992,* Washington, DC.

U.S. Initiative on Joint Implementation (1996) *Activities Implemented Jointly: First Report to the Secretariat of the United Nations Framework Convention on Climate Change:*

Accomplishments and Description of Projects Accepted Under the U.S. Initiative on Joint Implementation, Washington, DC.

U.S. Initiative on Joint Implementation (1997) Activities *Implemented Jointly: Second Report to the Secretariat of the United Nations Framework Convention on Climate Change: Accomplishments and Descriptions of Projects Accepted Under the U.S. Initiative on Joint Implementation,* Washington, DC.

Vine, E. and Sathaye, J. (1997) *The Monitoring, Evaluation, Reporting, and Verification of Climate Change Mitigation Projects: Discussion of Issues and Methodologies and Review of Existing Protocols and Guidelines.* Lawrence Berkeley National Laboratory, Berkeley.

Wisniewski, J., and Sampson, R.N. (Eds) (1993) *Terrestrial Biospheric Carbon Fluxes: Quantification of Sinks and Sources of* CO_2. Kluwer Academic Publishers, Dordrecht.

World Bank (1994) *Greenhouse Gas Abatement Investment Project Monitoring and Evaluation Guidelines,* Washington, DC.

World Resources Institute (1998) *Climate, Biodiversity, and Forests: Issues and Opportunities Emerging From the Kyoto Protocol,* Washington, DC.

World Wildlife Fund (1995) *Position Statement: Joint Implementation,* presented at the First Conference of the Parties to the U.N. Framework Convention on Climate Change, Berlin.

Chapter 8

DO AIJ PROJECTS SUPPORT SUSTAINABLE DEVELOPMENT GOALS OF THE HOST COUNTRY?

A HAMBLETON[1], C FIGUERES[1] and K CHATTERJEE[2]
[1]*Center for Sustainable Development in the Americas;* [2]*Development Alternatives*

Key words: sustainable development, Costa Rica, India, host country priorities, developing countries

Abstract: One of the criteria established for the UN Framework Convention on Climate Change (FCCC) Activities Implemented Jointly (AIJ) pilot phase projects was to support national priorities of host countries. Success in contributing to these countries' sustainable development was less a factor of the design of the pilot phase, and more the result of national capacity to define and implement policies, legal frameworks and financial instruments for sustainability. Developing countries will pursue greenhouse gas mitigation projects only if these promote national priorities. Costa Rica's successful AIJ program is one of the elements in a strategy to obtain sustainable financing for a clear national environmental agenda. India's demand side management project has specific community level benefits. As we move forward toward the design of the Clean Development Mechanism (CDM), a combination of national, as well as community level, sustainable development criteria, may be helpful in ensuring that international mitigation projects will, in fact, contribute to the sustainable development of host countries.

1. BACKGROUND

In this paper we shall argue that the responsibility for ensuring that AIJ pilot phase projects contributed to the sustainable development of the host country rested on the shoulders of the host country. Sustainable development was not an agreed upon criterion for the AIJ pilot phase. However, interested individual host countries were free to set country specific host country acceptance criteria which, in some cases, included sustainable development guidelines. Those countries that were pro-active in defining their national development goals, and set AIJ project acceptance criteria to reflect these priorities, were able to benefit from AIJ in a measurable and concrete manner. Sustainable development disappointments from the FCCC AIJ pilot phase were largely the result of differing expectations between hosts and investors.

2. DEFINITION OF SUSTAINABLE DEVELOPMENT

The World Commission on Environment and Development (WCED) in their report *Our Common Future* (1987) first defined sustainable development as meeting the needs of the present without compromising the ability of future generations to meet their own needs. The word sustainable has acquired tremendous currency in recent years, particularly after the UN Conference on Environment and Development (UNCED) in Rio de Janeiro, Brazil, 1992. Sustainable development is a critique of the concurrent growth-oriented, top-down development paradigm. It emphasizes a development framework, which is based on an appreciation of the finite human and natural resources available on planet Earth. However, the concept has not yet been fully integrated to the operational aspect of development strategies of industrialized and developing countries. It remains the sovereign right and responsibility of each country to define sustainability in its own terms and to ensure that sustainable development is being pursued.

3. THE AIJ PILOT PHASE AND SUSTAINABLE DEVELOPMENT

The 1995 Berlin Mandate, developed at the 1st Conference of the Parties (COP-1), established an AIJ pilot phase for joint implementation (JI) projects in reaction to fears that too many questions regarding the benefits of international mitigation projects remained unanswered. In addition to myriad technical uncertainties, the question of whether and how joint implementation project activities could avoid inequitable treatment of host (non-Annex 1) countries loomed large in the Group of 77 (G77) agenda at the first Conference of the Parties (COP).

This chapter will focus on one of the five criteria established for the AIJ pilot phase:

AIJ should be compatible with, and supportive of, national environment and development priorities and strategies, contribute to cost-effectiveness in achieving global benefits, and could be conducted in a comprehensive manner covering all relevant sources, sinks, and reservoirs of greenhouse gases (Decision 5, COP-1).

It is important to note that no mention of sustainable development is made in the Berlin Mandate. It was precisely the ensuing discussion about whether or not these projects actually contributed to the sustainable development of the host country, which lead to a change in wording for the CDM. The CDM defined under the Kyoto Protocol, clearly states as one of its purposes to assist Parties not included in Annex I in achieving sustainable development.

The following discussion centers on two developing countries', Costa Rica and India evaluation of whether and how AIJ projects undertaken in their countries were in fact compatible with and supportive of national environment and development priorities and strategies. Two perspectives are presented. Costa Rica stresses the compatibility with national sustainable development goals, whereas India emphasizes the importance of a community approach.

4. COSTA RICAN NATIONAL ENVIRONMENT AND DEVELOPMENT PRIORITIES AND STRATEGIES

In the mid 1990's the small, environmentally aware country was already signatory to both the FCCC and the UN Framework Convention on Biological Resources, and was considered an international center for biodiversity research and creative conservation approaches and projects. Costa Rica had also already been successful with innovative financial mechanisms such as debt-for-nature swaps and bio-prospecting, as well as nature-based ecotourism. There was recognition of the many different values of natural resources, as well as a comfort level with merging private and public sectors in the pursuit of environmental goals. Traditional sources of funding for conservation were diminishing, and the country recognized that any really effective attempt to grow sustainably would have to involve the private sector.

Beginning in 1994, Costa Rica based its national development strategy on the concept of sustainable development, which it defined as development which represents the balance provided by political stability, social equity, economic stability and development in harmony with nature (Figueres,

1996). There was a firm commitment to prove that intelligent environmental management can complement economic growth, and that civil society and the private sector can and should be fully engaged in the decision making process.

Some of the sustainable development priorities for 1994 – 1998 (the period during which AIJ projects were designed) were:

- Territorial ordainment, or consolidation and improved management of protected areas.
- Climate change and energy policies, the country was at the time already generating 90% of its power from renewable energy sources. The commitment was made to further increase clean energy generation as well as promote energy efficiency in the industrial sector.
- Biodiversity, Costa Rica has 5% of the world's biodiversity. An impressive 24% of its national territory is designated as national parks or protected areas. The goal was set to consolidate the park system so as to guarantee its existence for future generations.
- Macroeconomic Policies, restructuring the economy toward an open and more balanced one, reducing one of the highest deficits in Latin America and promoting internal savings.
- Incorporate civil society and the private sector into the decision making process

4.1 COSTA RICAN EXPERIENCE WITH THE AIJ PILOT PHASE

Costa Rica was the first developing country to recognize that the FCCC AIJ pilot phase could be designed to become a tool for financing national sustainable development priorities. Thus, Costa Rica has done more than any developing country to establish a comprehensive JI/AIJ regime as a strategy to promote its own sustainable development goals, while at the same time meeting the objectives of the climate treaty.

In 1994, a high level consultative committee on climate change was formed to shape JI policy and oversee the completion of the national greenhouse gas emissions inventory. In September 1994, Costa Rica and the USA signed a bilateral Statement of Intent on Sustainable Development and Joint Implementation (Castro, Tattenbach, 1998).

The Costa Rican Office for Joint Implementation (OCIC) was established in 1995 within the Ministry of Natural Resources, Energy and Mines. (Figueres, 1996). OCIC carries out the country's JI evaluation, acceptance and promotion activities. The office combines expertise from forestry, to energy, to marketing, and is housed in the private sector export promotion office, Coalición Costarricense de Iniciativas de Desarrollo (CINDE). OCIC is directed by a leading forestry non-governmental organization. It is therefore a tri-partite collaboration among the government, private and non-governmental (NGO) sectors. OCIC works with project developers, national policy makers, and other countries, in order to ensure that quality projects are designed and marketed abroad.

In 1995, OCIC established project approval criteria and procedures. Since then, OCIC has assisted in the development of more than 15 project proposals. Three of the seven projects approved in the first round of evaluations by the U.S. Initiative on Joint Implementation (USIJI) were from Costa Rica. These included a wind energy project, Plantas Eolicas, and two carbon (C) sequestration projects: ECOLAND and CARFIX. During the second round of USIJI, Costa Rica had five more projects approved: three renewable energy projects (Doña Julia Hydroelectric project, Aeroenergia wind project, and Tierras Morenas wind project) and two carbon sequestration projects (BIODIVERSIFIX and Klinki Forestry project).

More recently, the OCIC strategy has been to consolidate small projects into three national scale programs focusing on:

- consolidation of parks,
- natural forest management by private landowners, and ,
- renewable energy.

The development of national level AIJ programs reduces the per unit transaction costs associated with developing, evaluating and marketing projects. Thereby, one of the most important obstacles to developing country participation in AIJ project development is removed. The first two of these umbrella programs, known as the Protected Areas Project (PAP) and the Private Forestry Project, which focuses on land use, will eventually encompass most of the threatened forested areas of the country.

Carbon offsets generated by the PAP are certified annually by the Societe Generale de Surveillance (SGS) and are being sold to interested buyers in one Mg increments as Certified Tradeable Offsets (CTOs). Two of the previously approved projects, the CARFIX project and the BIODIVERSIFIX project, have subsequently been incorporated into the PAP, which received USIJI approval in the third evaluation round.

OCIC has recently made its 1999 report on the AIJ pilot phase to the FCCC Secretariat. The report summarizes Costa Rican projects as follows (Costa Rica, 1999):

TABLE 1: Land use change and forestry AIJ projects in Costa Rica.

Name of the project	Type of project	Area (ha)	Life of project (years)	Total cost (US$ millions)	Emission reductions (Mg C)
ECOLAND	Conservation	2.340	15	1	366.200
KLINKI	Reforestation	6.000	40ˇ	3,8	1.966.495
CR/Norway	Conservation	2.000	25	3,3	313.646
	Reforestation	1.000			
	Regeneration	1.000			
PAP	Conservation	422.800	25	180	18.000.000
	Regeneration	107.698			
TOTAL		542.838		188,1	20.646.341

TABLE 2: Energy sector AIJ projects in Costa Rica.

Name of project	Type of project	Installed capacity (MW)	Annual Production (GWhr/yr)	Total cost (US$ millions)	Emission reductions (Mg C)
Plantas Eólicas	Wind	20	98	30.4	506,720
Tierras Morenas	Wind	20	90	27	562,020
Aeroenergía	Wind	6.4	30	8.85	146,000
Doña Julia	Hydro	16	85	27	562,020
TOTAL		62.4	303	93.25	1,776,760

TABLE 3: Agriculture AIJ projects in Costa Rica.

Name of project	Type of project	Total cost (US$ million)	Life of project (years)	Emission reductions (tm C)	Emission reductions ($MgCO_2$)
ICAFE/ BTG	Waste water treatment	0.973	10	34,645	127,031

The renewable energy projects, which will all be operational by September 1999, represent a total of $93,000,000 in new investment and will generate 6.5% of Costa Rica's installed electricity capacity. Of the C sequestration projects, both ECOLAND and CNFL are fully funded. Klinki Forestry is partially funded and under implementation. The PAP is 6% executed through the sale of 200,000 CTOs at $10 Mg of C to the Government of Norway (Costa Rica, 1999).

Costa Rica has established a C fund which serves as a financing agent for national C offset projects by supplying funds both to purchase threatened lands in order to expand the government forestry incentives, and to support renewable energy projects. Individuals may receive compensation for ecosystem services, including C sequestration provided by their land under arrangements that cede their environmental services rights to the government for its use and possible resale. The C fund markets and sells those rights internationally, producing CTOs. Funds generated by the sale of CTOs are then distributed to landholders under ecosystem services contracts. (Castro and Tattenbach, 1998).

4.2 DO COSTA RICA'S AIJ PROJECTS CONTRIBUTE TO ITS NATIONAL PRIORITIES?

In order to meet sustainable development objectives, a country must formulate clear policies, develop effective legal frameworks, and implement efficient financial instruments. Costa Rica is a case study in this attempt. Over the last decade Costa Rica has developed a variety of legal, policy and financial instruments with the goal of supporting sustainable development goals. Costa Rica's AIJ program is one of the elements within the framework for the sustainable financing of a clear environmental agenda. It provides an avenue for generating foreign funds and allowing private sector participation, two important pillars of the national definition of sustainable development.

Costa Rica and Australia are the only two countries that specifically mention sustainable development in their AIJ project acceptance criteria, although many countries require evaluation of non-GHG costs and benefits including environmental, social and economic impacts. Costa Rica's AIJ project criteria specifically require that an approved project be compatible with and supportive of Costa Rican national environment and development priorities and strategies, including biodiversity conservation, reforestation and forest preservation, sustainable land use, watershed protection, air and water pollution reduction, reduction of fossil fuel consumption, increased utilization of renewable resources and enhanced energy efficiency.

The criteria also require that projects enhance the income opportunities and quality of life for rural people, transfer technological know-how and capacity building, and minimize adverse consequences. In order to be accepted for host country approval, project developers must prove to OCIC that they will meet these criteria.

Costa Rica's renewable energy projects have been reported as having the following benefits: (US EPA, 1999)

- Help support Cost Rica's goal of satisfying Costa Rica's current and future energy needs with a diverse portfolio of clean, renewable and environmentally sound energy sources
- Attract foreign capital investment that would not otherwise be invested

- Provide protection from rising fuel prices and dependence on imported fuel
- Transfer modern, clean and efficient technology
- Provide energy to meet Costa Rica's 7-9% annual increase in energy demand
- Local economy benefits from improved roads, access to electricity and increased local commerce
- Reduce non-GHG pollutants/improve air quality by replacing highly emitting fuels
- Provide employment during both construction and operation

Costa Rica's land use change, forestry and agriculture projects have been reported as providing the following benefits:

- Prevent deforestation of primary forests
- Protect biodiversity in critical unprotected habitats
- Maintain water quality
- Reduce soil erosion
- Reduce logging pressures on natural forest
- Reduce use of chemicals through substituting other low-yield crops with forest plantations
- Provide a stable source of income to small, medium and large local farmers through payments for plantings and C credits

All of these benefits are compatible with and supportive of (Costa Rica's) national environment and development priorities and strategies. Costa Rica has been able to further its sustainable development agenda through AIJ. Increased financial resources for investment in environmentally beneficial projects that are additional to official development assistance (ODA), and in most cases, from private sources, have begun to flow to Costa Rica. A monetary value is being given to environmental services provided by the country's natural resources, and Costa Rica has a greater control over the sustainable use of these natural resources. Incentives, such as contracts with individual landowners for their environmental service rights, provide competitive economic alternatives to destroying biodiversity rich primary forests.

5. INTEGRATED AGRICULTURAL DEMAND SITE MANAGEMENT (DSM) PROJECT IN ANDHRA PRADESH

Consumption of electricity in the agricultural sector of Andhra Pradesh presents a significant challenge for the Andhra Pradesh State Electricity Board (APSEB). This sector consumes much electricity and is growing rapidly. Energy consumption by the agricultural activities is 19 % of the total energy sales. At the same time, agricultural tariffs do not provide sufficient revenue to cover the cost of production. A number of studies have indicated that there are significant cost effective opportunities to increase the efficiency of electricity usage both in terms of improving the efficiency of the distribution system and in terms of increasing end-use efficiency.

The project combines the following four technical measures in an integrated package to generate electricity saving and greenhouse gas emission reductions:

- improvement of the distribution system efficiency by conversion of low voltage (LV) feeders to high voltage (HV) feeders
- reduction in system demand and line losses, as well as improved service to non agricultural customer through the use of automated load control
- the provision of customer meters to provide information on energy consumption
- improvements in end-use efficiency by replacement of pumpsets and associate pipe and valves with efficient products

This pilot project encompasses approximately 5800 pump sets on 8 feeders in two geographically separate areas of Andhra Pradesh. The cost of this project is estimated at $ US 4.60 million. The funding for the pilot project is being made available under the World Bank / Norway AIJ Program.

5.1 OBJECTIVES OF THE PROJECT

The objective of the pilot project is to demonstrate that an integrated rural distribution and agricultural energy efficiency project can save electricity in a way that is

- technically, economically and financially sound as an integrated project
- has the potential of reducing GHG emissions
- sustainable as a business proposition for the APSEB.
- replicable by other Indian electric utilities.

Furthermore, as funding for this activity is being made available under the World Bank AIJ Program, the project must meet the monitoring and verification requirements of AIJ projects.

5.2 PROJECT SITE DESCRIPTION

Two typical areas, one in the North and the other in the South of Andhra Pradesh, were chosen for the DSM project. The two sites are the Karimnagar District in the North, , and the Chittoor District in the South having. Both sites have high agricultural potential, but different ground water potential and different cropping patterns.

The total area of the Karimnagar District is 11823 square km, 19.6% of which is forest area. The cultivable area of the district is 60% of the total area as per census. Agriculture is the most important occupation in the Districts. Rice, jowar, bajra, maize, groundnut and fruits are the major crops. Sericulture is an agrobased major industry. A total of 1,75,323 pump sets are being operated in the District.

The total area of the Chittoor District is 15,152 square km, 30% of which is forest area. The cultivable land in the district is 9,71,700 ha. Agriculture is the most important occupation in the district. Rice, jowar, bajra, and maize are the major crops of the district. A total of 1,31,032 pump sets are being operated.

5.3 CONTRIBUTIONS OF THE AIJ PROJECT

The major economic contributions of the projects are:

- freeing of electricity in the system due to increased efficiency of the pumping systems and the reduction of line losses in the distribution system
- reduction in the need for rewinding pumpset motors, resulting in the saving of Indian Rupees (Rs.) 1000 annually in the rewinding of motors. Studies have estimated that farmers spend on an average Rs. 1500-2000 per year on pumpset maintenance
- reduction in the need to purchase replacement pumpsets. The expected life of a pumpset is about 20 years. The project will provide farmers with new pumpsets. This expenditure of purchasing new pumpsets can be deferred by the farmers
- boost for the economy of the pumpset manufacturers in the country.
- efficient pump irrigation system will result in better yields in food production brining additional economic returns from the sale of agricultural yields. This will also improve the farmers per capita income and purchase power
- economic analysis of the projects has shown that implementation of the project will provide excellent economic benefits for the region. Approximately 12% of the economic benefits accrue directly to the farmers who participate in the program due to reduced maintenance and capital costs
- reduced pilferage of service (it amounts to 20% of the consumption of the agricultural load). With HV distribution, it will be totally avoided.

The social contributions of the project are:

- increased flow of irrigation water to the farmers due to introduction of an efficient pumping process;
- better crop yields and food quality;
- better employment benefits. The project will result in the creation of short term jobs in and around the project sites for the implementation of the distribution system conversion to high voltage, the implementation of the distribution automation and the manufacture and installation of the pumpsets; in addition to the extent that the farmers are better off as a result of the project, there should be employment benefits resulting from the spending of their additional incomes; and,
- better purchase capacity of the people around the project area.

The environmental contributions of the project are both at the local and at the global levels:

- due to conservation of fossil fuels, particularly coal, fewer emissions of pollutants (e.g., NO_x, CO, SO_x) and suspended particulate matters.
- reduction in GHG emissions based on a preliminary feasibility study the DSM proposed in the project would result in the saving of 180,000 kWh a year. It is assumed that the electrical energy available for re-sale is equivalent to deferring building new infrastructure in the electricity sector. Assuming a 20-year lifetime for this DSM project, the expected C reduction is about 150,000 Mg abatement over the life of the project. The C abatement cost being paid to Andhra Pradesh, Government of India is about US $ 30 per Mg, while the cost to Andhra Pradesh works out to roughly US $ 6.4 per Mg.

Increased efficiency of pump sets may result in pumping of water in excess of the agricultural requirement. This may result in water logging, salination of top soil and providing breeding ground for mosquitoes, causing adverse impacts on health of the people due to increased incidence of malaria, dengue fever and other vector borne diseases. Important as the global environment is, for this and future generations, India's role in international efforts to avert climate change must be based on domestic development priorities. The pressing issues of alleviating poverty, providing for basic needs and economic development are the starting points for assessing the potential for AIJ. Only by addressing these needs and critical environmental problems will the concept of AIJ flourish in India.

6. ANTICIPATED FEARS AND ACTUAL SUSTAINABLE DEVELOPMENT DISAPPOINTMENTS OF AIJ

The early days of the JI negotiations and the FCCC AIJ pilot phase raised a lot of concerns on the part of developing countries (Aggarwal and Norain 1999). The anticipation that these concerns would appear under the CDM resulted in the requirement that sustainable development priorities be given equal footing with creating a means for lower cost international cooperation for reducing emissions. While most of the concerns countries expressed were not manifested during the AIJ pilot phase, it can be argued that once credit is granted for GHG emission reductions and increased levels of investment are realized under CDM, they may re-appear.

Initial concerns about JI and AIJ revolved primarily around historical inequities of investment and trade between developed and developing countries (Cisse and Diop 1999). The term, carbon colonies was coined to represent the potential perils for developing countries engaging in cooperative emission reductions with industrialized countries. (Maya, 1996). Among the many issues raised by this work and others were: loss of low-hanging fruit easy emissions reductions; apprehension about being shortchanged; projects would lead to caps that inhibit economic growth; fear of losing ODA; unintended adverse impacts; mismatching of host country priorities and technology needs; and; fear of a loss of sovereignty, particularly in cases where land is owned by foreign investors. A general feeling that developing countries ought not to take on what should be an industrialized country's obligation for reducing global emissions enhanced these fears.

The FCCC AIJ pilot phase did not see these fears come to fruition in any significant manner. Rather, sustainable development disappointments during the AIJ pilot phase were mostly the result of differing expectations on the parts of host and investor countries. These disappointments included:

- an inequitable geographic distribution of projects;
- differing expectations regarding price;
- low levels of investment/technology transfer/capacity building;
- a perceived preference for land use projects; and a mismatch with country development priorities (Jepma and van der Gaast 1999).

AIJ investment tended to concentrate in Eastern Europe and Latin America, with the countries of Africa and Asia lagging behind. These regions argued that this distribution was inequitable and that it prevented them from building capacity for CDM, as well as receiving sustainable development investment.

Prices negotiated for early paper credits, also, were problematic. While credits did not have a technical value during the AIJ pilot phase, investors sought, and were willing to pay for, reductions on speculation. As there was no established market, prices ranged dramatically, from $10.00 Mg C to $00.25. Early deals were struck with low prices before countries had a chance to realize the opportunity cost or real value of their reductions. As country capacity developed, several AIJ agreements were later re-negotiated with different prices and more equitable distribution of credits between hosts and investors. These experiences clearly illustrate the need for host countries to build capacity for negotiating and reviewing projects in advance of the start of the CDM (Jackson and Begg 1999).

Low levels of actual investment, technology transfer and capacity building also were disappointing to those countries eager to use AIJ as a vehicle for economic development (Lile et al, 1998). Many of the 130 AIJ pilot projects are not fully capitalized, or have undergone significant restructuring since their inception. Raised host country expectations as a result of exploratory initiatives on the part of investor countries have created a somewhat cynical perspective. For example, a 1996 investor organized JI workshop in Central America solicited 57 project proposals from the country in which it was held. None of the projects received investment.

A perceived investor preference for less expensive (and non-tech transferring) Land Use Change and Forestry (LUCF) projects also provoked criticism, particularly from those African and Alliance of Small Island States (AOSIS) countries who do not support C sequestration as a reduction option (Sokona et al 1999). This, criticism is not borne out by the AIJ pilot phase, since only sixteen out of one hundred twenty six current AIJ projects are C sequestration projects.

Finally, the mismatch of projects with development priorities resulted in several disappointments, although this could be attributed to the lack of clearly defined national development priorities or criteria for AIJ projects. It is important to remember that AIJ projects can only occur in GHG related sectors such as energy production, transportation and LUCF. In addition to improved GHG management, host countries can stipulate additional environmental, social and economic criteria. However an AIJ project cannot be judged on whether it reduces infant mortality, improves access to primary education, or reduces overfishing, for example, if it does not primarily contribute to emission reduction, avoidance or C sequestration.

7. CONCLUSIONS

Under the UN FCCC AIJ pilot the contribution to the sustainability of the host country was largely the responsibility of the host country. Under the CDM, this may not be dramatically different. The stated objective of the CDM covers the needs of the host countries as well as the needs of the industrialized countries. Everyone agrees that offset projects must contribute to sustainable development. But agreement on how to measure sustainable development, and whose responsibility it should be to insure it, is more difficult to achieve.

Three approaches are currently being discussed:

- let the individual countries determine their own sustainable development goals and approve projects based on them;
- establish a set of international sustainable development indicators that projects can be measured and certified against by operating entities or certifying bodies; and,
- a combined approach which would establish a general set of principles or indicators, and encourage countries to specify individual country goals/criteria beyond the general indicators (Jepma and van der Gaast 1999).

As we move toward the design of the CDM, we should learn the lessons of the AIJ pilot phase (Matsuo 1998). The combined general/specific approach may be helpful in order to offer both broad guidelines, as well as respect the sovereign right of each host country to define its own priorities. However, the responsibility for insuring the contribution to sustainable development cannot be placed on the international system, which is being created. This responsibility continues to rest on the shoulders of developing countries themselves. They must ensure that their criteria accurately reflect the nation's sustainable development goals and priorities. Projects must then be evaluated and approved according to their contribution to those objectives. Ultimately, the congruence between mitigation projects and national sustainable development goals is not only a sovereign right; it is an inescapable national obligation.

REFERENCES

Aggarwal, A., Narain, S. (1999) Whither Joint Implementation? Joint Implementation Needs to Consider Needs of Developing Countries. India's Centre for Science and Environment. Delhi.

Castro, R. and Tattenbach, F. (1998) The Costa Rican Experience with Market Instruments to Mitigate Climate Change and Conserve Biodiversity. FUNDECOR. San Jose.

Chatterjee, Kalipada (Ed) 1997 *Activities Implemented Jointly to Mitigate Climate Change : Developing Country Perspectives*, Published by Development Alternatives, New Delhi.

Chomitz, K., Brenes E. and Constantino, L. (1998) Financing Environmental Services: The Costa Rican Experience, Economic Notes, Number 10, The World Bank. Washington DC

Cissé, M.K, Diop, I.N. (1999) Comments on the Mechanism of North-South transfer in the UNFCCC. ENDA-TM Energy Programme. Dakar.

Climate Change Secretariat. (1997) Convention on Climate Change. UNEP/IUC Geneva Executive Center, Geneva.

Climate Change Secretariat. (1997) The Kyoto Protocol to the Convention on Climate Change. UNEP/IUC Geneva Executive Center, Geneva

Costa Rican Government (1999) Reporte Nacional sobre actividades conjuntas durante la fase piloto. República de Costa Rica. Report submitted to the Secretariat of the Framework Convention on Climate Change. San Jose.

Dutschke, M. (1998) Financing Sustainable Development. The Case of Costa Rica. HWWA-Institut fur Wirtschaftsforschung-Hamburg.

Figueres, C., Hambleton, A., Lay, L., MacDicken, K., Petricone, S., and Swisher, J. (1996) Implementing JI/AIJ: A Guide for Establishing Joint Implementation Programs. Center for Sustainable Development in the Americas, Lawrence Berkeley National Laboratory and U.S. Agency for International Development, Washington DC.

Figueres, J.M. (1996) Sustainable Development, *Sais Review, A Journal of International Affairs* 16: 187-203, Washington DC.

Government of Andhra Pradesh (1997) Proposal for Integrated Agricultural Demandside Management Project submitted to the AIJ Task Force of the Government of India. Ministry of Environment, New Delhi.

IEA/UNEP (1997) Summary Report. IEA/UNEP Workshop on Mutually Beneficial Incentives to Promote Activities Implemented Jointly, Paris.

Jackson, T. and Begg, K. (1999) Accounting and Accreditation of Activities Implemented Jointly, European Commission DG XVII, Brussels.

Jepma, C. and van der Gaast, W. (1999) On the Compatibility of Flexible Instruments, Kluwer Academic Publishers, Dosdrecht, in press.

Lee, R., Kahn, J., Marland, G., Russell, M., Shallcross, K., and Wilbanks, T. (1997) Understanding Concerns About Joint Implementation. Joint Institute for Energy & Environment. Washington, DC.

Lile, R., Powell, M. and Toman, M. (1998) Implementing the Clean Development Mechanism: Lessons from U.S. Private-Sector Participation in Activities Implemented Jointly. Resources for the Future. Washington, DC.

Matsuo, N. 1998 How is the CDM Compatible with Sustainable Development? The Institute for Global Environmental Strategies, Hayama.

Maya, R.S. and Gupta, J. (1996) Joint Implementation: Carbon Colonies or Business Opportunities? Weighing the odds in an information vacuum. Southern Centre for Energy & Environment. Harare.

Meeting Progress Towards Sustainablity: The Neemrana Deliberation (1993) *Development Alternatives Newsletter*, Vol. 3:12 New Delhi.

Sokona, Y., Humphreys, S., and Thomas JP. (1999) Sustainable Development: A Centrepiece of the Kyoto Protocol. An African Perspective. ENDA Tiers monde, Energy Programme, Dakar.

U.S. Environmental Protection Agency Third Report to the UNFCCC Secretariat. Accomplishments and Descriptions of Projects Accepted under the U.S. Initiative on Joint Implementation, 1999. Washington, DC.

Chapter 9

TECHNOLOGY TRANSFER AND AIJ PROJECTS

J. SATHAYE[1], R BRADLEY[2]
[1]Lawrence Berkeley National Laboratory, [2]U.S. Department of Energy

Key words: technology transfer, deployment, diffusion, energy efficiency, fuel substitution, bioenergy, land-use change and forestry, renewable energy

Abstract: Technology transfer is a central concept of the UN Framework Convention on Climate Change (FCCC). Technology is transferred both within and across countries and is supported by a variety of actors including private firms, governments, industry, and non-government organizations (NGOs). Many cost effective energy (eg, energy efficiency and fuel substitution) and forestry (forestation or forest conservation) sector greenhouse gas (GHG) mitigation technologies have been identified and are being applied within the activities implemented jointly (AIJ) pilot. AIJ technology transfer projects are distributed across five continents with the majority of projects in Eastern Europe and Latin America. Barriers, policies and incentives for technology transfer vary widely and examples of successes and failures are presented in this chapter. Experience with a limited number of AIJ projects suggests that technology transfer is being achieved in some regions. Successful projects should be replicated on a large-scale basis to reach all appropriate developing and transition countries Parties to the FCCC.

1. INTRODUCTION

Recognizing the differentiated responsibilities and respective capabilities of each country, the 1992 UN Framework Convention on Climate Change (FCCC) stressed the need for the transfer of technologies for mitigation of, and adaptation to, climate change. Article 4.5 of the FCCC requires the Parties to promote, cooperate, in the development, application, diffusion, including transfer of technologies, practices, and processes, that control, reduce, or prevent anthropogenic emissions of greenhouse gases (GHGs). The importance of technology transfer was also recognized in Article 10c of the Kyoto Protocol, which encourages all Parties to foster the transfer of publicly owned technologies, and to create an enabling environment that would support private sector efforts to transfer environmentally sound technologies.

What is the potential for the penetration of mitigation technologies? What can we learn from experience with AIJ projects in promoting these, or similar, technologies? What barriers exist to the increased market penetration of such technologies? Can these be overcome through the implementation of a mix of judicious policies, programs and other measures? We first discuss the potential for mitigation technologies in the energy and forestry sectors to set the stage for a discussion of technology transfer opportunities. We then present the experiences with AIJ projects in their ability to transfer technologies, and provide examples of barriers to technology transfer activities that have been overcome at three levels: macro or national, sector-specific and project-specific.

1.1 Technology Transfer Process

Technology transfer often carries many meanings. Technology is seen as a commodity, knowledge, or as a socioeconomic process by various scholars (Rosenberg, 1982). Some view its transfer as a cost-less process that merely involves the reproduction of the technology in a different socioeconomic setting. More recently scholars have recognized that technology transfer is a complex and costly process that involves learning from others. Learning would imply that the recipient of the technology is able to replicate its functioning in a new set of conditions.

Technology transfer occurs both within and across countries and is supported by a variety of actors, e.g., private firms, governments, industry, NGOs, etc., who are engaged in promoting the use of a particular technology. The market penetration of a technology may proceed from research to development, and demonstration (RD&D), adoption, adaptation, and replication. At a project-specific level, the elements of the transfer are different, and may proceed along project formulation, feasibility studies, loan appraisals, implementation, monitoring and evaluation and verification of carbon (C) and associated benefits. The transfer may include many actors, starting with laboratories for RD&D, manufacturers, financiers and project developers, and eventually the customer whose welfare is presumably enhanced through their use. The actors may make specific types of arrangements such as joint ventures, public companies, licensing, etc. that are mutually beneficial. These arrangements will determine the particular modalities chosen for technology transfer.

Governments at all levels, private industry, and/or local communities may drive the transfer. The transfer of a particular technology may occur through a variety of means and mechanisms, as it evolves from R&D towards commercial application. The importance of the pathway may change over time, as activities that were carried out earlier by governments are turned over to private industry or to communities. On the other hand, in times of crisis, the government role may become more prominent as national or international interests become the primary driver for taking action.

The spread of a technology may occur through transfer within a country and then transfer to other countries, both may occur simultaneously, or transfer across countries may precede that within a country. Generally, the spread of a technology is more likely to proceed along the first option rather than the other two, since the transfer of technologies to markets within a country is likely to be less expensive given the proximity to the market. Transfer of technology from one country to another will generally face trade and other barriers both in the initiating and recipient country, which may dissuade manufacturers and suppliers from such transfer.

Many market barriers prevent the transfer of technologies, including that of cost-effective GHG mitigation options. In the energy sector, these barriers include the high first cost of equipment, a lack of information on new technologies, the presence of subsidies for electricity and fuels and high tariffs on import of energy technologies. In the forestry sector, barriers include pressures on land availability for mitigation; absence of institutions to promote participation of local communities, farmers and industry; risk of drought, fire, pests; inadequate research and development capacity in countries; and poorly developed reforestation and sustainable forestry practices. Both sectors also suffer from an absence of appropriate methods and institutions to monitor and verify carbon flows (Watson, Zinyowera, and Moss, 1996).

1.1.1 Criteria for Determining Effectiveness of Technology Transfer

Evaluation of technology transfer requires criteria that are specific to the manufacture and use of technologies, and to the process of transferring these technologies. The IPCC discussed the criteria that could be used to evaluate technologies and measures (IPCC, 1996). These criteria are also useful to determine the effectiveness of technologies that are transferred. The criteria for evaluating the effectiveness of a technology are grouped into three categories as shown below:

- GHG and other environmental criteria (biodiversity, soil conservation, etc.)

- economic and social criteria (cost-effectiveness, investment needs, jobs created or lost, etc.)
- administrative, institutional and political criteria (information collection, monitoring, consistency with other public policies, etc.)

Some of these criteria, such as the GHG reduction potential, are amenable to quantitative evaluation, while others such as political considerations are at best evaluated qualitatively. An example of the application of these criteria would be the market penetration of wind power technology within a country and/or across countries. It would be possible for instance to quantitatively evaluate the transfer of this technology with respect to the criteria and qualitatively with respect to the third criterion. In addition, it would be important to identify and describe the policies, programs and measures that were used to overcome barriers to the transfer of the technology, and how these affected one or more of the aforementioned criteria. For example, the program may have resulted in a more equitable distribution of jobs, increased government subsidy, or replication of wind turbine projects.

In addition to the above criteria one needs to consider criteria to evaluate the effectiveness of the process of technology transfer. These process-related criteria include:

- rate and geographic extent of technology transfer
- long-term institutional capacity building
- monitoring and evaluation considerations
- leakage that reduces the impact of a program or measure

1.2 AIJ, Technology Transfer, FCCC and the Kyoto Protocol

The negotiations concluded in Rio de Janiero, Brazil 1992, that led to the FCCC, the Convention on Biodiversity and Agenda 21 found technology diffusion and transfer a central focus. In Agenda 21, for example, Principles 9 and 14 deal with technology and its transfer and Chapter 34 is dedicated to this topic. While technology transfer is highlighted in all three documents, we will center our discussion on the FCCC. Technology is referenced in many places in the FCCC beginning with its preamble. There it specifically recognizes, in its penultimate statement, that because developing countries need access to resources required to achieve sustainable development, their energy consumption will need to grow taking into account the possibilities for achieving greater energy efficiency and for controlling GHG emissions in general, including through the application of new technologies on terms which make such an application economically and socially beneficial.

The search for technology and its diffusion and transfer finds its way into a number of the articles that follow. In Article 4.1, the article that identifies commitments for all Parties, taking into account their common but differentiated responsibilities and their specific national and regional development priorities, objectives and circumstances, states in paragraphs (g) and (h) that all Parties shall promote and cooperate in technological research and in the full, open and prompt exchange of relevant technological, technical information. While this article refers to all Parties, most of the FCCC places the responsibility for undertaking and paying for the diffusion and transfer of technology on the developed country Parties.

Articles 4.3,4.5, 4.7, 4.8 and 4.9 all address technology transfer to some degree, primarily calling upon developed country Parties to undertake this transfer and at their expense. For example in Article 4.3, developed country Parties agreed that they shall also provide such financial resources, including for the transfer of technology, needed by developing country Parties to meet the agreed full incremental costs of implementing measures. Article 4.5 states, The developed country Parties and other developed Parties included in Annex II shall take all practicable steps to promote, facilitate and finance, as appropriate, the transfer of, or access to, environmentally sound technologies, and know-how to other Parties, particularly developing country Parties. While the language is open to a number of interpretations regarding exactly how much must be done, it clearly shows the degree to which technology was central to the debate that created the FCCC.

Articles 4.7, 4,8, and 4.9 squarely place the success of the FCCC on the contributions of developed country Parties to developing country Parties, including specifically in the area of technology. For example, Article 4.7 states, The extent to which developing country Parties will effectively implement their commitments under the FCCC will depend on the effective implementation by developed country Parties of their commitments under the FCCC related to financial resources and transfer of technology. The needs of special groups of countries are highlighted in Article 4.8 (a) through (i) with particular reference to actions related to funding, insurance and the transfer of technology. The needs and situations of the least developed countries are specifically referenced in a stand-alone article, Article 4.9, especially with regard to funding and transfer of technology. The Parties created the interim financial mechanisms, the Global Environmental Facility (GEF) to accomplish the purpose of these articles.

Technology and technology transfer is a theme in other parts of the FCCC as well. The Subsidiary Body for Scientific and Technological (SBSTA), one of the two organizations created by the FCCC to provide timely information and advice to the Parties, is mandated in Article 9.2(c) to identify innovative, efficient and state-of-the-art technologies and know-how and advise on the ways and means of promoting development and/or transferring such technologies. The financial mechanism defined in Article 11.1 provides for financial resources on a grant or concessional basis, including for the transfer of technology. Further, developing country Parties are invited, on a voluntary basis to propose projects for financing, including specific technologies needed to implement such projects in Article 12.4.

At the FCCC Third Conference of the Parties (COP-3) Kyoto, Japan in December 1997, three new mechanisms of cooperative implementation were established (Climate Change Secretariat, 1998). The new mechanisms include transactions among Annex I Parties, international emissions trading (IET) which provides for cooperation among the Annex B Parties, and the CDM which extends the scope of the cooperation to non-Annex I Parties. It is important to recognize that all of the above mechanisms, and AIJ will lead to the transfer of technologies to mitigate climate change. COP-4 in Buenos Aires, Argentina in November 1998 discussed the issue of technology transfer further and SBSTA made a set of specific recommendations with a special emphasis on capacity building and consultative processes.

1.3 Energy and forestry sector mitigation technologies in AIJ projects

Many cost-effective technical strategies for GHG abatement have been identified for the energy sector of developing nations. These options can be classified into two categories: improving energy efficiency and switching to less-C-intensive fuels. Improving energy efficiency reduces the energy used without reducing the level of service. Reduced energy use decreases associated environmental impacts, including emissions of GHGs. Switching to less-C-intensive fuels means moving away from fuels that have a higher C content per unit of energy to those that have lower C content. Among the most commonly used fuels, C content decreases from coal to oil to natural gas. Renewable energy sources such as wind, solar energy, and nuclear energy have no direct C content. Hydroelectric sources may release CO2 and methane CH_4 depending on the type of vegetation and soils beneath their reservoirs (Fearnside, 1997).

Opportunities to improve energy efficiency exist on both the supply and demand-side. Supply-side energy efficiency can be increased by improving equipment for power generation, and improving transmission and distribution; reducing the flaring of natural gas; plugging of pipeline leaks; and improving coal distribution. Improvements in electricity supply include rehabilitation and efficiency improvements for power plants and rehabilitation and upgrading of transmission and distribution systems, including district heating system, as being done under Swedish-funded AIJ projects in Eastern Europe (Tables 1 and 2). Older vintage power plants typically supply electricity in developing countries because investment capital to build new power plants is scarce. Rehabilitating such power plants and improving their efficiency is thus preferable to acquiring new sites and constructing power plants. Many developing countries have transmission and distribution losses that exceed 20% compared to approximately 6-8% for well-managed systems in developed countries (Meyers et al, 1993). Reducing transmission losses is cost effective and can lead to significant reductions in fuel use and carbon emissions (Jhirad, 1990; World Bank, 1993).

Demand-side energy-efficiency opportunities include improving the energy performance of equipment such as cookstoves, lamps, electric motors, appliances, boilers, buildings, vehicles, etc., and that of processes, particularly in the energy-intensive industries of aluminum, cement, chemicals, fertilizers, iron and steel, and paper. Cost-effective projects and programs to improve lighting efficiency are being pursued in Mexico and in Poland (Sathaye et al., 1994; International Finance Corporation, 1996). Appliance energy performance is being improved through standards and guidelines in Thailand and South Korea (Duffy, 1996), and industrial energy-efficiency programs have been successfully pursued in China (World Bank, 1993; Levine et al, 1992). Air-conditioner energy conservation programs have been proposed for an Australian-sponsored AIJ project in the Solomon Islands.

There are opportunities to switch to less-C-intensive fuels on both the demand and supply sides. Demand-side fuel-switching strategies to reduce C emissions include the use of compressed natural gas or ethanol in vehicles, and the use of renewable biomass sources for cooking, e.g., in the AIJ project in Burkina Faso. Biomass cookstoves, however, release CO_2 and other GHGs as well, so switching household cooking to biomass-fuel may not reduce overall GHG emissions (Smith et al, 1993; Smith, 1995). The use of alternative vehicle fuels has been tried with varying degrees of success in Brazil, Argentina, and India, among other countries (Sathaye et al., 1989; Michaelis et al, 1995), and has been proposed under a Dutch-Hungarian AIJ project to use compressed natural gas (CNG) in buses. Because CH_4 has a GWP that is 21 times higher than CO_2, small leaks in the process of supplying natural gas can negate the advantage of switching to this fuel; it is not clear that switching to methane will necessarily lead to reduced CO_2 emissions (DeLuchi, 1991).

Supply-side fuel switching includes the use of natural gas in place of coal or oil in power plants, and the use of renewable and nuclear electric power sources. Example strategies include using combined-cycle, natural-gas-based power plants in place of coal; creating bagasse/biomass cogeneration/bioenergy systems; and using small hydropower, wind, off-grid solar photovoltaics, fuel cells, solar thermal power generation, and other renewable energy sources. Several AIJ projects have been proposed in this category, e.g., the USA/Costa Rica and USA/Mexico wind energy projects, and the USA/Honduras biomass power project. Cost-effective opportunities have been demonstrated for wind, small hydro, and sugar cogeneration (Johansson et al., 1993); however, many renewable energy sources may not be cost effective when compared with conventional power plants.

1.3.1 Forestry Mitigation Options

Brown et al. (1996) have identified three broad categories of forest mitigation options:
- C sink conservation and management
- C storage management (expanding C sinks)
- Fossil-fuel C substitution management

C Sink Conservation and Management: The annual rates of deforestation in developing countries between 1981 and 1990 were estimated to be 16.27 Mha, of which 15.41 Mha occurred in tropical forests in developing countries (FAO, 1994). Causes of deforestation include land clearing for agriculture, mineral extraction, hydro reservoirs, and fuelwood and timber harvesting (Brown et al, 1995). The forest C sink conservation mitigation strategy involves halting or slowing deforestation and forest degradation (Brown et al., 1996). Forest conservation measures include legislative bans on forest clearing, accompanied by measures to provide alternate sources of biomass (fuelwood and timber), alternate livelihoods for forest-dependent communities, and increased agricultural productivity. Other mitigation measures are: introducing sustainable logging as in the proposed USA/Indonesia reduced impact logging (RIL) AIJ project in Indonesia; improving the efficiency of harvesting, processing, and use of forest products; and recycling of forest-wood-based products. Creating Protected Areas (Nature Reserves) is also an attractive mitigation measure, e.g. the biological reserves AIJ project to consolidate Costa Rican national parks. Converting forests to protected areas can conserve or even increase the C density in vegetation and soil. In developing countries, 486 Mha of forests are currently protected (FAO, 1995). Forest conservation and forest C sink conservation should be priority mitigation options for all tropical countries because, in addition to their CO_2 emissions reductions benefits, these strategies have significant implications for biodiversity conservation, watershed protection, and prevention of desertification and land degradation.

C Storage Management: Deforestation and degradation from over-extraction of fuelwood and timber, grazing, and fire all lead to loss of C in vegetation and soil to atmosphere. C sink management entails measures to increase C density in vegetation and soil in the existing degraded forests as well as measures to create new C sinks on non-forest lands or in areas where forests have been destroyed (Dixon et al., 1994). These options include promotion of natural regeneration (through protection of degraded forestlands), reforestation (on deforested lands), afforestation (on non-forest lands) and agroforestry (on cropland). The USA/Russia RUSAFOR project is an example of this type of AIJ project. C sequestration stops when a forest matures; thus, mitigation potential is finite. The period required for forest maturity depends on forest type, species of trees, soil, and climate conditions. C pools can also be created through production of durable wood products, (assuming the demand for wood products will expand at a faster rate than the decay of wood), and by extension of the life of wood products through means such as timber treatment and the production of long-lasting particle board (Elliot, 1985).

Fossil Fuel C Substitution Management: Substitution management views forest plantations as renewable resources (Ravindranath and Hall, 1996) and has the largest long-run mitigation potential. Substitution management involves conversion of biomass into products that substitute for or reduce the use of fossil fuels. Measures include substituting bioenergy for fossil fuels or fossil fuel electricity as well as replacing non-wood products that require fossil fuels for their manufacture (e.g., steel and cement) with sustainable wood products (Brown et al., 1996). The AIJ Project Salicornia will promote halophyte cultivation in Sonora, Mexico, and one use of the product will be as a substitute for particleboard.

With C storage management, C accumulation stops at maturity; with sustainably managed substitution options, perpetual C emission reductions could be achieved (Hall and Hemstock, 1996). Substituting biomass-fueled for coal-fueled electricity can avoid four times more C emissions than the amount of C that would be sequestered in plantations over a period of 100 years (Ravindranath and Hall, 1996). Primary forests should not be used for harvesting fuelwood for energy production but should be conserved for their contributions to biodiversity and watershed maintenance as well as for their role as C sinks. Degraded forest and non-forest land can be used for sustainable production of biomass for energy (Ravindranath and Hall, 1995).

Short-rotation woody crops have the potential, with advances in energy conversion and yield, to reduce global fossil fuel emissions by up to 20% (Hall et al, 1991; Sampson et al., 1993; Graham et al., 1992). A study in India has shown that biomass-based, decentralized, small-scale electricity systems could offset nearly a quarter of India's fossil fuel emissions (Ravindranath and Hall, 1995).

2. TECHNOLOGY TRANSFER DURING THE AIJ PILOT PHASE

Discussions of project-based mechanisms during the negotiation of the FCCC and the Kyoto Protocol identified a number of issues concerning the relationship of such systems to technology transfer.

- Would JI/AIJ transfer technology to developing countries and what kinds of technology would be transferred? There were concerns that older, more polluting technologies would be transferred rather than state-of-the-art technologies, or the technologies would be inconsistent with the development objectives of the host countries.
- Would only some sectors benefit from AIJ/JI? In particular, it was feared that forestry projects, because of their presumed lower costs, would dominate the pilot phase in particular, since without crediting there would be little incentive for truly environmentally additional mitigation projects.
- Would the technologies generate local environmental and socioeconomic benefits in addition to the global benefit associated with the reduction in greenhouse gas emissions?
- What role would governments play in the technology transfer?

These are the questions that this section explores using the evidence from the AIJ projects.

2.1 Distribution of AIJ Projects by Region

Various sources have reported on AIJ projects, some claiming that there are nearly 130 projects (Joint Implementation Quarterly, 1999). However, only 95 projects have been reported to the FCCC Secretariat as part of the annual reporting on AIJ. The FCCC database of reported projects was last updated October 13, 1998. Since it is a database of the officially reported projects, those using the uniform reporting format (URF), there is some assurance that the AIJ criteria have been met. Also, since there is not a complete set of comparable information on other projects, we decided to focus on the FCCC database of projects.

In looking at these questions, it is important to remember that the AIJ pilot phase did not include crediting for the GHG emissions reductions. This is an important limitation on the generality of the conclusions that can be drawn about project-based crediting systems from this chapter. As will be seen below, the location and types of projects are, in many cases, at odds with what one might expect with crediting. Certainly the absence of crediting affected the number of projects, but we believe it also influenced the extent of private sector involvement in the AIJ pilot phase. In addition, in general, projects are developed where the profitable opportunities are greatest, i.e., where the business environment is favorable and sales and revenue growth can be expected. During most of the pilot phase, Asia had the fastest growth among regions in the developing world. It would be reasonable to expect that a large percent of the projects would have been in this region

TABLE 1.Allocation of AIJ projects by region.

Region	Percent of Total
Latin America	21%
Asia	6%
Africa	1%
Eastern Europe and Russia	68%

Instead, as Table 1 indicates, the majority of the projects occurred in Eastern Europe where economic decline has largely characterized this decade. In fact, the largest number of AIJ projects occur in countries like Latvia, Estonia, and Lithuania as can be seen in Table 2.

TABLE 2. Allocation of AIJ projects to selected countries.

Country	Percent of Total
Latvia	25%
Estonia	20%
Lithuania	9%
Costa Rica	8%
Russia	6%

What explains this odd distribution of the AIJ projects? Fifty-four percent of all projects in the pilot phase were in these three small countries. There are probably several factors at work.

While the data in the FCCC Secretariat's compilation of projects is inadequate to know for sure, it appears that the Eastern European projects were undertaken for non-climate change reasons, namely the reduction of transboundary fluxes of acid rain precursors. This picture becomes clearer when one looks at other aspects of the technology transfer picture, e.g., the types of technologies and the sectors of their application. The Ziegzdriai Boiler Conversion and Energy Efficiency project in Lithuania is a typical example. The project involved conversion of a small boiler plant to combined biofuel and oil firing, rebuilding of the distribution network and closing down of steam projection. Most of the Swedish projects were retrofits of district heating systems to substitute fossil fuels for biofuels.

Sweden supported all but two of the projects done in Latvia, Estonia, and Lithuania. In fact Sweden, which was the investor nation for 49 (over half of the total) of the projects, only invested in those three countries. The Netherlands and Germany funded the two projects in those three countries that were not funded by Sweden. Norway is also an investor nation in AIJ projects in Eastern Europe, although their AIJ program has broader geographic distribution than Sweden's.

It appears that the desire to contribute to a global solution has been integrated into the Swedish national policy objective to preserve their local environment. Sweden and other Nordic countries have for over a decade promoted the notion of investing in activities that reduce emissions of acid rain precursors, even when the sources are in other countries. It will be important to remember this reality in interpreting the other aspects of technology transfer. However, it is a particularly noteworthy lesson learned from the pilot phase.

What the Swedish AIJ experience illustrates is government's incentive to invest in AIJ experience is in part for the local environmental benefits that flow from these projects in the form of reduced transboundary fluxes. At the same time, the host country benefited from reduced local pollution as well. The existence of crediting, we believe, would only magnify the number of these opportunities.

2.2 The Roles of Government, the Private Sector, and Other Organizations

There have been two broad views expressed about how project-based systems should function. The USA, Canada and others favored a private sector based system. Government funding would generally not be used to implement the activities of JI/AIJ projects. Other countries, including Japan, many of the European Union (EU) countries and developing countries see a more explicit funding role for governments. These two views are manifest in the approaches taken during the AIJ pilot phase.

While the data from the URF is frequently sketchy on the nature of participants in projects, it appears that between 30-35% of the projects involved private sector entities on the host government side. They played a variety of roles including developer, financier, utility/generator, consultant, and project administrator. On the investor-nation side, about 50 percent of the projects involved the private sector in roles similar to that played by the host private sector. With the available data it is impossible to make judgements about which model worked best in transferring technology.

Since the AIJ pilot phase does not include marketable credit for the investors in the projects, the incentive for the private sector to assume the additional transactions costs is quite weak. Several explanations have been suggested including:

- opportunity to influence the future rules that will govern project-based crediting systems,
- gain experience with such projects
- public relations value of conducting a green project, and
- potential for lower sovereign risk associated with a project, since one AIJ criteria was host and investor country government approval of the project.

This reality may also explain why many of the non-C sink projects were relatively small. Participation in the AIJ pilot phase did involve additional transaction costs in the form of submissions to host and investor governments, the establishment of monitoring and verification systems, etc. Large projects, particularly those in active markets and with strong profit and market share acquisition characteristics, could not be bothered with the few motivations noted above.

Nevertheless, 95 projects were listed by last year and as noted above the number of projects after this year may be in excess of 130. It appears that most projects currently in the database, governments played an active financing and implementation role.

In addition to private sector business enterprises participating in the AIJ pilot phase, other non-profit organizations also participated. Included in that group were environmental, developmental, educational, and research non-governmental organizations, such as The Nature Conservancy (TNC). A unique feature of project-based crediting systems envisaged by the FCCC and the Kyoto Protocol is the encouragement for these other segments of NGOs to participate. Again data limitations due to imprecise responses to the URF, limited precision in estimating NGO participation is possible, but it appears that 30 to 35% of host country NGOs participated in projects. About 20-25% of investor country NGOs participated. In both the case of investor and the host, the NGO participation was primarily in areas of project evaluation and monitoring.

The actual role played by governments is also difficult to discern from the URF. The types of agencies involved, however, can be a rough indicator of the expertise they bring to bear to a project and perhaps also indicate their responsibility for climate change within their national governments. About 70% of the projects had some participation by a host government agency. Table 3 indicates the relative participation of host government agencies in AIJ projects.

TABLE 3. Participation of host government agencies in AIJ projects.

Host Government Agency	Percentage Participation
Environment	29%
Agriculture/Forestry	4%
Energy	8%

Finance	2%
Commercial/Technology Development	0%
Local Government	39%

It is impossible to deduce from the URF the expertise of these agencies in promoting technology transfer or evaluation. The most likely candidate for such expertise would be among agencies promoting commercial interests of countries or technology development and none of these agencies appears to have participated in an AIJ project. It also appears from the URF that countries were not consistent in reporting the roles of host governments. There may be some cases where an agency responsible for approving the project as an AIJ project was listed as a participant and in other cases such agencies were not listed.

One particular oddity is the relatively large percentage of local governments involved. It is tempting to view this as a success of AIJ, since encouraging local government involvement in projects and technology choices would be a desirable feature of AIJ. High local government participation could imply, for example, that local environmental and social benefits from the project were being given due consideration. Unfortunately, no such conclusion is warranted. Most of these cases of local government participation are due to the Eastern European projects which almost exclusively dealt with retrofitting boilers and energy efficiency improvements in municipal district heating plants. Local governments therefore appear qua utility rather than as articulator of local environmental, economic and social interests.

2.3 Technology Transfer by Sectors and Type

Table 4 indicates the sectors covered by the AIJ projects.

TABLE 4. AIJ projects by economic sector.

Sector	Number of Projects
Residential/Commercial	57
Manufacturing	3
Electricity	19
Transportation	2
Forestry	12
Agriculture	5

Several of the projects involved multiple sectors. Cogeneration is represented by inclusion in multiple sectors for example. Again, the large number of residential projects reflects the influence of Sweden's AIJ projects in Eastern Europe. These projects were almost exclusively retrofitting of district heating plants. While they are represented in the residential sector in this analysis, they in most cases would also provide heating to commercial buildings. Rather than combine the two sectors however, we separated them so that the four projects that dealt exclusively with the commercial sector could be noted.

Excluding the residential sector however, does show those supply side changes in the electricity sector, either efficiency improvements or fuel switching, dominate the number of energy efficiency improvements at the end use. The two transportation projects are on natural gas pipelines. There are no projects addressing traffic management, mode switching, or efficiency improvements in the transportation sector.

Table 5 indicates the AIJ projects by major type.

TABLE 5. AIJ pilot phase projects by type.

Project Type	**Number of Projects**
Forest Preservation/Management	8
Afforestation/Reforestation	6
Retrofit-Fuel Switching	37
New Generation – Fuel Switching	15
Supply-Side Efficiency Improvements	16
Demand-Side Efficiency Improvements	11
Gas Capture	2
Recycling	1

The effect of the Swedish Eastern European projects can again be seen in the large number of fuel switching and supply-side efficiency improvements. A large number of those projects involved new boilers and improvements in the system. Removing the retrofit-fuel switching from the mix however shows a reasonable spread among different project types. Of particular note, are the 14 forestry and land-use change (LUCF) projects that include agricultural land use switching to agroforestry projects under forest preservation/management category. Many critics of AIJ had feared that forest projects would dominate in number. Such projects are generally thought to have the lowest marginal cost with some estimating the cost at $1 per Mg C sequestered. There is however a large number of energy projects. The picture remains cloudy nevertheless since all but two of the LUCF projects involved the private sector and not governments. The two other LUCF projects were government-to-government deals. It might be expected that the private sector was more sensitive to project costs than governments. In the absence of crediting, as we noted the motivation for the private sector to participate in the pilot phase is less compelling. The limited number of purely private sector projects seems to have gone disproportionately to LUCF projects.

2.4 Technologies and Projects

Table 6 presents the list of supply-side technologies that were included in the AIJ projects. The energy efficiency technologies were not easily determined from the URF. However, energy efficiency projects did include the installation of more efficient lighting, air conditioners, insulation and demand-side management (DSM). Because of the large number of Swedish district heating projects, many of these projects involved new heat exchangers, controls, metering, and system optimization efforts.

TABLE 6. AIJ pilot projects by technology.

Technology	Number of Projects
Biofuel	35
Pipelines	4
Natural gas generation	2
Wind	4
Solar	5
Hydro	2
Geothermal	1

The large number of biofuel projects reflects the Swedish projects. However, there are also two Bio-Gen projects in Honduras included in the U.S. Initiative on Joint Implementation (USIJI) portfolio of projects. It does appear that given the limited number of projects for which clear technology information could be gleamed, a nice spread of generation technology choices were used.

Finally, it is important to note that technology is not just equipment. It includes to a considerable extent the knowledge required to make appropriate technology choices, operate and maintain equipment, and address the external effects of technologies. While it was not possible to assess the capacity building aspects of the AIJ projects, it is worth noting that almost all projects reported a capacity building aspect to the project.

3. TECHNOLOGY TRANSFER PROJECTS: BARRIERS, POLICIES AND INCENTIVES

Many mitigation studies highlight the fact that developing countries could pursue additional, cost-effective GHG mitigation options in the energy and LUCF sectors. Pursuit of these cost-effective options would reduce the rate of increase in C emissions from developing countries without jeopardizing, and in some cases enhancing, the countries' economic growth. Although the estimated GHG emissions reduction from a mitigation option depends on the baseline scenario that is somewhat subjectively defined for each country, experts from 14 of the countries participating in the U.S. Country Studies Program (CSP) have nonetheless identified feasible negative -cost mitigation options.

The mix of these options and their impact on emissions reduction differs across countries. Energy-efficiency and renewable energy options that displace grid connected electricity will not reduce emissions as much in Brazil where electricity is largely generated using hydropower as in China or India where coal-based electricity generation predominates. Further, studies show a higher potential for renewable energy options in countries rich in natural resources. In the LUCF sector, the studies identify large tracts of land that are suitable for forestation. Land availability may be a constraint, however, if the certain legal and socio-cultural barriers are not overcome.

Many barriers prevent the transfer of technologies across and within countries. Trade barriers can prevent the transfer of efficiency technologies while subsidizing the transfer of inefficient power plants and related equipment. Price subsidy can be a disincentive to the adoption of efficiency measures, and irrational fuel pricing can prevent the introduction of natural gas or limit its use to one sector. Lack of knowledge about growing trees dissuades farmers from planting trees whose products may supplant those made from unsustainably harvested roundwood. A substantial body of literature discusses market barriers to the implementation of energy options (Golove and Eto, 1996; Huntington et al, 1994; Johansson et al., 1993) but, not as much has been written about barriers to options in the LUCF sector.

The International Energy Agency (IEA) in its review found several barriers associated with Swedish AIJ projects in Eastern Europe and FSU. Barriers to project development (in the baseline scenario) relate to the unfamiliarity of local banks with such projects, the lack of know-how to manage and implement projects involving bank loans on the part of district heating owners and operators, the perceived risk related to the financial viability of these projects on the part of Nordic equipment suppliers and, sometimes, lack of securities because of unclear ownership of district heating facilities. The involvement of the Swedish National Board for Industrial and Technical Development (NUTEK) mitigated these barriers effectively by providing consultant assistance throughout project implementation (design of call for tenders, evaluation of offers, procurement of equipment, supervision of installations and construction works) as well as the training of local experts to facilitate the increasing use of domestic capacity for future projects.

AIJ, GEF and other projects can be, and have been, greatly facilitated in countries where barriers are being removed. The International Energy Agency (IEA) Climate Technology Initiative (CTI) and bilateral programs of individual OECD countries (eg, Denmark, Germany and USA all have extensive ODA programs aimed at transferring GHG mitigation technologies) help address barriers to technology transfer. NGOs such as the World Business Council for Sustainable Development (WBCSD) are active leaders in identifying and transferring climate friendly technology to developing and transition countries (Anonymous 1999; Dixon, 1997; Dixon 1998). Below we discuss the barriers at three levels: macro, sector, and project-specific, and cite examples of AIJ and other projects, where appropriate policies and programs to remove or diminish the barriers have fostered successful technology transfer.

Macro Conditions: Macro conditions relate to the entire socioeconomic structure of a country, particularly investments in which energy and forestry options are ignored, undervalued, or considered too risky by economic actors. Examples of macro barriers include a low level of competition among firms resulting from regulation of the domestic market and/or from policies that constrain entry of imported products, high tariffs on imported goods, and a low level of capital market development.

At its broadest level, the collapse of the former Soviet Union (FSU) created many opportunities to redress the gross neglect of efficiency in energy use in countries in Eastern Europe and elsewhere in the FSU. One outcome of the economic and geopolitical restructuring was that Sweden was able to establish 49 AIJ projects in Eastern Europe, and AIJ projects such as RUSAGAS and RUSAFOR were started in Russia.

Another example of a macro barrier is government regulation prohibiting foreign firms from bidding on the construction of new industrial factories or power plants, which can limit a country's access to advanced foreign technology. Conditions that constrain the entry of imported products can lead to the use of obsolete technology. The history of government intervention to address a severe paper shortage in India during the early 1970s illustrates this problem. To address the shortage, the Indian government promoted the establishment of small paper mills that could be quickly set up (Datt and Sundharam, 1998). This led to the import of inexpensive second-hand paper mills that were set up in many regions of the country. The inefficient mills grew to account for 50% of the country's paper production. Then, in 1988, the government removed the protection it had accorded the paper industry, which led to the shutdown of many of these small, inefficient plants. This elimination of government protection will in the long run increase energy efficiency and total productivity.

An example of the limitations created by government regulation was a high import duty imposed on compact fluorescent lamps (CFLs) in Pakistan. When this duty was reduced from 125% to 25% in 1990, the price of CFLs dropped by almost half, and sales started to rise, leading to improved energy efficiency (USAID, 1996).

Capital for investment from domestic sources is scarce in many developing countries, particularly where foreign exchange is required. AIJ projects can be particularly effective here in providing this source of capital. Virtually all AIJ projects have occurred in countries where capital is too expensive and available only for the short term. Thus AIJ projects have played a key role in bringing in new capital to sectors that otherwise would have been neglected.

Sector Conditions: There are several energy and LUCF sector-specific barriers to the implementation of mitigation options in transitioning and developing countries. These include subsidized pricing of fuel, electricity, and other products; government policies regulating energy- and forestry-related activities; reduced government budgets and financing for energy and forestry projects; weakness in a country's institutional and legal framework; the uncertain status of private firms in the energy and forestry sectors, lack of information on mitigation options, and limited access to financing.

Rationalization of electricity and fuel prices is a key feature of reforms being carried out in developing countries where subsidies for LPG and kerosene are being reduced or removed, and electricity tariffs are being adjusted to reflect the cost of production. Fuel subsidies have been reduced or removed in India, Pakistan, Thailand, Brazil, and Mexico, to name a few countries.

There is significant potential to reduce GHG emissions by substituting bioenergy for fossil fuels. Often, bioenergy may have to compete with subsidized fossil fuels and fossil-fuel-based electricity. For example, electricity for pumping irrigation water (a major electricity-using activity) in rural areas of India is supplied free or at a highly subsidized rate. For bioelectricity to compete with fossil-fuel electricity in this situation, it may be necessary to adopt rational electricity pricing for fossil fuel electricity (Hall and Hemstock, 1996; Ravindranath and Hall, 1995; Brown, Sathaye et al, 1995). In the Honduran AIJ project, however, this barrier does not exist and the project could go ahead by selling electricity to the grid.

Many financial, administrative, and policy reforms are necessary to promote LUCF measures to reduce GHGs and to enhance C sinks (Sathaye and Ravindranath, 1997). Strong forest policies are required to regulate or ban forest clearing. India, for example, enacted a Forest Conservation Act in 1980 that regulates all conversion of forest land to non-forest uses. This policy has significantly reduced deforestation rates (Ravindranath and Hall, 1994). Vietnam is in the process of enacting a policy to regulate extraction from forests for export. Brazil has issued a decree suspending the granting of fiscal incentives to new ranching projects in Amazon forest area, in order to decrease the rate of at which the forest is cleared for ranches (Sathaye and Ravindranath, 1997). A policy is still needed to stop clearing of Brazilian forests for cattle pasture, however. One of Costa Rica's AIJ projects, the Private Forestry Project (PFP), was initiated in combination with a policy mandate to tax gasoline and use the revenue to pay private land owners to maintain forest cover, or convert land from pasture to forests (Subak, 1998).

Forest-based industries in many countries often obtain large concessions from the government forest department for extracting timber from forests for which they pay very low royalty charges (Repetto and Gillis, 1988). There is a need for withdrawal of such subsidies (Kadekodi and Ravindranath, 1997). In addition to eliminating subsidies, efforts are needed in order to prevent timber harvest from natural forests by persuading forestry firms to acquire raw materials from non-forest areas such as farm lands or degraded lands. In place of subsidies for harvesting from natural forests, financial incentives such as tax concessions could be provided to enterprises that source wood from non-forest lands. Farmers could be provided with seedling material, low-cost credit, and an assured market and price to encourage farm forestry program for supplying wood to the industry. The Scolel Te AIJ Project in Chiapas, Mexico is designed, and being implemented, to accomplish these objectives (Imaz, Gay, Friedmann and Goldberg, 1998). The project will develop initially 1200 ha of agroforestry and another 100 has of natural forest management within an area of 13000 ha where about 3500 persons live. Farmers will receive an annual income from the project, and will receive technical assistance in the aforementioned activities. An interesting development in India along these lines has been the recent planting of teak by private entrepreneurs, with capital raised in private capital markets. An AIJ project for planting teak trees has been proposed and recommended for acceptance by the Indian government. There are also examples of paper mills in India that provide credit and seedlings and enter into agreements with farmers to purchase eucalyptus wood at harvest (Ravindranath and Hall, 1995).

Government budgets are stretched tight as the energy sector expands faster than economic growth, and the government budget is inadequate to finance new power, natural gas and petroleum, and renewable energy projects. Deregulation to allow private producers into an area, which hitherto has been dominated by government-owned companies, is one solution to this problem. The governments of Argentina, Chile, and Brazil have privatized the generation and supply of electricity and fuels. Several renewable-energy AIJ projects in Central American countries are owned by private companies and are designed to generate electricity and sell it to the central grid.

The development of companies that could provide energy services that reduce GHG emissions growth can be limited in countries where there is no existing legal framework for contracts with energy service companies (ESCOs), such as in China. Although ESCOs are being formed in many countries, their status is often uncertain; the prices that they will be paid for saving electricity are being negotiated on a utility by utility basis, for example, in India.

Analysts and governments are not fully aware of the variety of mitigation options that can be pursued in a sector. The bilateral programs, such as the US CSP, or the German and Dutch programs were created, in part, to fill this need by training analysts and providing information to developing-country governments about the consequences of pursuing mitigation options. The programs have trained hundreds if not thousands of experts in developing and transition countries on mitigation methods and analytical techniques (Dixon et al., 1996). AIJ projects are also fulfilling this need for knowledge. The Australia/Solomon Islands AIJ project will provide occupant education on proper air conditioner use, while also installing timer switches and regular air conditioner maintenance.

Project conditions: Most of the climate change projects in developing countries are being funded by foreign investors or governments under the AIJ or JI concept. AIJ/JI allows for full or partial financial support from a potential investor country. The IEA in its analysis of AIJ (International Energy Agency, 1997) found government's playing an active role in barrier removal. For example, NUTEK played an active role in many of the Swedish AIJ projects. NUTEK provided financing for many of the Swedish projects in the Baltics.

Several concerns have been voiced regarding AIJ/JI projects. These include the transfer of high cost and/or obsolete technology, negative local impacts, additionality of funds and carbon emissions reduction, lack of capability to monitor carbon flows of projects, sharing of carbon credits, and macroeconomic impacts. A recent study of current and proposed energy, LUCF, and bioenergy projects in Brazil, India, Mexico, and South Africa shows that most of these concerns are unfounded. The review of projects showed that these are, or will be, contributing to rural employment, reducing air pollutants, in addition to GHGs, increasing manufacturing capacity, conserving biodiversity, reclaiming degraded lands, and protecting watersheds (Sathaye, Makundi, and Andrasko, 1998). The additionality of funding, however, could not be assured under the reviewed projects, and the equitable sharing of C credits remained a concern. The macroeconomic benefits of these projects were also positive since they provided new jobs and reduced oil and timber imports.

Another study of impacts of mitigation options for India showed significant positive impacts on economic output from forests, as well as a three-fold increase in the contribution of the LUCF sector to GDP, a large increase in employment, and a net contribution to foreign exchange earnings (Kadekodi and Ravindranath, 1997). These energy and forestry studies clearly show that there need not be any conflict between the global benefits and the local or national socio-economic and environmental benefits for climate mitigation projects.

One of the features of LUCF mitigation projects is the long period before carbon mitigation is realized, which makes it necessary to ensure long-term community rights to control land and forest products (Ravindranath and Hall, 1995; Kadekodi and Ravindranath, 1997). Strong local institutions with legal backing are necessary to enable communities to participate in forestry programs over the long term (Poffenberger and Banerjee, 1996). Government policy is fickle, however, as the Costa Rica PFP AIJ project demonstrates, the use of gasoline tax revenue to pay land-owners has now been suspended, and it is unclear as to how the revenue to pay landholders will be generated.

4. CONCLUSIONS

The historical GHG emissions from industrialized countries have been much higher than those from the rest of the world. The 1995 IPCC report pointed out that the former contributed about 80% of the C emissions from fossil fuel combustion from 1860 to 1990. Consistent with their economic and population growth rates, the growth rate of future GHG emissions is projected to be higher from the developing world. The transfer of environmentally sound technology has the potential to slow emissions growth at a lower cost in the latter countries. AIJ projects provide one mechanism to transfer technologies in order to achieve emissions reductions in the developing and transitioning countries.

The review of AIJ projects above suggests that almost half the projects were on improving energy efficiency in Eastern Europe and countries of the FSU. The Swedish government sponsored most of these projects since they had the added benefit of reducing acid rain from the Nordic countries. About 15 projects were on forestry and agriculture sector mitigation options. The AIJ projects include technologies to improve energy efficiency in district heating systems, lighting, boilers, air conditioning, renewable energy sources, including bioelectricity, that might not been readily available in the host countries. The AIJ projects will result in the transfer of knowledge, in addition to hardware, in most cases, as developers and communities in the host countries learn to use the hardware and/or new management systems. The forestry and agricultural projects will provide ways to manage tree farms and conduct logging in a sustainable manner, provide monetary benefits to local communities, promote ecotourism, etc. Many of the AIJ energy projects have built capacity building components built into the project design.

The AIJ projects have demonstrated the benefits of removing barriers for the transfer of mitigation technologies at all levels. Institutional changes in the national electric utility systems have provided an opportunity to pursue renewable electricity AIJ projects in the Czech Republic, Costa Rica and Mexico. Of course, none of the projects in the FSU and Eastern Europe would have occurred without the massive socio-political changes in these countries.

This review of AIJ projects demonstrates that the technologies transferred to reduce GHG emissions and atmospheric C concentration can simultaneously reduce deleterious local environmental impacts, and provide local jobs and other socio-economic benefits. The magnitude and quality of benefits varies between forestry and energy, and rural and urban projects. The experience with the limited number and duration of AIJ projects suggests that the transfer of hardware and/or knowledge to recipient countries is has been successfully implemented in many projects, and their implementation has the potential to create conditions for replicability within the country and abroad.

ACKNOWLEDGEMENTS

Abe Haspel contributed to the development of this chapter.

REFERENCES

Anonymous (1999) Workshop on Baseline for CDM, NEDO, Tokyo.

Brown, S., et al. (1996) Forests and the Global Carbon Cycle: Past, Present, and Future Role, in The Role of Forest Ecosystems and Forest Management in the Global Carbon Cycle, Springer Verlag, New York.

Brown, S., Sathaye, J, Cannell, M, and Kauppi, P. (1995) Management of Forests for Mitigation of Greenhouse Gas Emissions, Climate Change 1995, Cambridge University Press, Cambridge.

Climate Change Secretariat (1992) United Nations Framework Convention on Climate Change, UN Environment Programme, Information Unit for Conventions, Geneva.

Climate Change Secretariat (1998) The Kyoto Protocol to the Convention on Climate Change, UN Environment Program, Geneva.

Datt, R. and Sundharam, K.P.M. (1998) Indian economy, S. Chand and Company Ltd., New Delhi.

DeLuchi, M.A. (1991) Emissions of Greenhouse Gases from the Use of Transportation Fuels and Electricity, Vol. 1., Argonne National Laboratory, Argonne.

Dixon, R.K. (1998) The U.S. Initiative on Joint Implementation: An Asia-Pacific Perspective. Asian Perspective 22:5-19.

Dixon, R.K. (1997) The U.S. Initiative on Joint Implementation. Int. J. Environment and Pollution 8:1-18.

Dixon, R. et al. (1996) Greenhouse Gas Mitigation Strategies: Preliminary Results from the US Country Studies Program. Ambio 25:1-8

Dixon, R.K., Brown, S., Houghton, R.A., Solomon, A.M., Trexler, M.C., and Wisniewski, J. (1994) Carbon Pools and Flux of Global Forest Ecosystems, Science, 263:185 - 190.

Duffy, J. (1996) Energy Labeling, Standards and Building Codes: A Global Survey and Assessment for Selected Developing Countries, International Institute for Energy Conservation, Washington D.C.

Elliot, G.K. (1985) Wood properties, and future requirements for wood products, in Attributes of Trees as Crop Plants, M.G.R. Cannell, J.E. Jackson (eds.) Institute of Terrestrial Ecology, Edinburgh.

Fearnside, P.M. (1997) Greenhouse Gas Emissions from Amazonian Hydroelectric Reservoirs: The Example of Brazil's Tucuruí Dam, Hydropower Plants and Greenhouse Gas Emissions, (Eds) L P Rosa, M A dos Santos, Rio de Janeiro.

FAO. (1995) Forest Resources Assessment 1990, Global Synthesis. FAO, Rome.

Golove, W., and Eto, J. (1996) Market Barriers to Energy Efficiency: A Critical Reappraisal of the Rationale for Public Policies to Promote Energy Efficiency, LBNL- 38059, Lawrence Berkeley National Laboratory, Berkeley.

Graham et al. (1992) Climatic Change. 22:223.

Hall, D.O., and Hemstock, S.L. (1996) Biomass Energy Flows in Kenya and Zimbabwe: Indicators of CO_2 Mitigation Strategies. The Environmental Professional 18: 69-79.

Hall, D., Mynick H., Williams R. (1991) cooling the Greenhouse with Bioenergy, Nature 353: 11-12.

Huntington, H.L., Schipper, L., and Sanstad, A. (Eds) (1994) Energy Policy: Special Issue: Markets for Energy Efficiency **22**: 10.

Imaz, M., Gay, C., Friedmann, R., and Goldberg, B. (1998) Mexico Joins the Venture: Joint Implementation and Greenhouse Gas Emissions Reduction, LBNL-42000, Lawrence Berkeley National Laboratory, Berkeley.

International Energy Agency (1997) Activities Implemented Jointly: Partnerships for Climate and Development, IEA, Paris.

International Finance Corporation (1996) Republic of Poland, Poland Efficient Lighting Project, International Finance Corporation, Washington, D.C.

Intergovernmental Panel on Climate Change (1996) Technologies, Policies and Measures for Mitigating Climate Change, in IPCC Working Group II Technical Paper, R.T. Watson, M.C. Zinyowera, and R.H. Moss (eds.), Cambridge Press, Cambridge.

Jhirad, D. (1990) Power Sector Innovation in Developing Countries: Implementing Multifaceted Solutions. Annual Review of Energy 15: 335-363.

Joint Implementation Quarterly (1999) Joint Implementation Quarterly, **5**, 1:14.

Johansson, T.B. et al., (Eds.) (1993) Renewable Energy, Island Press, Washington D.C.

Kadekodi, G. and Ravindranath, N. (1997) Macroeconomic Analysis of Forestry Options on Carbon Sequestration in India. Ecological Economics **23**:201-223.

Lashof, D. and Tirpak, D. (Eds) (1990) Policy Options for Stabilizing Global Climate: Report to Congress, U.S. Environmental Protection Agency, Washington, DC.

Levine, M., Liu, F., and Sinton, J. (1992) China's Energy System: Historical Evolution, Current Issues and Prospects. Annual Review of Energy **17**:405-435.

Meyers, S., Goldman, N., Martin, N., and Friedman, R. (1993) Prospects for the Power Sector in Nine Developing Countries. Energy Policy.

Michaelis, L., Bleviss, D., Orfeuil, J.P., and Pischinger, R. (1995) Mitigation Options in the Transportation Sector, Climate Change 1995, Cambridge University Press, Cambridge.

Mintzer, I. (1994) Institutional options and operational challenges in the management of a joint implementation regime, in K. Ramakrishna (ed.), Criteria for Joint Implementation under the UNFCCC. The Woods Hole Research Center, Woods Hole.

Poffenberger, M. and Banerjee, A. (1996) Village Voices, Forest Choices- Joint Forest Management in India, Oxford University Press, New Delhi.

Ravindranath, N.H. and Hall, D.O. (1996) sustainable forestry for bioenergy versus forestry for carbon sequestration as climate change mitigation options. The Environmental Professional 18:119-124.

Ravindranath, N.H. and Hall D.O. (1995) Biomass, Energy and Environment – A Developing Country Perspective From India, Oxford University Press, Oxford.

Ravindranath, N.H. and Hall, D.O. (1994) Indian forest conservation and tropical deforestation. Ambio **23**:521-523.

Reddy, A. (1991) Barriers to improvement in energy efficiency. Energy Policy **10**: 10.

Repetto, R. and Gillis, M. (1988) Public policies and the misuse of forest resources, Cambridge University Press, Cambridge.

Rosenberg, N. (1982) Inside the Black Box: Technology and Economics, Cambridge University Press, New York.

Sampson, R.N. et al. (1993) Biomass Management and Energy. Water, Air, Soil Pollution **70**:139.

Sathaye, J. et al. (1994) Economic Analysis of Ilumex. Energy Policy.

Sathaye, J. et al. (1989) Promoting Alternative Transportation Fuels: The Role of Government in New Zealand, Brazil and Canada. Energy **14**: 575-584.

Sathaye, J., Makundi, W., and Andrasko, K. (Eds.). (1998) Concerns About Climate Change Mitigation Projects: Summary of Finding from Case Studies in Brazil, India, Mexico, and South Africa, LBNL-41403, Lawrence Berkeley National Laboratory, Berkeley.

Sathaye, J. and Ravindranath, N. (1997) Policies, Measures and the Monitoring Needs of Forest Sector Carbon Mitigation, Mitigation and Adaptation Strategies for Global Change, **2**: 101-115.

Smith, K., Khalil, M., and Rasmusen R., et al. (1993) Greenhouse Gases from Biomass and Fossil Fuel Stoves in Developing Countries: A Manila Pilot Study. Chemosphere **26**: 479-505.

Smith, K. (1995) Health, Energy, and Greenhouse Gas Impacts of Biomass Combustion. Energy for Sustainable Development **1**:23-29.

Subak, S. (1998) Forest Protection and Reforestation in Costa Rica: Evaluation of a Clean Development Mechanism Prototype. Environmental Management, in press.

USAID (1996) Strategies for financing energy efficiency, US AID, Washington, D.C.

World Bank. (1993) China Energy Conservation Study, World Bank.

Watson, R.T., Zinyowera, M.C., and Moss, R.H. (Eds) (1996). Technologies, Policies and Measures for Mitigating Climate Change, IPCC Working Group II Technical Paper I, Cambridge University Press, Cambridge.

Chapter 10

CLIMATE CHANGE, CAPACITY BUILDING AND THE AIJ EXPERIENCE

J. LEONARD[1], I. MINTZER[1] and D. MICHEL[2]

[1]*Pacific Institute for Studies in Development, Environment and Security;* [2]*Johns Hopkins University School of Advanced International Studies*

Key words: capacity building, climate change, enabling activities, institutional strengthening, policy reform, public awareness, technology transfer

Abstract: Participant experiences during the activities implemented jointly (AIJ) pilot have been rich and varied. The experiences of this period offer a number of useful lessons, revealing the critical importance of technology cooperation to international efforts aimed at achieving the objective of the UN Framework Convention on Climate Change (FCCC). Events since the FCCC First Conference of the Parties (COP-1) have demonstrated that reducing the risks of rapid climate change and achieving the FCCC's objective will require the introduction and diffusion of many new, low-emissions technologies. Innovative AIJ pilot projects have demonstrated that a mix of capacity-building activities are needed to facilitate successful technology cooperation. Simultaneously, the AIJ pilot phase has highlighted the inherent difficulties of technology cooperation for government leaders, private companies, international organizations, inter-governmental organizations (IGOs) and non-governmental organizations (NGOs). Specifically, the AIJ pilot has shown that the full potential can only be realized through careful design of projects and procedures. These designs must incorporate a mix of capacity building activities that are appropriate to the specific circumstances, local institutions, and cultural context of the industrial and developing country partners. The purpose of this chapter is to highlight the lessons learned from capacity building during the AIJ pilot and explore their implications for the Clean Development Mechanism (CDM) and Joint Implementation (JI) under the Kyoto Protocol.

1. BACKGROUND

Recent international efforts to address global climate change have included the formulation of new mechanisms to encourage private sector investments and public-private partnerships. These new mechanisms are designed to promote sustainable development and to encourage investment in new, more efficient technologies which reduce greenhouse gas (GHG) emissions or that sequester carbon dioxide (CO_2) in natural sinks. For the purposes of this paper, technology is understood to refer to a dynamic combination of hardware, software, know-how, and organizational structure needed to manage the production and distribution of a service or a good. Beginning with the AIJ pilot, continuing with JI and the CDM under the Kyoto Protocol, these efforts have led many to conclude that capacity building is key to successful partnerships and will play a critical role in these cooperative international regimes.

2. WHAT IS CAPACITY BUILDING?

If it means so much to so many, does capacity building really mean anything to anyone? Do people build their own capacities, or do they need to contract the job out to professional builders? ... Is capacity building a pre-condition for, or a by-product of, international cooperation? Is it synonymous with development? A means or an end, or both? Or is capacity building just another piece of unwieldy jargon whose very imprecision disguises its emptiness? (Eade and Williams, 1995)

2.1 Evolution of the Concept of Capacity Building

Capacity building is a term familiar to many, but with a meaning that is rarely agreed upon. It surfaces frequently in policy discussions covering a wide range of fields. The same term is heard in contexts ranging from public education to good governance to market transformation to technology diffusion. In many of these discussions, capacity building refers narrowly to the efforts of industrial countries and their foreign assistance agencies to help developing countries create independent, efficient, and effective institutions (Trostle, 1997). In the context of today's (and tomorrow's) climate change debate, however, it must be viewed as a much wider concept, involving both South-North and South-South cooperation as well as traditional North-South, donor-host relationships.

Since the end of World War II, the operating agencies of the UN and a number of industrial country governments have run programs of assistance to developing countries. In principal, one of the aims of these programs has been to build, reinforce, and strengthen indigenous institutions. The way in which these programs addressed institutional issues has evolved over time. The pattern of change in these programs reflects, to some extent, changing demands by host governments and shifting perceptions of development cooperation among donor countries. In the last decade, the concept of capacity building has evolved to include private sector and public-private partnerships outside the development assistance context.

Peter Morgan traces this thematic evolution back to the 1950s and 1960s, when development assistance focused on institution building (Morgan, 1993). This early effort was designed to assist post-colonial governments to create a basic inventory of public-sector institutions needed to deliver the services provided to citizens by a nation-state (Grindle, 1997). In the 1960s and 1970s, emphasis shifted to institutional strengthening, i.e., the reinforcement of existing institutions rather than the creation of new ones (Morgan, 1993). The goal in this phase was to smooth the transition to full independence, weaning recipient countries from the need for future donor assistance. By the late 1970s, attention centered on development management (Grindle, 1997). The concept of development management was designed to emphasize the responsibilities of host governments for the success of national development strategies and, in particular, for the successful delivery of essential services to meet the basic human needs of the population's poorest groups. In the 1980s, the concept of institutional development re-emerged, now focusing attention on the delivery of goods and services beyond those meeting basic human needs. This involved a broader range of entities and processes that deliver added value to consumers and citizens. This approach emphasized the contribution of NGOs and the private sector (as well as public institutions) to successful and sustainable development (Moore, 1995).

During the 1990s, the concept of change management has gained great currency. Change management brings attention to issues of system dynamics, institutional economics, governance, and the empowerment of individuals and communities. It focuses on inter-related hierarchies of institutional responses and organizational behaviors that are required to make lasting and self-sustaining changes in societies. The success of such processes can be measured by the extent to which individuals and recipient communities institutionalize the knowledge, methodologies, and skills necessary to achieve local goals. In this process, the meaning of capacity building has evolved into something that is both synonymous with and essential to the process of sustainable development (Grindle, 1997).

2.2 Current Views of Capacity Building

What you see as capacity building often depends on where you sit. In today's debate, international development assistance agencies, IGOs and multi-lateral development banks seem to have distinct but more-or-less parallel perspectives of the capacity-building process. The views held by many donor country governments differ in focus and emphasis from those of national governments in developing countries. In the broadest dimensions, developing country governments share roughly similar views but these can differ markedly from the approach of both international institutions and local governments on the ground at project locations. Academics and NGOs have their visions, sometimes shared, often divergent. And private sector entities, including joint venture managers and project developers, have views which differ markedly from all of the above.

The following definitions are drawn from reports by UN organizations and agencies, NGOs, and private sector entities. These definitions illustrate some of the meanings given to the concept of capacity building and its synonyms:

- Agenda 21, agreed by 160+ governments at the UN Conference on Environment and Development (UNCED) (Rio de Janeiro, Brazil, June 1992) sets the discussion in a broad and balanced context. This perspective focuses on the state of human capacities and institutional capabilities in a country:

  The ability of a country to follow sustainable development paths is determined to a large extent by the capacity of its people and its institutions as well as by its ecological and geographical conditions. Specifically, capacity building encompasses the country's human, scientific, technological, organizational, institutional and resource capabilities. A fundamental goal of capacity-building is to enhance the ability to evaluate and address the crucial questions related to policy choices and modes of implementation among development options, based on an understanding of environmental potentials and limits and of needs as perceived by the people of the country concerned. As a result, the need to strengthen capacities is shared by all countries. (United Nations, 1992a)

- The United Nations Development Program (UNDP), the UN agency mandated to focus primary attention on capacity building, views it as a systemic and enduring process of social change that builds on existing capabilities. It defines the concept as follows:

  Capacity development is the process by which individuals, groups, organizations, institutions and countries develop their abilities, individually and collectively, to perform functions, solve problems and achieve objectives.

To break that down: development suggests that capacities already exist within the country, and that UNDP's aim is to build upon them and sustain them. The term development also implies that maintaining capacities is a long-term learning process. Process describes the way individuals and organizations interact, learn, assess information, gauge opportunities, solve problems and make decisions to reach their goals. A process is integrated, multi-faceted, and systemic. Individuals, groups, organizations, institutions and countries shows that capacity development requires the full involvement of the people and groups whose capacities are to be developed, and stresses the relationships between the different actors. It also implies that capacity development reaches beyond government institutions to include all levels of society and the overall environment in which the capacity is being developed. Perform functions, solve problems and achieve objectives implies that capacity development involves more than human resource development. It aims to foster specific changes by working through the processes, organizations and institutional context of a given system. (UNDP, 1997)

- The United Nations Environment Program (UNEP) links the definition of capacity building to an understanding of the FCCC and to the technology transfer process:

Technology transfer must be accompanied by capacity building. The delivery of new hardware alone rarely leads to 'real, measurable and long-term environmental benefits' in the host country. In many cases it is absolutely essential to strengthen existing local institutions. This includes building managerial and technical skills and transferring the know-how for operating and replicating new technological systems on a sustainable basis. Without such preparation, advanced technologies may fail to penetrate the market. Capacity building also has a role to play in ensuring that new technologies are, in the words of the Convention, compatible with and supportive of national environment and development priorities and strategies, (and) contribute to cost-effectiveness in achieving global benefits. (UNEP, 1997)

- Environnement et Developpement – Tiers Monde (ENDA-TM), a major regional NGO based in Dakar, Senegal, defines capacity building in concrete terms related to the process of project development and implementation. Drawing on a rich experience of on-going project-based activities and on recognition of the need to create local conditions that are conducive to sustaining the gains achieved in these activities, ENDA defines capacity building as follows:

> Capacity building here means providing frameworks for project identification, formulation and implementation, making the maximum use of existing skills and resources and, crucially, bearing in mind that implementation of the project is not the end of the road. At a more basic level than funding or the transfer of technology and know-how, the primary task in the effective implementation of the FCCC in sub-Saharan Africa is the creation of an adequate institutional framework, grounded in social relations between competent bodies. (Cissé, Sokona, and Thomas, 1998)

- The Asian Institute of Technology, a Bangkok, Thailand-based institution emphasizes the multi-dimensionality of the capacity building process. Their vision links human resources, organizational strengthening, and development of an appropriate policy environment as integral elements of the capacity building process:

> Capacity Building is much more than training and includes the following:
>
> Human resource development: the process of equipping individuals with the understanding, skills and access to information, knowledge and training that enables them to perform effectively.
>
> Organizational development: the elaboration of management structures, processes and procedures, not only within organizations but also the management of relationships between the different organizations and sectors (public, private and community).
>
> Institutional and legal framework development: making legal and regulatory changes to enable organizations, institutions and agencies at all levels and in all sectors to enhance their capacities. (Sheng and Mohit, 1997)

- It is worth mentioning that developing country governments have a particular perspective on capacity building. These countries tend to frame their view of capacity building in terms of key national priorities that include poverty alleviation, employment creation, development of basic infrastructure, and the delivery of services to meet basic human needs. They are concerned that unless capacity building addresses these priorities, the new cooperative mechanisms under the FCCC will repeat the old, failed experiences, describing delivery of obsolete hardware as technology transfer.

– A recent joint report by UNEP and the International Petroleum Industry Environmental Conservation Association (IPIECA) defines capacity building as:

> A process of constructive interaction between countries and the private sector designed to develop the capability and skills to achieve environmentally sound forms of economic development through the use of modern technologies and management systems, a competent workforce and appropriate laws and regulations. (UNEP and IPIECA, 1995)

The UNEP/IPIECA report emphasizes the critical links between technology cooperation and capacity building. The report observes that:

> Hardware alone will not deliver improved environmental performance. Unless technology is accompanied by all the elements needed to support it – such as appropriate infrastructure, management skills and systems, and a trained, competent and environmentally aware work force – it will not deliver the maximum benefits. The benefits of technology are, in reality, the benefits of putting a whole system in place – a system that includes technology at its heart, and management techniques, infrastructure and skilled human resources as its body.

The private sector perspective illustrated by this report highlights the notion that capacity building is a two-way street and the key to sustainable development. The UNEP/IPIECA report asserts:

> Today the process involves much more than a one-way transfer of technology from rich countries to poor ones. Modern technology cooperation takes place between and within all types of countries, with a common goal to contribute towards sustainable development.

The UNEP/IPIECA study contains a number of case examples of capacity building activities associated with oil and gas exploration programs. In an effort to develop and exploit indigenous hydrocarbon resources, the IPIECA companies undertook programs of technical training, institutional strengthening, environmental assessment, and technology development. The partnerships emerging from these cooperative undertakings were better prepared to develop local resources while minimizing damage to the environment and, in some cases, were able to advance national development priorities in ways that might not have been possible otherwise.

Box 1. Examples of private sector capacity building.

While investigating the potential for oil and gas development in the Casanare foothills of Colombia, British Petroleum (now BP Amoco) developed an innovative set of environmental monitoring and assessment techniques. Local residents have been trained in the use of these techniques and will now be better able to track and evaluate the damage to tropical river systems and to local ecosystems that may result from oil and gas exploration in the region. As a result of this joint work, local consultants are now able to train other Colombians as well as citizens of other countries in the techniques required to monitor environmental damage to fragile tropical ecosystems. For its part, BP Amoco has acquired a technique and a scientifically sound methodology for environmental monitoring and assessment that can be applied in other countries that have no accepted protocol for environmental monitoring.

Canadian Occidental developed a partnership with the government of Yemen in order to assess and develop Yemen's Masila hydrocarbon resources. The Masila region is a sparsely populated desert region with villages scattered among a number of dry valleys. The Canadian Occidental project required advanced process control technology and a skilled workforce to operate the local facilities. Because most villages in the region had limited access to clean, potable water supplies and only the simplest educational facilities, Canadian Oxy undertook to develop a Community Affairs Program that provides seed capital for local infrastructure investments, underwrites school construction and provides technical training for local workers. Canadian Oxy is also establishing partnerships with local NGOs and international development assistance agencies (e.g., UNICEF) to initiate programs combining plastic recycling and revegetation.

2.3 Capacity Building in the Context of Climate Change and AIJ: a Simple Framework

In the climate change context, capacity building refers to a set of processes capable of fortifying human assets, strengthening institutional capabilities, and developing strategic resources of information and data. In the context of AIJ, these processes can facilitate the introduction of new technologies (where the term technology refers to a dynamic combination of hardware, software, know-how, and social organization). By building partnerships that operate as learning organizations, AIJ projects increase the ability of local communities to understand the scientific basis of the climate problem and to adapt to global environmental change. These projects also help governments, businesses, and civil society to understand, modify, and adopt new technologies. Thus, capacity building is an essential component of international efforts to protect the global environment and of national efforts to promote sustainable economic development. Capacity building is both a means to an end and an important end in itself.

We have identified several general categories of capacity building that are relevant to climate change and AIJ. These include:

- Activities enhancing human resources and raising public awareness;
- Activities strengthening existing institutions at the international, regional, national and local level;
- Activities generating data and information of strategic importance to technology cooperation and diffusion;
- Activities highlighting for officials of all levels the utility and relevance of AIJ, CDM or JI projects to achieving national and local economic development goals; and
- Activities stimulating policy reform and creating conditions conducive to technology diffusion.

2.4 Enhancing Human Resources and Raising Public Awareness

Human resource development is the foundation of capacity building and the most important ingredient of any successful program of sustainable development. For many developing countries, the number of trained technicians, scientists, and other professionals available to address environmental problems is severely limited. Responding to the challenges of sustainable development and global environmental change may mean reallocating these limited resources or training additional individuals in critical skill areas. It may also be necessary to increase the knowledge base of existing specialists in order for them to address these new challenges.

Developing and enhancing human resources involves a lengthy and continuing commitment to education, training, evaluation, and communication with local communities. The goal of these activities is to enhance public awareness, train technical specialists, and to develop local and national leadership. Although some combination of these activities is a necessary component of all successful project-based activities, it is unlikely that these tasks can be completed in any single project cycle.

The AIJ pilot phase has shown that building national cadres of individuals capable of dealing with the challenges of climate change will require long-term programs for human resources development at the national and local level. Sustained efforts will be required to improve local skills in scientific observation and the monitoring of global change, engineering analysis and systems integration, risk assessment and risk management, project finance, project evaluation, and policy analysis. Such programs will be needed in all countries, but are especially important in developing countries.

The key question raised by the need to enhance human resources is, of course, Who will pay for this investment? In some cases, the scale of capacity building necessary to ensure the success of a joint venture dwarfs the resources of the private partners in a cooperative project. In these cases, unless there is a commitment of public financing to the capacity building effort, the transaction cost of the project will be unacceptably high. As a result, the private partners are likely to withdraw and the project is likely to fail. On the whole, the optimal balance between private financing and public financing for capacity building remains to be determined.

2.5 Strengthening Existing Institutions

Although programs to enhance human resources are necessary, they are not sufficient to guarantee the uptake and diffusion of new technologies or the success of national sustainable development efforts. Training of personnel must be linked tightly to the utilization of new skills (an issue of organizational dynamics and management) (Grindle, 1997). Several studies suggest that the economic success of a number of Asian developing countries in the 1980s and early 1990s was partially attributable to capacity building programs that combined institutional strengthening with efforts to enhance human resources (Morris, 1997). By contrast, these studies point out that failure to attend to issues of institutional strengthening contributed to failures of planning, policy-making, and implementation in a number of African and Latin American countries (Cohen and Wheeler, 1997). These studies also suggest that, in Africa for example, some of the benefits of earlier capacity building activities have been lost as a result of the inability of institutions to retain trained staff. The problem is compounded by the effects of financial and political pressures (including structural adjustment programs) that have weakened support for many institutions, especially in the public sector.

In addition to providing technical training to enhance human resources, capacity building is necessary to ensure adequate institutional arrangements for coping with the existing shortage of trained staff. Such efforts are needed to reinforce the management and administrative infrastructure of many developing country institutions. The international climate debate has begun to address these issues through the concept of Centers of Excellence, arguing that leading NGOs and academic institutions can provide training and networking to strengthen other institutions within developing countries.

2.6 Expanding National and Regional Resources of Strategic Data and Information

In order to exploit the opportunities presented by new technologies and the capabilities of newly trained personnel, most governments need expanded access to strategic resources of data and information. For example, in preparing to introduce renewable energy technologies, energy efficiency systems, or sink-enhancement projects, many countries will need additional data concerning the availability and behavior of these resources and the corresponding energy demands. Systematic capacity building activities can disseminate information about how to locate existing information that is either country-specific or global in scope. In addition, capacity-building programs can increase the ability of local experts to monitor and collect environmental and economic data. Analysis of this resource data can create new information to facilitate market analysis and project design. Expanded networking among local and regional institutions can be useful in supporting, the longer-term processes of technology adaptation and technology diffusion and ensuring attention is paid to sustainable development.

2.7 Recognizing Multiple Benefits

Capacity building initiatives are particularly needed in order to focus the attention of business and political decision-makers in both developing and industrialized countries on the potential of AIJ projects and climate response strategies to deliver multiple benefits. From the perspective of industrial country leaders, AIJ projects can yield not only cost-effective emissions reductions but these projects can also work to strengthen trade relations, stimulate the evolution of new markets and accelerate the development of advanced technologies. From the perspective of the private sector in industrial countries, participation in AIJ projects can provide access to and experience in developing countries and economies in transition (EITs). These projects can also help to open new markets, facilitate new partnerships, and promote the co-development of technologies well suited to host-country markets. From the perspective of developing country governments, capacity building associated with AIJ projects can reveal how these projects may simultaneously advance other, more immediate national development goals. Participation in AIJ projects, for example, would not only bring new technologies into these countries but could help them leverage additional funds, expanding foreign direct investment in their economies. From the perspective of the private sector and of host communities in developing countries, active participation in AIJ projects and engagement in the associated capacity-building activities can provide important local benefits. The capacity building associated with these projects can enhance local human resources and strengthen existing institutions while the projects provide access to new business partners, support the introduction of new technologies, and help local leaders to attract new investment. On balance, capacity building initiatives are needed to help decision-makers in both industrial and developing countries to recognize the potential of these projects to advance multiple national economic development priorities.

2.8 Stimulating Policy Reform to Create Enabling Environments

Ultimately, success in addressing the risks of rapid climate change hinges on the formation and implementation of efficient, equitable, and environmentally sound policies of economic development. These policies will promote the development, demonstration, adaptation, and diffusion of technologies that are capable of reducing the rate of growth in national GHG emissions while accelerating progress toward national goals of sustainable development. Rapid diffusion of new, more energy efficient and less polluting technologies can only occur in a policy environment that rewards entrepreneurs and promotes new partnerships. In many industrial and developing countries, systematic programs of capacity building are needed to stimulate the necessary policy reform and to create the enabling environments for rapid technology development and diffusion.

The following sections explore the special characteristics of the climate change problem and will link this simple conceptual framework to the new cooperative mechanisms developed in the context of the FCCC and its Kyoto Protocol.

3. CLIMATE CHANGE AND THE NEED FOR CAPACITY BUILDING

Current scientific understanding of the risks of human-induced climate change suggests that persistent scientific and economic uncertainty, long lag times, and unpredictable impacts at the local and regional level will continue to characterize this issue for decades to come. Thus, making sound and profitable decisions concerning public policy and private investment will require an integrated, multi-disciplinary approach. Indeed, climate change is an issue that can only be successfully addressed if analyzed in the larger context of sustainable development. The complex nature of the problem, the long analytic time horizon needed for sound decision-making, and the extent to which stakeholders are unfamiliar with the challenges facing affected institutions indicates that the achievement of an equitable and efficient resolution of this issue will require extensive capacity building for all.

The defining characteristics of this complex problem have been highlighted in the reports of the Intergovernmental Panel on Climate Change (IPCC). The IPCC has carefully reviewed the scientific evidence and the economics research on climate change due to the greenhouse effect (UNEP, 1990; UNEP, 1995). In 1995, a part of IPCC Working Group III led by Nobel Laureate Kenneth Arrow evaluated the range of decision-making frameworks that can be employed to address the challenges of climate change (Arrow, et al. 1995). Arrow and his colleagues observed that decision-makers face unavoidable choices as they consider possible responses to the risks of human-induced climate change, and that any proactive choice would have potentially significant but uncertain consequences. Even inaction is a choice carrying risks and consequences.

Arrow et al. (1995) found that despite persistent and irresolvable uncertainties concerning the physical, ecological and socio-economic impacts of climate change, public officials and private investors must continue to make decisions about public policies and private investments. These factors complicate the decision-making process and reduce the confidence among decision-makers that they can identify and select superior options using conventional economic methods and decision rules.

Extraordinarily long time lags characterize the period between emissions of greenhouse gases and the experience of climate change impacts. Similar lags exist between policy or investment decisions, the diffusion of new technology, and changes in the trajectory of future emissions. These inherently long time scales also affect the uncertainties present in estimates of future social impacts and the valuation of the effects of climate change on economic societies (Arrow, et al., 1995).

The geographic scale of future climate changes identified by the IPCC creates the expectation for many that all human societies are vulnerable to the impacts of GHG buildup. Although it is likely that not all societies will be affected equally, scientific uncertainty concerning the regional and temporal distribution of future impacts makes it impossible to predict today with confidence which areas will be negatively affected and which will enjoy the benefits of a warmer, wetter world.

International organizations, national governments, local institutions, private enterprises, and NGOs are separately and collectively faced with a nearly insuperable challenge. The expected risks span the gamut from low-probability or high-consequence weather events capable of decimating whole cities, to rapidly shifting relative prices and international terms of trade, to confounding uncertainty in regulatory approaches to key economic sectors. These large and seemingly irreducible uncertainties, the long lag times, and the wide geographic and temporal scope of potential impacts make conventional decision tools appear to be almost irrelevant.

If the climate problem contained only these aspects, the current situation would appear to be perilous and bleak. However, there is an upside to this situation, For example, UNEP notes that, despite the difficulties and complexities of the climate problem, the situation offers substantial opportunity for capturing national and global benefits. UNEP's Information Unit on Conventions notes that, when viewed in the integrating context of sustainable development, actions to address the risks of climate change can promote both socio-economic development and environmental protection (UNEP/IUC, 1997a). UNEP's analysis highlights the conclusion of several recent studies suggesting that many climate response policies can generate multiple and collateral benefits, including more cost-effective energy systems and more rapid technological innovation (UNEP/IUC, 1997b).

The UNEP analysis concludes, the prudent response to climate change is to adopt a portfolio of actions aimed at mitigation, adaptation, and research. The optimal combination of elements in such a strategy is likely to vary among countries and will, in all likelihood, change over time. To be effective, such policies will require a high degree of public awareness and substantial support from the full range of stakeholders (UNEP/IUC, 1997b).

But how will countries identify the optimal combination of policy elements? How will governments, private sector enterprises and NGOs build a shared vision of sustainable strategies? How will these stakeholders find each other? Will they be able to initiate coalitions that are politically and economically empowered to develop and diffuse the technologies needed to achieve their objectives? Few countries are today prepared to take on these challenges. In most developing countries, the existing institutions are not prepared to cope with or address the complex, multi-disciplinary challenges associated with building a shared vision or establishing a portfolio of policies in support of sustainable strategies for development. Similarly, in most industrial countries, neither the awareness nor the political will exists to reduce consumption levels systematically and to support sustainable patterns of post-industrial development. Official institutional arrangements need to be expanded in both regions, emphasizing North-South dialogue and South-South networking of institutions with complementary strengths.

In both industrial and developing countries, systematic and lasting programs of capacity building will be needed to create the enabling conditions in which such policies can be articulated and the necessary technologies developed.

4. WHAT IS THE ROLE FOR CAPACITY BUILDING IN THE AIJ REGIME?

The FCCC entered into force in March 1994. This was followed in December 1997 by the Kyoto Protocol, which has not yet entered legally into force. Together, these two instruments establish an internationally agreed regime for dealing with the climate issue. They include general and specific commitments for both industrial and developing country Parties. They establish a process for advancing toward the overall objective of the FCCC (United Nations, 1992b). The FCCC links an international environmental goal, for the first time, with the economic objective of sustainable development. And the Kyoto Protocol establishes specific, legally binding targets for emissions reductions by industrial countries.

4.1 Technology, Capacity Building and the AIJ Regime

When the FCCC entered into force, very few countries, industrialized or developing, had the human and institutional capacity to design, develop, monitor, evaluate and verify AIJ projects. It was thus necessary to encourage capacity building to enhance human resources and increase public awareness of climate change. Targeted and systematic programs were needed for strengthening existing institutions. All governments needed to build the capability for selecting policies to advance national economic priorities while slowing the rate of emissions growth. Private enterprises and other entities needed technical assistance in learning to identify and assess the potential profitability of new, climate-friendly technologies. Additional resources of strategic information and data were required to evaluate policy and investment options that might reduce emissions growth rates while advancing national economic development priorities in both industrial and developing countries. International cooperative undertakings along with intra- and inter-regional dialogue among stakeholders were necessary to stimulate policy reform. Simply setting up a program to facilitate capacity building for AIJ projects, whether in developing or industrialized countries, would prove to be complex and time-consuming.

Some of the biggest challenges related to the initiation of AIJ projects lay on the technology side of the problem. For the purposes of this discussion, we will continue to define technology as a dynamic combination of hardware, software, know-how, and social organization capable of filling a particular economic demand (e.g., warm buildings, efficient lighting, sustainable and affordable forest products, or convenient and reliable transportation).

The AIJ regime covers two types of technologies:

- Technologies that mitigate, reduce or prevent emissions while advancing sustainable national economic development priorities; and
- Technologies that improve land use management practices in order to expand, enhance, or sustain biological sinks that are capable of sequestering CO_2.

The processes of designing, developing, monitoring, evaluating, and verifying AIJ projects that involve these two technology types and then diffusing the lessons learned throughout the host economy poses a complex challenge. In general, facilitating these processes in developing countries involves a series of capacity building activities, some of which can be conducted concurrently.

Initially it is necessary to take steps to enhance local human resources, both in government and in the larger civil society. Public awareness programs are needed to sensitize key stakeholder groups to the dimensions of the climate problem and to the national benefits of climate-friendly technologies. Leadership development programs are needed to enhance the capabilities of government policy-makers to formulate and implement policies that create the appropriate enabling conditions for new technology partnerships. Technical training programs are needed to expand the pool of skilled professionals (scientists, engineers, financiers, managers, lawyers, etc.) that are familiar with the climate problem. Such training programs require designs that can encourage the retention of those who obtain these skills and stimulate a sustained demand in country for the skills that result from these programs.

In addition to capacity building to enhance human resources, exploiting the full opportunities presented by climate change and the AIJ regime requires carefully targeted programs of institutional strengthening in developing countries and EITs. These programs are needed to improve retention rates for skilled incumbents and to increase the efficiency, effectiveness, and transparency of existing organizations. New efforts to improve information transfer and professional exchanges between institutions and among countries are useful. Programs to reinforce and expand the capability of local institutions in host communities are helpful in delivering a sense of empowerment to the participants in some AIJ projects.

Identifying and developing AIJ projects is very difficult in the absence of good data about technology options, local resources, and opportunities for acquiring financial support from international or other outside agencies. Capacity building activities are needed to develop and disseminate this strategic data. In addition, careful and targeted efforts are needed to synthesize the data into useful information and to make the data accessible to a wide range of stakeholders and potential project participants.

In many potential developing and EIT countries, the business climate is not attractive to potential investments in AIJ. For countries that wished to attract AIJ projects, it has proven useful to evaluate existing policies, and to reform those that generated barriers to investment and to technology cooperation. Similarly, for investing countries, it is necessary to evaluate existing policies and to reform those policies in order to encourage investments that promote international cooperation on the development of climate-friendly technologies. Where both sets of activities are needed, capacity-building activities could provide support for the necessary policy analysis and, where applied, help national decision-makers to evaluate the costs and benefits of policy reform.

Capacity building in support of AIJ project development is not only needed in developing countries and EITs. Most stakeholders and potential project participants in investing countries have only limited sensitivity to the climate issue and have an incomplete understanding of the AIJ regime. Thus, capacity building activities in support of technology cooperation and AIJ project development are also needed in industrial countries which could be the source of AIJ investments.

The mix of capacity building activities needed to promote AIJ project development in industrial countries has some similarities to the needs experienced in host countries. Public awareness raising is necessary to sensitize civil society and stakeholder groups to the climate issue and to the potential benefits of climate-friendly technologies.

The broad availability of technically trained personnel notwithstanding, many industrial countries have a large (and largely unmet) need for capacity building to stimulate policy reform. Such reforms enhance market access and improve market efficiency in support of technology cooperation and AIJ project development. Current policies in many countries force new climate-friendly technologies to compete with their conventional alternatives in markets that have been distorted by subsidies and other governmental interventions. The environmental costs of conventional technologies are often treated as non-market externalities in industrial countries, skewing relative prices. Outdated standards and traditional practices generate barriers to the introduction and diffusion of new technologies in both industrial and developing countries. Duties and non-tariff barriers limit access to domestic markets in industrial countries for products that may have been manufactured using climate-friendly technologies in AIJ host countries. As a result of all these obstacles, industrial country markets give preferential opportunities to conventional technologies, particularly in a world of low fuel prices. Without capacity building to highlight these market failures and encourage policy reform, opportunities to promote technology cooperation through AIJ projects will remain unexploited in many industrial countries.

Capacity building is needed not only for governments in industrial countries. Capacity building is also needed in the private sector of industrial countries and in the NGO community. Institutional strengthening activities are particularly important for these stakeholder groups. In many cases, business entities and NGOs that could participate in AIJ projects have faced a steep learning curve concerning the potential of the new mechanisms. Industrial country companies were (and still are) often unaware of the ways in which climate change and the international policy response to climate change, may alter the business climate for key sectors of the domestic and international economy. A handful of NGOs are aware of the ways in which participation in AIJ activities can be structured to advance other national policy goals related to sustainable economic development, and have taken on the roles of matchmaker, broker, and facilitator. To be successful in achieving the objective of the FCCC, many more NGOs and businesses must grow to understand and participate in the cooperative mechanisms under the FCCC. Without capacity building to raise the awareness and understanding of these linked issues, industrial country participation in the AIJ and related regimes will remain only a small fraction of what it might otherwise be.

5. ILLUSTRATIONS OF CAPACITY BUILDING IN THE AIJ CONTEXT

As of mid-1999, 130 AIJ pilot projects have been reported to the FCCC Secretariat using the Uniform Reporting Framework (URF). These activities have been approved as AIJ projects by both the host and investing country governments. Although many of these projects are not yet operational, they are planned for a variety of regions and national circumstances, with the largest concentrations proposed for Central America and Eastern Europe. Africa has had the least experience with AIJ, accounting for just 2% of the total number of projects. Some have suggested that the small number of AIJ projects implemented in Africa reflects relatively low capacity levels. Conversely, the African experience with AIJ highlights the importance of capacity building activities. The following section offers a series of illustrations of capacity building activities associated with AIJ projects. These experiences are drawn from AIJ activities in Costa Rica, Chile, the Philippines, and the Czech Republic.

5.1 Creating Strong Institutions and Building an Enabling Enviornment in Costa Rica: how the AIJ Regime can Stregthen National Institutions and Stimulate Policy Reform

In recent years, the government of Costa Rica has been an active and enthusiastic participant in the AIJ regime. As of December 1998, Costa Rica had registered 10 AIJ pilot projects with the FCCC Secretariat and had attracted over US$150 million in AIJ financing (Joint Implementation Quarterly, 1998). Costa Rica has made a national commitment to AIJ at the highest political levels. With strong backing from its President, Costa Rica has been a forceful advocate and a respected leader in international debates on JI and AIJ.

The power and success of the Costa Rican AIJ program shows the results that can be obtained from a determined national effort at capacity building. The Costa Rican government made an early and strong political commitment to strengthen national institutions that could facilitate AIJ project development, to stimulate policy and legal reforms, and to create a policy environment conducive to international cooperative ventures.

Costa Rica created the first national JI program with the formation in 1994 of its Joint Implementation Program within the Ministry of Natural Resources, Energy and Mines. The Costa Rican JI program initiated bilateral negotiations with the US, leading to the first bilateral JI agreement. In 1995, the Costa Rican government established the national Office of Joint Implementation (OCIC) and endowed it with unusually broad ranging decision-making authority. OCIC provided guidance to investors concerning host country priorities, project guidelines, and criteria for acceptance. These decisions increased investor confidence and led to approval of three US AIJ projects as well as investments by Norway and the Netherlands (Aslam, 1997).

Innovative work by OCIC led to the development of a new investment concept and another round of AIJ pilot projects. Costa Rica, under OCIC leadership, developed the concept of the Certifiable Tradable Offset (CTO). Each CTO represents a Mg of carbon (C) emissions offset or sequestered. The CTOs are backed by a reserve of unsold emissions credits and are guaranteed by the Costa Rican government for a period of 20 years. The first trades of CTOs were consummated in 1998 (Aslam, 1997).

Costa Rica continued to strengthen national institutions by reforming the electric power sector, establishing a renewable energy portfolio standard (equal to 15% of national electricity generation), and increasing the allowed share of foreign participation in energy sector investments. In addition, a new forestry law passed in 1996 granted legal recognition to the concept of payment for environmental services. According to M. A. Aslam, this law provided the legal basis for the land use and C sequestration component of the CTOs (Aslam, 1997).

Aslam (1997) concludes that the strengthening of the legal and institutional framework in Costa Rica combined with the enthusiastic participation of public, private, and NGO actors in Costa Rican society to give great momentum to the AIJ pilot phase in this small country. Initial supporters of OCIC included the Ministries of Energy and Environment, FUNDECOR (an NGO), CINDE (a not-for-profit foundation) and a number of private firms. The willingness of these stakeholders to take ownership of the concept and to commit financial and human resources to OCIC substantially increased the likelihood of success.

5.2 Introducing an Innovative Wind Project in Chile: an Example of AIJ Strengthening National Institutions and Private Enterprises

The Wind Energy Project in Northern Chile is a renewable energy project certified by the US Initiative on Joint Implementation (USIJI) (Gordon, 1999a). The project was developed by the International Institute for Energy Conservation (IIEC) of Washington, DC, USA (an international NGO) and Corporacion Nacional del Cobre de Chile (CODELCO) of Santiago, Chile. The project involves a proposed 37.5 MW wind energy power plant to be located in the desert region of northern Chile, near Antofagasta. The wind plant will supplement the northern power grid, which serves CODELCO's Chuquicamata mine as well as other mining and industrial operations in the northern region. As the first large-scale wind energy installation in Chile, the project is expected to play a capacity-building role in many respects, providing technical training, technology demonstration, and innovative financing. The project will also strengthen existing institutions in the Chilean electric sector, reinforcing their ability to manage and dispatch an intermittent renewable energy source by providing strategic resources of information and data. These capacity-building aspects will become evident over the next several years as the project is structured, built and operated. But the USIJI application and approval process itself has also had important capacity-building effects in Chile.

First, the process of identifying a local partner afforded multiple opportunities to educate Chilean companies, NGOs and government agencies about climate change. The project developer, IIEC used these opportunities to raise public awareness about the FCCC and JI. IIEC first learned of the potential wind energy project at the northern Chile site in 1997. A US-based wind developer had previously identified the site's potential through a resource assessment conducted cooperatively with a Chilean university. The original project was never developed due to a lack of financing.

Given the dependence of Chile's northern region on coal, IIEC recognized the GHG mitigation potential of a wind-driven power plant in this region. The challenge was to interest a Chilean company as a local partner in the application to the U.S. Initiative on Joint Implementation (USIJI). IIEC engaged in extensive discussions with several electricity generation and distribution utilities, encountering both cautious interest in the concept and skepticism regarding its GHG emission reduction potential. Ultimately, CODELCO's growing understanding of the climate problem benefited from and reflected the prior experiences of a key staff member who had worked with a USA NGO. The company became persuaded of the potential future value of GHG emission reduction credits and decided to become the local partner on the USIJI application.

The USIJI application process required the project partners to conduct an emissions baseline study and to estimate the avoided GHG emissions that would result from the project. The USIJI program sponsored a technical assistance mission to Chile in late 1997. The mission sent two wind energy experts from a USA national laboratory to Chile to review the wind measurement data for the site, assess ways to overcome challenges posed by the intermittence of the wind regime, and to investigate the operations and dispatch procedures of the northern power grid.

Following the mission, IIEC was able to estimate the baseline emissions using carbon emission factors for the existing coal-fired generation capacity and to take account of the imminent introduction of natural gas supplies into the region. IIEC drew on IPCC and U.S. Environmental Protection Agency (EPA) methodologies for estimating GHG emissions, establishing a project-specific procedure that was reviewed and accepted by both the Chilean government and the USIJI Evaluation Panel.

The process of gaining host country acceptance in Chile helped to firm up institutional roles and procedures for review and evaluation of AIJ projects. The Chilean National Climate Committee was charged with reviewing the proposed project in light of Chile's national energy, environmental and policy objectives. The Committee's mandate included determining whether and under what terms the project could be approved. Ultimately, the project was reviewed by three federal agencies represented on the Chilean national climate change committee: the National Environment Commission (CONAMA), the National Energy Commission (CNE) and the Ministry of Foreign Relations. As the first GHG mitigation project approved by the Chilean government under the AIJ pilot phase, the

Wind Energy Project helped lay the foundation for assessment, evaluation and approval of future AIJ projects in Chile.

5.3 Providing Energy Efficiency Street Lighting n the Philippines: an Example of Institutional Strengthening and Policy Reform

The Energy-Efficient Street lighting Project proposed by the International Institute for Energy Conservation (IIEC) and the Cagayan Electric Power and Light Company (CEPALCO) was the first AIJ project approved by the Philippines Inter-Agency Committee on Climate Change (IACCC). The project served as a test case for the committee, strengthening institutional capability to establish criteria and set policies for review and approval of GHG mitigation and reduction projects (Gordon, 1999b). The IACCC required explicit capacity building and technology transfer components to be integrated into the project as a condition for host country acceptance. This indicates the growing capability and willingness of the Philippines government to establish clear terms of engagement within the context of the FCCC. The results of this capacity building activity will substantially decrease the transaction costs and increase the likelihood of replication in Philippine follow-on projects.

5.4 Fuel Switching and Cogeneration in the Czech Republic: an Example of AIJ Strengthening National and Municipal Institutions

The Center for Clean Air Policy (CCAP) worked with the municipal government of Decin in the Czech Republic to develop one of the first AIJ pilot projects. The project calls for fuel switching and the introduction of a cogeneration plant at the city's district heating facility. The Czech government approved it in April 1997 (Aslam, 1997).

At the start of the Decin project there was very limited understanding of the climate problem or of the potential for AIJ pilot projects in the Czech Republic. Even the issue of jurisdiction over the project was unresolved. Throughout the project development cycle, leading Czech and USA-based NGOs worked with the national government and the municipal authorities to illuminate the process. The NGOs also helped the government and municipal officials to explore the range of institutional and technical options available for structuring the project.

As part of the follow-up to the Decin experience, the Czech government has clearly delineated institutional responsibilities for development and approval of AIJ pilot projects. In contrast to the Costa Rican case discussed above, the Czech government has decided to partition these responsibilities among different bureaucracies. The national JI office is responsible for working with project developers on issues of approval, monitoring, and verification of AIJ pilot projects. The Ministry of Environment is responsible for establishing the criteria and guidelines under which the national JI office must operate. Aslam (1997) concludes that although this division of responsibility for AIJ activities has not been as successful as the Costa Rican experience, the evolution of clear institutional roles and responsibilities has led to the considerable achievements of the Czech Republic in attracting AIJ investments.

6. EFFECTS OF CAPACITY BUILDING ON THE NET COSTS OF AIJ PROJECTS

None of the capacity building activities outlined in the preceding section are cheap. The success of these capacity-building efforts depends upon a long-term commitment to these activities, a commitment that will often extend beyond the life cycle of individual projects. How will capacity building affect the costs of AIJ projects and the benefits of the AIJ regime?

In the short term, the costs of capacity building appear to increase the costs of AIJ projects. If required to implement capacity building activities, AIJ project developers will perceive demands for capacity building as additional transaction costs associated with these projects. These incremental transaction costs will make AIJ projects less financially attractive in the short term than other, more conventional joint venture opportunities.

Ironically, capacity building activities are likely to have the opposite effect on project costs in the long term. Efforts made in the early period to enhance local human resources are likely to expand the local pool of skilled labor. Institutional strengthening activities will lead to indigenous institutions that are better positioned to integrate new information, to adopt new technologies and to cope with environmental change. Efforts to develop strategic resources of data and information will provide project developers with more penetrating insights about the availability of local resources, the opportunities for garnering outside support, and the likely needs of local populations. Efforts to stimulate policy reforms in industrial countries that level the playing field for climate-friendly technologies can create a business climate in the future that is far friendlier to technology cooperation and technology diffusion. By increasing the number of trained professionals and workers, sensitizing the civil society to the benefits of new, climate friendly technologies, and generating strategic resources of data and information, capacity-building activities will create public goods. Taken together, these activities will increase the efficiency of many productive factors in the national economy.

Managing the introduction and maintenance of capacity-building activities is not easy. One of the challenges is to determine how the success of these activities can be measured and how the value added to the economy can be monetized. Some benefits of capacity building may be easy to measure. Most governments, for example, now monitor employment rates both by geographic area and by sector. Perhaps, the impact of capacity building activities on local job markets and on wage rates could be monitored to assess the cost-effectiveness of capacity-building activities. Another way of measuring the aggregate impact of capacity building activities is given by the UN's Human Development Index. The details of how this index can be linked to specific project investment decisions are not yet clear, but should be investigated.

Other impacts of capacity building will be harder to measure and nearly impossible to monetize. The financial value of new data and information about local resources will be very hard to assess with confidence. The significance to the economy of more efficient policies will always be controversial. The public benefit from stronger, more resilient institutions may be largely in the eye of the beholder. Understanding may always be limited concerning exactly how these conditions affect future rates of technology innovation and technology diffusion. Given the irresolvable uncertainties, measurements of the success of capacity building activities and their value to the affected societies may necessarily remain incomplete.

Some companies, especially in the extractive industries and those who expect to have a long-term relationship with a host country, already finance capacity building beyond training of the cadre of technical specialists minimally required by the project. But for small projects and small companies, the issue is not so clear. In the context of the cooperative mechanisms under the FCCC, the question that remains to be resolved is: How should the costs of capacity building be divided between project developers and national governments or international agencies (or others)? Some observers maintain that project developers should contribute to the cost of capacity building -- that it should be considered an integral, perhaps growing, component of a project, not an externality. But how much of these costs can project developers realistically be expected to bear under the Kyoto mechanisms when they can choose other investments that do not demand the same (or equally expensive) activities?

There are indications that international agencies in the UN system (with a mandate for capacity building) are preparing to support capacity building on CDM project activities. Evidence is emerging from the AIJ pilot phase to suggest that bilateral assistance agencies will move in similar directions. It would be useful if these international efforts were sufficiently coordinated and non-duplicative so as to provide the means for capacity building that is beyond the resources of private project developers.

7. CONCLUSIONS, UNRESOLVED ISSUES AND LESSONS LEARNED FROM THE AIJ PILOT PHASE

Participant experiences during the AIJ pilot phase have been rich and varied. The experiences of this period offer a number of useful lessons, revealing the critical importance of technology cooperation to international efforts aimed at achieving the objective of the FCCC. Events since COP-1 have demonstrated that reducing the risks of rapid climate change and achieving the FCCC's objective will require the introduction and diffusion of many new, low-emissions technologies. Innovative AIJ pilot projects have demonstrated that a mix of capacity-building activities are needed to facilitate successful technology cooperation. Simultaneously, the AIJ pilot phase has highlighted the inherent difficulties of technology cooperation for government leaders, private companies, international organizations and NGOs. Specifically, the AIJ pilot has shown that the full potential of JI can only be realized through careful design of projects and the regime. These designs must incorporate a mix of capacity building activities that are appropriate to the specific circumstances, local institutions, and cultural context of the industrial and developing country partners.

The identification and development of AIJ pilot projects has demonstrated that capacity building is needed in both the host and investing countries. This learning process is necessarily a two-way street. The investing organization and the investing country can learn, as much from the implementation of a well-designed project as do their counterparts in the host country. Capacity building can help investors to be more sensitive to the cultural context of the host country. In addition, political leaders and potential investors in investing countries could benefit from capacity building experiences that make them more aware of the potential benefits of the mechanisms under the FCCC and the Kyoto Protocol. Conversely, it is critical that political leaders and project participants in host countries also develop an understanding of the potential benefits of AIJ or CDM for achieving larger national development priorities.

Capacity building activities associated with the design of the AIJ regime and the implementation of AIJ pilot projects can take many forms. The most important of these focus on building public awareness, enhancing local human resources, strengthening indigenous institutions, expanding strategic resources of data and information, and stimulating policy reform. All activities are not needed in all countries, nor are they all necessary for the success of any particular project or technology. Determining which ones are appropriate to a given pilot project requires systematic efforts to understand the character of local conditions, the roles and responsibilities of involved institutions, the availability of accurate compilations of data, and the presence of an enabling environment.

Experience with conventional technology transfer shows that investment in capacity building leads to more robust and successful project outcomes. Project developers and investors need to understand that lack of such investment leads to unsustainable projects in which hardware and infrastructure are not maintained. In such circumstances, project achievements usually can not be replicated.

Capacity building activities can strengthen and empower host communities, but they have real costs. In the short term, project developers may view such costs as a tax that raises project transaction costs. Over the longer term, however, successful capacity building activities will reduce transaction costs for AIJ, JI, and CDM projects, absorbing the value added by a skilled labor force, an efficient, competent bureaucracy, and a well-informed citizenry.

Successful experiences with AIJ pilot projects in Costa Rica, Chile, the Philippines, and the Czech Republic demonstrate the importance for AIJ pilot projects of a national commitment to technology cooperation. These experiences also highlight the fact that development of a policy environment conducive to investments in innovative joint ventures can increase investor confidence. This type of policy environment goes a long way toward mitigating the scientific, economic, and political risks that are generally associated with any international environmental regime.

One of the benefits of capacity building for host countries may be to sensitize political leaders to the fact that the cooperative mechanisms under the FCCC can provide useful leverage for attracting investment that supports national development priorities. National leaders who recognize this new influence on foreign direct investment may choose to stimulate discussion of policy reforms leading to enabling environments that amplify the flow of funds through future CDM and JI projects. Experience in the AIJ pilot phase has shown that countries that designate a national focal point to facilitate development of projects congruent with national needs are more likely to become the site for such investments.

Experiences during the AIJ pilot phase raised a number of issues that remain unresolved. These include the following:

- who should provide technical training in the best methods for calculating base line emissions and for estimating the incremental environmental benefits derived from AIJ projects?
- how can the size of the optimum investment in capacity building be determined for any proposed AIJ project?
- who will pay for capacity building activities? Will it be the host government, the project developer, international financial institutions or bilateral aid agencies?
- how can the effects of capacity building in AIJ pilot projects be measured?
- what determines the ability to replicate an AIJ project or technology?
- what is the most appropriate geographic and temporal scale for conducting capacity building programs in support of AIJ?

The AIJ pilot has also been an important vehicle for individual and collective learning about capacity building. The capacity building activities associated with AIJ pilot projects are bringing real, measurable and lasting benefits to both project participants and to the larger civil society in host and investing countries. Some of the most important lessons learned include the following:

- capacity building is an on-going, long-term, learning process;
- capacity building is needed in both the host and investing countries;
- capacity building is absolutely critical to the successful transfer and diffusion of technology;
- capacity building must be viewed as an integral component of AIJ/JI/CDM projects;
- capacity building implies different combinations of activities for different stakeholder groups;
- capacity building activities include: building awareness (general public and officials), enhancing local human resources, strengthening indigenous institutions (including encouraging networking and regional centers of excellence in developing countries), expanding strategic resources of data and information, and stimulating policy reform;
- investor countries need to implement policies that will encourage and facilitate private investment through the AIJ/JI/CDM mechanisms;
- capacity building initiatives are needed to assist local and national leaders in host countries in recognizing that AIJ/JI/CDM projects can advance national economic development priorities;
- developed and developing country governments as well as international institutions must work together to identify the optimal mix of public and private funding for capacity building in support AIJ or CDM projects;

The lessons learned from the AIJ pilot phase can usefully inform the planning and implementation of the CDM (for cooperative projects in developing countries) and of JI (for cooperative projects in countries with economies in transition).

ACKNOWLEDGEMENTS

Ms. Kelly Gordon of the International Institute for Energy Conservation (IIEC) and Dr. Lilia Abron, President, PEER Consultants contributed to this chapter.

REFERENCES:

Arrow, K.J., Parikh, J. and Pillet, G. (1995) Decision-Making Frameworks for Addressing Climate Change, in: Bruce, J., H. Lee, and E.F. Haites, Climate Change 1995: Economic and Social Dimensions of Climate Change, Cambridge University Press, Cambridge.

Aslam, M.A. (1997) Endogenous Capacity Building for AIJ: Developing Country Needs, ENVORK, Islamabad.

Cissé, M. K., Sokona, Y., and Thomas, J.P. (1998) Capacity Building: Lessons from Sub-Saharan Africa, Environnement et Développement du Tiers-Monde (ENDA-TM), Dakar.

Cohen, J.M. and Wheeler, J.R. (1997) Training and Retention in African Public Sectors, in: M.S. Grindle, (Ed) Getting Good Government: Capacity Building in the Public Sector of Developing Countries, Harvard University Press, Cambridge.

Eade, D. and Williams, S. (1995) The Oxfam Handbook of Development and Relief, Oxfam, Oxford.

Grindle, M.S. (1997) The Good Government Imperative, in M.S. Grindle, ed., Getting Good Government: Capacity Building in the Public Sector of Developing Countries, Harvard University Press, Cambridge.

Joint Implementation Quarterly (1998) Vol. 4, no. 4, Foundation JIQ, Paterswolde.

Moore, M. (1995) Institution Building as a Development Assistance Method: A Review of Literature and Ideas, Swedish International Development Authority, Stockholm.

Morgan, P. (1993) Capacity Building: An Overview, in: Workshop on Capacity Development at the Institute of Governance, Ottawa.

Morris, J. (1997) The Transferability of Western Management Concepts and Programs: An East African Perspective, in: W.D. L.Stifel, J.E. Black, and J.S. Coleman, eds., Education and Training for Public Sector Management in Developing Countries, The Rockefeller Foundation, New York.

Parikh, J. (1994) North-South Cooperation in Climate Change through Joint Implementation, Indira Ghandi Institute for Development Research, Bombay.

Sheng, Y.K. and Mohit, R.S. (1997) Re-inventing Local Government for Sustainable Cities, in: International Expert Group Meeting on Capacity Building for Sustainable Cities in Asia, Human Settlements Development Program, Asian Institute of Technology, Bangkok.

Trostle, J.A., Sommerfeld, J.U. and Simon, J.L. (1997) Strengthening Human Resource Capacity in Developing Countries, in: M.S. Grindle, ed., Getting Good Government: Capacity Building in the Public Sector of Developing Countries, Harvard University Press, Cambridge.

UNDP (1997) Capacity Development Assessment Guidelines, BPPS/MDGD, UNDP, New York.

UNEP (1990) IPCC First Assessment Report, UN Environment Program and World Meteorological Organization, Cambridge University Press, Cambridge.

UNEP and IPIECA (1995) Technology Cooperation and Capacity Building: The Oil Industry Experience, UNEP, London.

UNEP (1995) IPCC Second Assessment Report, UN Environment Program and World Meteorological Organization, Cambridge University Press, Cambridge.

UNEP/IUC (1997a) Limiting Emissions: The Challenge for Policymakers, in: Climate Change Information Kit, Information Unit on Conventions, UNEP, Geneva.

UNEP/IUC (1997b) Global Cooperation on Technology, in: Climate Change Information Kit, Information Unit for Conventions, UNEP, Geneva.

United Nations (1992a) Agenda 21: National mechanisms and international cooperation for capacity building in developing countries, UN, New York.

United Nations (1992b) Framework Convention on Climate Change, Article 2, UN, Geneva.

United Nations (1992c) Framework Convention on Climate Change, Article 4.1(c), UN, Geneva.

United Nations (1992d) Framework Convention on Climate Change, Article 4.2(a), UN, Geneva.

United Nations (1992e) Framework Convention on Climate Change, Article 4.5, UN, Geneva.

United Nations (1995) Decision 5/CP.1, Climate Change Secretariat, UN, Geneva.

Chapter 11

THE WORLD BANK'S EXPERIENCE WITH THE ACTIVITIES IMPLEMENTED JOINTLY PILOT PHASE

J. HEISTER[1], P. KARANI[1], K. POORE[1], C. SINHA[1] and R. SELROD[2]
[1]*Environment Department, The World Bank;* [2]*Bureau for Environmental Analysis*

Key words: activities implemented jointly, Mexico, Poland, India, Burkina Faso, baselines, verification, certification, demand-side management (DSM)

Abstract: The Parties to the UN Framework Convention on Climate Change (FCCC) established the Activities Implemented Jointly (AIJ) pilot phase during the First Conference to the Parties in 1995 as a learning exercise to gain practical experience in such project-based mechanisms. The World Bank, in collaboration with the Government of Norway, has been participating in the AIJ pilot phase and earlier demonstration joint implementation projects since 1993. The chapter provides an analysis of these projects both on a quantitative and qualitative level emphasizing the unique characteristics of the individual projects and the resulting lessons learned. The overall message of this analysis is the significant contribution of the AIJ pilot phase to the practical understanding of how such mechanisms can function. The pilot phase contributes knowledge to the ongoing negotiations related to the operationalization of the Kyoto Mechanisms, namely joint implementation (JI) and the CDM.

1. THE WORLD BANK – NORWAY COLLABORATION ON AIJ

In 1992, following the signature of the FCCC, the World Bank and the Government of Norway signed a US $4.8 million co-financing agreement to implement two demonstration joint implementation projects as it was referred to then. The objectives of the agreement were to analyze the methodological and practical issues related to the concept of JI, inter alia through experience gained from two projects which included the Poland Coal-to-Gas Boiler Conversion Project and the Mexico ILUMEX High Efficiency Lighting Project. The agreement provided for an additional contribution of funds from the Government of Norway to the existing project financing which included financing from the Global Environment Facility (GEF). At the time, both the World Bank and Norway were eager to take advantage of the demonstration to increase the understanding of how such projects might function.

In 1995, COP-1 formally established the official AIJ pilot phase (UN FCCC Decision 5/CP.1). Again, as a result of continued discussions with the Government of Norway, and in the hopes of increasing the experience gained from the AIJ pilot phase, the World Bank and Norway expanded the scope of their activities and established the AIJ Program formally in April 1996.

The overall objective of the World Bank AIJ Program is the maximization of participation in, and learning value from, the AIJ pilot phase, which is critical for establishing future environmental trading schemes, such JI and CDM. Additionally, as one of the main challenges of the AIJ pilot phase has been the operationalization of the AIJ projects, the program also aims to improve client countries' means for estimating emission abatement and for addressing AIJ-specific risks, as well as monitoring, verification, reporting and institutional requirements.

The World Bank is currently implementing four AIJ projects as a result of Norwegian financing. In addition to the Poland and Mexico projects, the Bank is implementing a sustainable energy project in Burkina Faso, the only AIJ project in the Africa region, and an agricultural DSM project in India.

In order to meet its objective of maximizing the learning value, the Bank has emphasized dissemination of the results both of the program and of its projects. All of the projects have been reported on an annual basis to the FCCC Secretariat. Several workshops have been held during the FCCC negotiations and within the developing countries. Additionally, each of the projects has undergone or will undergo an evaluation process. One evaluation in particular is a pilot verification and certification exercise, whereby the ILUMEX project was subjected to a verification and certification of its emission reductions by an independent third party environmental auditor.

The following paragraphs provide the reader with an analysis of each of these four AIJ projects. Each section provides relevant information as pertains to an economic and quantitative cost/benefit analysis. However, the analysis of the projects below will also represent a more qualitative and political description of the projects and their unique characteristics. An emphasis on this sort of analysis has been taken for the following reasons:

- Difficulty in attaining a true economic analysis results when trying to disengage the results of the AIJ project from the overall project – namely the Poland and Mexico projects, with GEF co-financing.
- There is a great deal of value-added in identifying the unique characteristics and issues that have arisen in the four projects. It should be emphasized here that each of the four projects is unique not only in their regional spread and stage of implementation, but also in the problems that have arisen, the solutions attained and the levels of analysis that have been undertaken thus far.

Therefore, for instance, more time is spent discussing the ILUMEX project as a result of the various analyses that have been undertaken, including a pilot verification and certification exercise. Each section opens with a brief discussion of the project's characteristics. Following this overview, the sections already begin to diverge but will each discuss to some degree an economic analysis of the project in quantitative terms and the baseline determination.

2. THE ILUMEX HIGH EFFICIENCY LIGHTING PROJECT

2.1 Project History as a Reflection of the FCCC Negotiations

The ILUMEX Project is a unique project representing the entire cycle of the international negotiations. The project was born out of the early and very controversial discussions surrounding JI in the early days of the FCCC negotiations.

The ILUMEX project was conceived and developed by Mexico with assistance from the World Bank for funding through the GEF. During the project preparation stage in 1993, the ILUMEX project was modified to include a financial contribution from Norway through which JI, an idea included in FCCC Article 4 would be demonstrated. The project served subsequently as one of the first AIJ projects resulting from the decisions taken at COP-1in 1995. The project formed the basis for a strengthened and formalized relationship between the World Bank and Norway in the form of a formal AIJ program begun in 1996.

In 1992, the demand for electric energy in Mexico was as high as 100 TWh with a projected growth of 5.3% per year. It was expected that an additional 14,000 MW of capacity would be needed over the next 10 years, which implied very large investment requirements on the order of USA $3 billion per year for generation, transmission and distribution. Apart from these huge investment projections, there was also concern over the gaseous emissions from thermo-electric power plants and their negative effect on the environment. Since Mexico's installed generating capacity was about 80% thermal, estimated CO_2 emissions from thermal power plants amounted to 57 million Mg annually and included significant emissions of sulfur dioxide (SO_2) and nitrogen oxide (NOx).

In 1990, the federally owned electric power utility, Comisión Federal de Electricidad (CFE), had already established a trust fund (FIDE) with funds provided by several national sources to support and finance energy conservation programs in the energy sector. These initiatives and others underscored Mexico's commitment to environmental protection in line with its programs to mitigate climate change as a signatory to the FCCC.

A large amount of learning and activity has taken place since 1995 in the international arena. The Kyoto Protocol, adopted in December 1997, acknowledges commitments on the part of the Annex I countries. The Protocol also outlines three flexibility mechanisms including JI and the CDM. The flexibility mechanisms help developing countries meet their sustainable development goals while helping Annex I countries meet their obligations. However, a great deal remains uncertain in terms of supporting a market for GHG emission reductions including the principles, modalities, rules and guidelines of the Kyoto mechanisms.

The ILUMEX project is able to provide valuable insight for the determination of these requirements. As a continued part of the World Bank AIJ Program, ILUMEX has undergone the first pilot verification and certification exercise. The intention of this exercise was to provide a practical, hands-on verification and certification of the emissions reductions resulting from the project's activities. Though no credits can be claimed by this exercise, the lessons learned as a result will be invaluable as the international negotiations continue. The text below provides more detail as to the results of this exercise. An additional section presents an economic analysis of the project.

The following project description coincides with this background, and the project itself has continued to illustrate CFE's and Mexico's dedication to improving energy efficiency and to generate the corresponding benefits to the global environment.

2.2 Project Description

The ILUMEX Project was implemented by CFE and replaced approximately 1.7 million standard incandescent light bulbs with as many compact fluorescent light bulbs (CFLs) in the two Mexican cities of Monterrey (region Novo Leon) and Guadalajara (region Jalisco). While CFLs provide the same or better quality lighting, they require only about 25% of the energy, lasting up to 10,000 hours, 13 times longer than ordinary light bulbs. AIJ financing for this project resulted in the replacement of approximately 200,000 bulbs. The financing package for the entire project totaled US$ 23 million, to which the GEF and CFE each contributed US$ 10 million and Norway US$ 3 million through its AIJ trust fund.

CFE, through implementing units in Guadalajara and Monterrey, administered the project by purchasing the CFLs and selling them, on average, at 37% of the cost, including project overhead and administration. The implementing units established sales counters in locations where customers pay electricity bills and also sold the bulbs in the workplace to employees of large companies. Customers bought the bulbs on credit terms of up to 2 years. Customer payments for the bulbs have been used to establish a trust fund. Finances from the trust fund are used to subsidize additional light bulb sales. Therefore, though the project has since been completed, additional CFLs have been purchased with the proceeds from selling the bulbs, bringing the total amount of CFLs sold to about 2.4 million.

The bulb replacement was expected to reduce energy consumption and its associated GHG emissions by more than 700,000 Mg CO_2 over the lifetime of the CFL's with a conservative estimate of six years. Guadalajara and Monterrey are Mexico's second and third largest cities with approximately 550,000 residential electricity consumers each. They have, however, different climates and economic structures and have, therefore, provided valuable information for the plans to replicate the program.

The objectives of the ILUMEX project were as follows:

- demonstrate the technical and financial feasibility of reducing GHG's and simultaneously to reduce local environmental contamination through the widespread installation of high efficiency lighting;
- build the institutional capacity for technological change and energy conservation;
- provide a replicable model for DSM in Mexico and elsewhere in the developing world;
- strengthen the capacity of CFE to practice DSM on a sustainable basis; and
- serve as a demonstration vehicle for elements of AIJ schemes.

2.3 Baseline and Project Emissions

The calculation of anticipated and achieved emission reductions requires a comparison between emissions in a business-as-usual or baseline scenario and in a project scenario. In the case of electricity saving projects of the ILUMEX type, the baseline must first be defined in terms of household energy use and second in terms of emissions from power generation. The project's expected emission reductions were initially calculated from an engineering point of view without taking into account the behavioral factors that may reduce the effectiveness of the program in terms of net greenhouse gas abatement.

Since CFLs were practically not available in Mexico prior to the project, the energy use baseline was set by assuming a continued use of incandescent light bulbs. The likelihood of a spontaneous development of a CFL market in Mexico or of the initiation of an efficient lighting project by the CFE without foreign support was regarded as small during the first five years of the project. Surveys were conducted prior to the project to assess bulb usage, in particular, type of bulbs and the length and time of usage for the various household types. The project's expected direct energy savings were then determined on the basis of the wattage difference between the incandescent and CFL bulbs. The baseline for GHG emissions, differentiated by season and time of day, was set on the basis of the fuel mix used for the electricity produced in neighboring plants and deemed necessary to meet the residential consumption without the project. Emissions factors were then applied to the kWh savings resulting from a switch by households to the CFLs in order to project the emission reductions expected from the project.

The determination of energy and emissions savings in demand-side management projects is often uncertain due to difficult to observe changes in consumer behavior. The ILUMEX project was intended to create a market for CFLs in Mexico so that households would buy CFLs outside of the project (free drivers). This positive effect on savings and emissions could not be quantified ex ante, but was confirmed to some extent by market observations. Similarly, consumers who would have purchased CFLs without the project may have taken advantage of the project by purchasing the subsidized CFLs (free riders). Households may also have burned the CFLs longer than the incandescent bulbs or may have increased their use of electric light or, in fact, other energy services (rebound effect). Both negative effects have not been quantified in the ILUMEX case and would lead to lower actual electricity savings and emission reductions.

Many utilities in the USA apply a free rider share of 30-50% in the evaluation of DSM measures. There are, however, some factors that may point towards a smaller free rider share in the ILUMEX project. In the first place, Mexico has a shortage of investment funds that preclude investors and consumers from making profitable investments. Secondly, investors who are actively seeking government support for their already planned investments often implement subsidized energy conservation investments in industrialized countries. This results in a large free rider share of such programs.

But there are also some factors that may suggest a larger possible free rider effect for projects like ILUMEX, in particular future stringent environmental regulations and possible price reforms:

- Forthcoming environmental regulations may contribute to reducing GHG emissions in a way the project has already done and may thus undermine the project's effect.
- If grant assistance is expected, the government may delay enacting environmental regulation that would in effect result in the same global and national environmental benefits as the grant financed project. Certainly, this perverse incentive has not played a role in Mexico.
- Residential energy prices in Mexico are below the economic costs of energy supplies. Price reforms would reduce local air pollution and GHG emissions and need be taken into account in the baseline assumptions and the calculations of the net abatement effect.

Another possible leakage in net abatement stems from the so-called snapback in energy use or the conservation rebound effect mentioned above. Rebound means that the predicted reduction in consumption is not fully achieved, because the efficiency improvement triggers behavioral changes, which partly offset the saving. A change in relative prices can change behavior in the desired direction, for example, the substitution of subsidized CFLs for incandescent bulbs. A simultaneous increase in available income, for example, caused by the lower·electricity bill of ILUMEX customers, can have the undesired effects of increasing consumption and demand for energy services (more lighting, heating etc.). The resulting emissions would then have to be counted against the emission savings from the project. An American econometric study of a residential conservation program found that approximately two-thirds of the initial saving due to engineering effects were eventually eroded by the rebound effect (Khazzoom, 1986). The rebound effect can be as high as 100%, i.e. the entire theoretical saving is eroded through additional consumption, if conservation measures alleviate or end a situation in which energy consumers were restrained from having their demand meet, which is not a rare condition in many developing countries.

Some other factors that occurred during the project's life were also not included in predicting energy and emissions saving at the planning stage of the project, because information was not available or because these factors were not anticipated.

- The introduction of new environmental standards for power installations, partly in response to NAFTA, resulted in a different fuel mix (conversion to natural gas) in the power plants.
- The introduction of daylight savings time in Mexico reduced the need for electric light and stretched the savings as the CFLs will burn out later.
- The stockpiling of CFLs by customers for later use was not anticipated.
- Changes in the differentiation of consumer tariffs may have had a negative impact on project participation by subsidized low consumption customers.
- The neighboring plant approach to emission factors seems somewhat simplistic, because both cities draw from a nationwide integrated power grid. However, the chosen approach had a minor effect on calculated emission reductions as proven by the verification exercise reported below.

From the above analysis it can be concluded that both from a conceptual and practical point of view the net abatement calculations at project level are very complex and require a good deal of information not necessarily available. These makes detailed planning and project preparation for all aspects related to the generation of valid GHG emission reductions all the more necessary.

It is also worth mentioning that assessing net emissions effects of particular projects with high accuracy may result in high transaction costs, which can outweigh the net benefits of potentially beneficial projects. One way, and perhaps the only way in which this problem can be overcome, is to define classes of projects for which more general guidelines can be established. In that case, the rationale should be that the estimated emissions reductions are correct on average but not necessarily in each specific case. Establishing general baseline methods is, however, not a small task either. It will require more experience with successful project-level baseline assessment than is available to date. And any generalized approach to baselines and to the calculation of net emission reductions is likely to require consensus of the FCCC Parties, if the achieved reductions are to be counted against legally binding reduction obligations in the Kyoto Protocol.

2.4 Benefits and Costs of ILUMEX

Due to its pilot nature, the ILUMEX project has resulted in several important and sometimes unusual benefits for a divers group of interested organizations and individuals. Table 1 below presents an overview over the project's direct benefits.

TABLE 1. Benefits of ILUMEX AIJ project for different participants.

Beneficiary Benefit	House-holds	CFE	Mexico	Global climate	UNFCCC process	World Bank	Gov. of Mexico	Gov. of Norway	Com-panies, Con-sultants	Research, NGOs etc.
Savings (fin. & econ.)	X	?	x							
Local environment	X	x	x							
Social impact	X									
Technology transfer		x	x							
Greenhouse gas savings				x	x					
Market trans-formation	X		x	x						
Learning value		x			x	x	x	x	x	x
Capacity building		x	x		x	x	x	x	x	x
Public image		x			x	x	x	x	x	

The GEF and AIJ co-financed ILUMEX project has been highly successful on many accounts and for many constituencies. Its institutional implementation proved to be highly adequate and it repeatedly achieved a high consumer satisfaction count. The project's organization and implementing has proven extremely flexible in responding to an ever changing national and international environment, in particular to always new demands which the turns of the FCCC process created. This has not been a small contribution to many of the benefits achieved.

- Most obviously, the ILUMEX project (including the GEF component) has saved GHG emissions during its first four years (over 170,000 Mg CO_2) and will possibly save the global environment almost 1 million Mg of CO_2 over its lifetime. It appears that ILUMEX has also led to a transformation of the lighting market in the project cities, where CFLs became gradually available in stores during the project's implementation.
- ILUMEX has also resulted in substantial economic savings for Mexico as a country and for households in Monterrey and Guadalajara. Households also benefited from employment training opportunities mostly for women. Whether the CFE would have profited by implementing the project itself is less clear and depends on how the reduction in the power sales is valued. The company is, however, continuing the project with the reflows from the CFL sales and plans to replicate and undertake similar conservation projects in the entire country with own funds. Finally, if the

emissions reductions could already be marketed as GHG credits, they would provide a sizeable source of income for the CFE.

- The project resulted in a reduction of all pollutants and saved resources involved in power generation and lighting with one exception. Mercury emissions are likely to increase slightly due to its used in CFLs and the risk that the CFLs will be disposed off as household waste at the end of their lifetime.
- ILUMEX has become a model for complex DSM projects of this kind. It has pioneered a highly participatory approach in its design, organization, implementation and management with the involvement of many people and institutions in the two project areas and in the entire country. This fact, combined with a professional project administration by the CFE, has greatly contributed to the project's success.
- The project has contributed greatly to learning and capacity building in and outside of Mexico. It has removed barriers for DSM projects in Mexico and has created human capacity for future GHG projects in Mexico both on the business and government level. Mexican experts are now capable of assisting in the transfer of this knowledge asset to other interested companies and countries.
- ILUMEX has been invaluable as a catalyst for learning and cooperation. It has served as a demonstration project. It has been widely studied for a variety of methodological and institutional questions related to the implementation of the FCCC and the project-based mechanisms of the Kyoto Protocol. It is a useful model because it brings together many organizations such as the World Bank, the Governments of Mexico and Norway, the FCCC Secretariat, FCCC member states and negotiators, CFE and other participating companies and consultants, various research organizations and non-government organizations (NGOs). The project has also helped to shape World Bank plans for the Prototype Carbon Fund (PCF).

This success has come at a cost. The total project cost was about 23 million of which approximately 75% was used for the acquisition of the CFLs. If the project costs were entirely attributed to GHG reductions, one Mg of CO_2 would cost approximately $30 USA. The cost per unit of CO_2 becomes negative, however, if the value of the avoided electricity generation is considered. Despite this fact it can be observed that the market for CFLs is still underdeveloped in many developing and industrial countries alike, which testifies to the fact that substantial barriers still exist and impede switching to more efficient lighting devices.

2.5 The ILUMEX Pilot Verification and Certification Exercise

A key component in determining the value and credibility of emission reductions generated by AIJ projects and by investment projects under Article 6 (JI) and Article 12 (CDM) of the Kyoto Protocol (KP) is their verification and certification. Bona-fide GHG emission reductions over business as usual emissions, or the baseline project case require certification and verification. The acceptance of emission reduction (ER) units as defined in Article 6 and of certified emission reductions (CERs) as defined in Article 12 can be expected to depend on proof of compliance through defined verification. Certification procedures with the requirements contained in the relevant articles of the KP and/or subsequent decisions of the Parties to the KP.

2.5.1 The objectives of the exercise

The first objective of the ILUMEX Verification Pilot was therefore to demonstrate that ILUMEX had in fact achieved ERs in the verification period (1995-98) in compliance with all relevant criteria for AIJ. Verification also certified the projects were consistent with criteria for JI and CDM projects as contained in KP Articles 6 and 12, respectively. This objective required:

- professional audit of the relevant records and data collected throughout the project's preparation and operation;
- review of the baseline and a recalculation of the emission reductions using appropriate models and conservative estimates where data were unavailable; and,
- establishment of indicators for compliance with the remaining criteria, in particular the project's assistance with achieving sustainable development.

A second objective was related to the fact that the ILUMEX Verification Pilot was designed as a learning exercise. It was therefore the task of the verification team to develop a format for this and future verification and certification activities. This objective required the team to:

- develop a concept and procedures for auditing and verification of emission reductions in line with established industry practices where appropriate;
- propose a format for the verification report and for a GHG emission reduction certificate;
- use the experience for drafting a model for a monitoring and verification protocol for future projects; and,

- report about the lessons learned for use by the World Bank, the FCCC and other interested parties.

These objectives were successfully met by the verification team consisting of Det Norske Veritas (DNV), a Norwegian verification and certification company, ICF Kaiser Inc., a technical consultant company, and CICERO, an Oslo-based climate change research institute. The verification team submitted four reports, which can be obtained from the World Bank.

In addition, the World Bank expected that as a result of the ILUMEX Verification Pilot also the following secondary benefits could be achieved:

- A contribution to the evaluation of the FCCC AIJ pilot;
- Assistance with the development of a public knowledge base for the verification and certification of emission reductions;
- Input into the international decision making process regarding modalities for verification and certification;
- The building of capacity in the host country (Mexico) and in the host company (the CFE) that would facilitate their involvement in CDM projects;
- The promotion of a market in cost-effective high quality verification and certification services that will assure the integrity, the credibility and the quality of project-based transferable emission reductions; and,
- Support for the preparations for future World Bank JI and CDM projects.

The ILUMEX Verification Pilot has laid the groundwork for achieving those benefits. But the full impact of the exercise will only materialize with time as interested parties draw lessons from the ILUMEX experience and use them in designing future activities.

2.5.2 Findings

The scope of the audit was limited to the ILUMEX project impacts over the period May 1995 to December 1998. The auditor's mandate was to audit only aspects relevant for GHG emissions. The audit covered only existing data from the project operations and, where such was not available, used conservative estimates to determine energy savings and GHG reductions.

The independent verification of GHG emission reductions concluded that, in the verification period, a minimum of 171,169 Mg of CO_2 equivalent emissions were saved and that this was achieved in compliance with the relevant AIJ, JI and CDM criteria including goals for sustainable development. This amount was subsequently certified by DNV. Of the amount, 18,828 Mg relate to the Norwegian financed AIJ component.

These figures represent conservative values for the ERs based on verifiable information. Following industry standards, the verification team always adopted a conservative approach so as to meet established international auditing rules. The team might have been able to verify a higher level of ERs if the audit and the verification had been anticipated and appropriately planned for in the design of the project by Mexico and the World Bank. If the CFE had been given more support to survey the important indirect emissions effects of free riding and market transformation (free driving) and to track power plant dispatching patterns and other factors that effect the verified ERs.

The verification team recommended to certify approximately 28% lower GHG emission reductions than CFE had originally determined for the verification period, the discrepancy being due to the verifier's conservative approach, insufficient data availability and limited auditing resources. Some of the GHG emission savings were shifted from the current project period to future periods due to the stockpiling of CFLs by households for future use.

The ILUMEX Verification Pilot exercise has resulted in a number of lessons. The most critical lesson for future AIJ, JI, and CDM projects is that a project-tailored monitoring and verification protocols (MVP) must be developed with sufficient funding and technical support before the project is implemented. The verification team was lucky that they could draw on the unusually fastidious data collection, record keeping, data processing and generally professional excellence demonstrated by CFE's ILUMEX teams, which permitted to verify a significant portion of the reductions reported by CFE. However, the limitations stemming from the lack of an MVP were all too apparent in the course of the verification exercise. Defined pre-project expectations and goals and their operationalization through the issuance of an MVP will go a long way to ensure that emissions reductions in future projects can be fully verified.

2.5.2.1 Project energy savings and emission reductions

GHG ERs were calculated by first determining the verifiable direct and indirect energy savings from ILUMEX. Direct energy savings were calculated based on the wattage difference between CFLs and incandescent light bulbs sold during the verification period. Results of wattage tests of the CFLs, which had been conducted by an independent electrical laboratory, were conservatively adjusted to account for a small sample size.

The verification team identified three important indirect effects that have an impact on energy savings due to the project, namely stockpiling, free driver and free rider effects. The verification team could only verify the stockpiling and free rider effect, which have a negative impact on energy savings. Due to severe data limitations, the verification team was unable to verify the positive free driver impact and therefore concluded to adjust energy savings only for stockpiling. The team felt comfortable that the fully adjusted savings would be significantly higher, since the market for CFLs began emerging in the project area by 1996 and any free driver impacts would more than offset the free rider impacts. It is worthwhile to note that such data limitation need to be avoided in future projects by already planning properly for the audit when designing the project.

To capture the full ERs impacts of ILUMEX, future energy savings over the lifetime of the bulbs were also estimated by conservatively assuming that the CFLs last until 2013 if current use patterns persist. It turns out that the vast majority of savings and therefore emission reductions will only occur from 1999 to 2013.

TABLE 2. Verified project and future GHG emissions reductions from ILUMEX.

Project GHG Reductions Including Transmission & Distribution Losses (Tons CO_2 equivalent)*				
Year ($CO_2 + CH_4$)	Nuevo Leon	Jalisco	Sales based GHG emissions**	Stockpile adjusted GHG emissions***
1995	4,621	2,719	7,340	6,745
1996	15,723	16,111	31,834	29,253
1997	27,142	32,902	60,044	55,175
1998	36,183	50,874	87,057	79,997
Project (95 - 98) GHG Reductions	**83,671**	**102,606**	**186,276**	**171,169**
Total Project (95 - 98) CO_2 Reductions	83,654	102,585	186,238	171,134
Total Project (95 - 98) CH_4 Reductions	17	21	38	25
Future (99 - 13**) GHG Reductions**	**328,691**	**424,134**	**752,825**	**767,932**
Future (99 - 13) CO_2 Reductions	328,623	424,046	752,769	752,869
Future (99 - 13) CH_4 Reductions	68	88	56	63
Bulb Lifetime GHG Reductions	**412,362**	**526,740**	**939,102**	**939,102**
Bulb Lifetime CO_2 Reductions	412,276	526,631	938,907	938,907
Bulb Lifetime CH_4 Reductions	86	109	95	95

* Numbers may not add due to rounding. ** Assumes that 2,436,627 bulbs were distributed, each having an average bulb-life of 9,362.5 hours. *** A stockpiling percentage of 8.11% was used based on a survey conducted in Nuevo Leon by ILUMEX. The stockpiling values derived from this survey were applied to all bulbs distributed under the project. **** Based on the expected burn-out of the least used bulb distributed in December 1998.

Finally, the verification team used the energy savings to calculate the GHG emission reductions during the verification period and to forecast the future ERs. This was done using weighted average marginal emissions rates, which were derived from CFE dispatching data, rather than using the model plant approach, which had originally been approved by the World Bank for ILUMEX. The marginal rate approach, while conservative, allows for a closer tracking of the actual emissions displaced at the grid based on the generation plants on the margin over time during the project period. Table 2 shows the GHG reductions for the verification period as well as projected reductions.

TABLE 3. Compliance of ILUMEX with AIJ, JI and CDM criteria.

Criteria*	AIJ	JI	CDM	Verification
Approval by involved parties:				
1. Voluntary participation of non-Annex I countries	√			OK
2. Acceptance/approval/endorsement by host countries	√			OK
3. Projects must have the approval of involved Parties		√		OK
4. Voluntary participation approved by each Party involved			√	OK
Compatibility with national priorities:				
5. Compatible with and supportive of national environmental and developmental priorities and strategies	√			OK
Cost-effectiveness:				
6. Contribution to cost-effectiveness in achieving global benefits	√			OK
Additional to financial obligations and ODA:				
7. Financing additional to Annex I Parties' financial obligations, and to official development assistance (ODA)	√			(OK)
Compliance with Kyoto Protocol (KP) commitments:				
8. The party acquiring emission reductions must be in compliance with obligations under Art. 5 and 7 KP		√		(Not applicable)
9. Acquisition of emission reductions must be supplemental to domestic actions for meeting Art. 3 KP		√		OK
10. Assist Annex I Parties in achieving compliance with part of their emission reduction commitment under Art. 3 KP			√	(OK)
Assistance to sustainable development:				
11. Project must assist non-Annex I parties in achieving sustainble development contributing to UNFCCC objectives			√	OK
Additional climate change mitigation:				
12. Supplementary and subsidiary means of achieving UNFCCC objectives	√			(OK)
13. Real, measurable, and long-term benefits related to climate change mitigation that would not have occurred in the absence of the project	√			(OK)
14. Reduction in emissions by source, or enhancement of removals by sinks, must be additional to any that would otherwise occur		√		OK
15. Real, measurable and long-term benefits related to climate change mitigation			√	OK
16. Reduction in emission additional to any that would occur in the absence of the project			√	OK

* Check marks mean that the criterion is an AIJ, JI or CDM requirement. OK means that the criterion has been verified and found fulfilled. (OK) means that the verification team had questions regarding the criterion's applicability, interpretation and/or compliance with the criterion.

2.5.3 AIJ and Kyoto Protocol criteria

Verified ERs can only be certified if they have been achieved through a project that meets the relevant project criteria. Therefore, the verification team undertook to verify that the ILUMEX project meets all the criteria for FCCC AIJ projects. The team also sought to verify, where possible, whether ILUMEX would meet the criteria for JI projects contained in KP Articles 6 (JI) and 12 (CDM). In addition, the team verified compliance with additional specific requirements of the host country (Mexico) and of the donor country (Norway), as reflected in their agreement as well as compliance with any relevant host country legislation. A summary of verified criteria is contained in Table 3.

In order to verify project compliance, the team had to develop suitable indicators for each criterion. This proved to be particularly challenging for the sustainability criterion in Art. 12 KP (No. 11 in Table 3), which says that it is the purpose of the CDM to assist Parties not included in Annex I of the UNFCCC in achieving sustainable development. The verification team met the challenge by formulating a set of indicators with the help of the relevant Mexican authority Instituto Nacional de Ecologia (INE). INE then developed weightings for the indicators on a scale of 1 to 5, whereas the verification team rated the project's compliance with each indicator on a scale of -2 to 2 with 0 indicating no progress. Since the sum of the indicators was positive, the verification team concluded that ILUMEX meets the CDM's sustainability criterion.

The verification team concluded that ILUMEX satisfies the 16 project criteria with a few minor reservations where some uncertainty is attached due to questions regarding the criterion's applicability, interpretation or compliance with it (shown as (OK) in Table 3). The team's reservations related to their inability to verify that ILUMEX will have a long-term effect on consumer behavior, and to the possibility that, given the high rate of return of CFL use for households, the CFL market may take off without the project.

2.6 Lessons learned

The ILUMEX Verification Pilot exercise has resulted in a number of lessons and recommendations as well as the identification of questions that remain to be resolved:

- *Established practice*: The ILUMEX audit shows that established industry practices for audits of environmental management systems can be used for GHG emissions projects. It is a clear recommendation from the auditors that the established industry practice is fully deployed. However, the acceptability of CERs hinges on the development and establishment of rules or standards that will guide the verification and certification process.

- *Accreditation and conflicts of interest*: Rules need to be decided for the accreditation of operational entities that offer validation, verification and certification services, and an accreditation authority should be established. Necessary are also rules that avoid possible conflicts of interests for organizations involved in GHG verification and certification. Such conflicts can occur if the same body is simultaneously involved in the functions of project party and project validator; project validator and GHG verifier/certifier; project party and verifier/certifier; and, certifiers and GHG reduction trader. Finally, rules should be developed for the registration of emission reductions under KP Articles 6 and 12 and the handling of non-conformance situations.
- *Monitoring and verification plan*: The exercise has clearly highlighted the need for the development of an elaborate monitoring and verification plan as part of a GHG project's design. The plan should be laid down in a project-specific MVP which would identify the relevant indicators to be monitored and include requirements for data recording and document control, for training of monitoring personnel as well as rules for the handling of non-conforming situations in the project. The MVP should be made binding for the parties involved in monitoring and verification in order to support an optimum preparation and execution of the monitoring, auditing and verification processes and to prevent disputes between the auditor and the auditee about unclear terms and obligations.
- *Generic criteria*: From the verifier's/certifier's perspective, there is a need for generic criteria or standards for project development, implementation and operation. Such criteria must at least include requirements for the development and documentation of baselines as well as for project-specific MVPs. The generic criteria should then be used for the design of each specific GHG project and its project-specific MVP.

The ILUMEX verification exercise shows that learning by doing can be one of the means to provide assistance to those who will resolve the above questions. Drawing upon established industry practice will be another. The long history of the globally operating certification industry suggests that this experience and the necessary decision by the relevant regulatory authorities, in particular the FCCC Parties, can over time establish a robust and suitable framework for verification and certification of GHG ERs.

3. THE POLAND COAL-TO-GAS BOILER CONVERSION PROJECT

3.1 Project Description

The Poland Coal-to-Gas Boiler Conversion Project is the second project that began as a demonstration JI project in 1993 and it is also a project within the GEF climate change portfolio. Co-financing from the Government of Norway allowed an expansion of the scope of the project and its use as a demonstration vehicle. The project has the following three components:

- The coal-to-gas conversion component invests in about 30 non-industrial small to medium sized heat plants (boilers) for their conversion from coal to natural gas. The boilers supply energy to residential houses and public buildings. The new technologies include gas-fired co-generation of heat and electricity and the use of condensing boilers as well as conventional gas-fired boilers.
- An energy efficiency component invests in the installation of energy efficient equipment in some hundreds of new residential units.
- A technical assistance component for project participants includes institutional training, assistance with project management, environmental monitoring and nation-wide marketing.

The project will, through lesser emissions of GHGs and pollutants, capture global environmental benefits and reduce local pollution. Additionally, the project seeks to support the objectives of the FCCC through the stimulation of technological and institutional changes and the demonstration of inter-fuel substitution and energy efficiency improvement, the encouragement of a more rapid transition from coal to gas, and the introduction of more energy efficiency in residential buildings.

3.2 Project costs and risks

The project cost was US$ 48.32 million with a grant proportion of 26.1 million, of which 25 million came from the GEF and Norway through the AIJ Program with the World Bank contributed 1.1 million. The remainder came from various sources in Poland, most notably from the Polish National Fund for Environmental Protection and Water Management, from the Polish Environmental Protection Bank and from the Polish private sector. The GEF and AIJ funds were not allocated to specific project activities, which resulted in a highly mixed project with activities and results that are not separable as to the sources of funding.

A prerequisite for financial support through the project was that the applicant boiler owner could not achieve an acceptable internal rate of return (IRR) without concessional funding. Project preparation calculations showed that, without grant funding, the IRR for typical coal-to-gas projects ranged between 2 and 8%. To achieve a chosen IRR of 25%, which is the boiler owners' cost of capital, the concessional financing would have to cover between 40 and 70% of the total project cost, depending on the technology choice. The incremental costs of investment and operation were figured to be between US$ 15 and 70 per Mg of removed CO_2.

A preliminary analysis by the implementing agency of the pipeline of 22 boiler projects, for which the grant agreement was signed, showed that the grant contribution will be USA $19,757,000, which is approximately 75% of the total US$ 26 million grant funding for the project. The grant funding will cover, on average, approximately 55% of the total project cost. The related CO_2 ERs are estimated by the implementing agency to amount to 195,189 Mg annually compared to the reference baseline. The marginal cost for all 22 boiler projects was calculated to be US$ 12.70 per Mg CO_2 at a 12% annuity rate.

Many factors may influence the success of the project. Poland's economic development, the availability of investment to boiler owners, changes in the price for coal relative to gas, new environmental regulations, sociological and behavior changes, the policies of national and local environmental funds, new technologies, a lack of institutional capacity are all factors. The supply of gas was considered sufficiently secure, with some limitations arising in the winter peak period, because of a lack in seasonal flexibility of gas imports from Russia and a limited national storage capacity. Polish production was considered a buffer for seasonal variations in demand. This assessment is still valid.

Project proponents tried to minimize identified risks by initiating a nation-wide marketing campaign and by recruiting a technical consultant early in the project. These efforts were, however, insufficient and significant problems emerged during project implementation. The project's implementation was also slowed down by a delayed approval of the grant agreement, delays in the appointment of an advisory panel and last, but not least, changes in the Ministry of Environment regarding responsibilities for the supervision of the project.

The institutional set-up for this project seemed overly bureaucratic. Too many entities were involved or established to advise, consider and approve individual project elements, which resulted in an ineffective project performance. The project suffered also from an initial lack of capacity and experience in project development. Support in this area should be a part of a

JI or CDM package; this will increase the costs, but decrease the risks. The institutional qualifications of the designated host country institution is key for the investor.

3.3 Baseline and indirect emission effects

At project appraisal, and for the immediate future, conversions from coal-to-gas fired boilers were not expected to be financially attractive without taking into account climate change and/or local pollution effects. Continued use of existing coal-fired boilers and a gradual shift to new coal-fired boilers was considered the most probable baseline. Regardless of the age of existing boilers at the time of their replacement, the baseline chosen was that all boilers were new coal-fired boilers with a lifetime of 17 years. The project financing would thus provide a grant equivalent to the additional life-cycle costs of converting existing coal-fired boilers to new more efficient gas-fired boilers over the cost of replacing the old boilers with new coal-fired boilers. As many old coal-fired boilers had many years of remaining lifetime, this was a very conservative choice.

During project implementation it was revealed that local authorities had initiated similar conversion projects, with support from national environmental funds as well as foreign institutions. As new environmental standards, taxes and fees are phased in to meet expected standards within the European Union (EU), and the costs of using polluting fuels are discouraged through fees and penalties, this conversion process is likely to be accelerated. Additionally, the expected further liberalization of coal prices and increasing labor costs are working against coal as a fuel for small and medium sized boilers. A more realistic baseline would therefore be a scenario of continued use of existing coal-fired boilers with improved operation and management procedures and slight modernization, along with a gradual shift from coal-to-gas fired boilers. In a recent AIJ evaluation report it was therefore suggested to calculate the investments and differences in discounted operations and maintenance costs against a 10-year lifetime of existing coal-fired boilers (Selrod et al., 1999).

Indirect effects on emission reductions are also a concern in the project. It is possible that some boiler owners would have made the conversion without support from the project and they must therefore be considered free riders. The proportion of free riders is difficult to estimate, but may represent large costs without real benefits. The owner of the first converted boiler could be considered a free rider in a 17-year perspective. However, it is believed that the extent of the energy efficiency measures that took place simultaneously with the conversion would not have happened without the project support.

The negative effect on real emission reductions due to the possibility of free riders is likely to be offset by the transformation process that is taking place in the Polish boiler market due to the project. It is expected that some boiler owners that are not participating in the project are inspired or influenced by it to convert to gas without support of the project and they must therefore be considered free drivers. The free driver effect represents additional indirect emission savings that would not have happened at this time without the project. The ultimate aim of the project and the marketing plan was to make every not-participating boiler owner a free driver. The project has probably had a limited effect. It might have had a stronger effect had the marketing plan been developed as planned.

Several other factors render project GHG reductions uncertain. Uncontrolled GHG emissions from black coal mines were assumed to be 20-25 m^3 per Mg of coal produced. These emissions will be reduced as conversions to oil and/or gas gain momentum. The related transport of coal will also reduce GHG emissions. New gas pipelines to converted boilers were expected to have negligible gas leakage. The precise figures regarding these factors will have to be confirmed, but in sum these factors reduce GHG emission indirectly as a result of the project.

Finally, it is worth noting that the Poland AIJ project faced significant practical and methodological difficulties, since the relatively small AIJ component was fully integrated with the much bigger overall grant support for the project from GEF and other sources. Since AIJ funds were not allocated to specific project activities, AIJ-specific methodologies for baseline assessment and the calculation of emission reductions cannot easily be applied and the project's results cannot easily be attributed to the sources of funding. This fact rendered the calculation of emission reductions for the AIJ component and an adequate reporting on the project to the FCCC Secretariat much more difficult.

3.4 Environmental benefits

Based on figures from the first operating conversion project, which uses one condensing boiler for the base load and two direct gas-fired boilers for the peak load, the annual ER was calculated to be 550 Mg CO_2 compared to the baseline of new coal-fired boilers. The actual CO_2 emissions after conversion are 427 Mg CO_2 per year. In Table 4 below, ERs and corresponding marginal costs are shown for the two baseline scenarios compared to the World Bank project estimate.

TABLE 4. CO_2 abatement (Mg CO_2) and marginal cost per Mg CO_2 in the Poland Boiler Conversion project.

Baseline	CO_2 abated (per year)	CO_2 abated (project life)	Marginal cost** (US$/Mg)
Estimate for chosen baseline	494	8,398	37.00
Real figures for chosen baseline	550	9,350	31.70
New suggested baseline	1,210	12,100*	18.20

*10 years lifetime. **Marginal cost related to conversion project, excluding administrative costs etc.

Compared to the expected 65% CO_2 reduction for the project, the figures from the first boiler conversion show a reduction from the pre-project situation of about 70%, while compared to a baseline of new coal fired boilers the reduction is about 42%.

The project will also have significant local and regional environmental benefits in particular improvements in local air quality. Compared to the existing boilers, the fuel switch will reduce emissions of SO_2 and particulates to nearly zero and NOx by 85%. These ER are of particular importance in Poland since the converted boilers are usually located in heavily populated areas. The project will also improve the air quality throughout Europe through the reduction of transboundary SO_2 and NOx pollution. Reduced mining of coal will reduce CH_4 from mining operations as well as the discharge of saline water from the mines causing considerable problems for downstream water consumers.

ERs from coal combustion have important positive health impacts both on a local and a regional scale. A short-term negative effect will be a loss of jobs in the coal mining industry, the transport sector and in the handling of coal at the boiler site. But conversions to environmental friendly and energy efficient technologies will need new investments, which may create new jobs and economic growth. Lower health risks and lower energy cost will, in a longer perspective, increase the social and economic standards for the population at large.

3.5 Other benefits and lessons learned

From a Polish perspective, the project yielded a number of lessons and benefits: Small-scale gas burning condensing boilers and combined heat and power (CHP) installations are new technologies in Poland. The project revealed that these technologies are not yet profitable in Poland compared to traditional direct gas-fired boilers given the current economic conditions. The project had comprehensive plans for dissemination of results, training and information, which was seen as a prerequisite for replication of the project concept. This element contributed to national capacity building, but has not been as successful as expected.

The project's organization and the procedures for processing applications for individual boiler conversions were designed to meet the political, economical, technical and environmental concerns and requirements for implementing similar projects. As the project progressed, however, its administrative structure turned out to be quite complicated and bureaucratic. Participants directly and indirectly involved in the project have, nevertheless, expressed satisfaction with the experience they have gained. But the project, due to the many difficulties encountered, has resulted in greater learning opportunities and may have yielded more lessons than might have been the case with a problem-free project.

Poland is likely to be a host country in future JI projects, since the unit cost of reducing GHG emissions is far lower than in many other industrial countries. The present AIJ project has been very important for health and environmental quality in Poland, with the global environment and technology transfer as second and third in importance (Selrod et al., 1999). The project has thus demonstrated that future JI projects would be highly advantageous for Poland. They would encourage sustainable development in energy utilization and reduce local and transboundary air pollution substantially.

The project has been instrumental in the formulation of Polish GHG policies. It has been a valuable learning experience for Polish authorities on how to deal with future projects of this kind. Among the most interesting areas for JI projects in Poland are energy saving projects, utilization of alternative energy sources such as biofuel, and reforestation. Polish authorities expect to resolve remaining legal and administrative issues and permit the implementation of JI projects beginning in the year 2000. From the perspective of the global environment, this project would, with the initial baseline, have yielded more emission savings than what would have been awarded in the form of emission credits. From the perspective of an investor in GHG reductions, the project might, with the initial baseline, have been too expensive, and the host country might not have been able to sell the project.

From the perspective of Norway as the donor country, the main value of the project has been its learning experience where a number of methodological and practical issues have been scrutinized. The project was started as a co-financing arrangement with the GEF at a time when the JI mechanism was controversial and unclear. With the guidelines for AIJ evolving over time, it proved to be a major weakness of the project that the AIJ components could not clearly be distinguished from the rest of the project, a lesson that should clearly be heeded in drafting guidelines for future GHG projects. It is believed that this project, along with other demonstration projects in the World Bank-Norway AIJ portfolio, has assisted Parties in their understanding of how JI might work, and has helped to bring forward the agreements for introducing JI and the CDM into the Kyoto Protocol.

From the perspective of the World Bank, the project has provided valuable learning experience to the Bank and other stakeholders in the climate constituency. It has helped to guide the debate and the process of learning about project-based mechanisms under the FCCC and later the Kyoto Protocol. The project has given rise to a number of methodological questions that have been addressed by the World Bank and researchers in many parts of the world. It has served as a reference project for a number of international seminars and workshops and helped in the formulation of draft monitoring and reporting systems as well as in addressing issues of how JI and the CDM might work in the future. As an AIJ pilot project, it has proved its value in paving the way for the development of the Kyoto mechanisms and has served its initial purpose well.

4. THE BURKINA FASO SUSTAINABLE ENERGY PROJECT

4.1 Project Description

In 1997, Burkina Faso had a population of 11 million, an annual growth rate of 2.8% and a per capita income of US$ 220. The country has limited energy resources in a semi-arid climate. Only a third of the 154,000 km^2 of forest land is accessible. The population's low-income level has limited fuel substitution efforts, with woodfuels accounting for over 91% of total national energy consumption and 98% of household energy use. The devaluation of the local currency slowed down the penetration of liquified petroleum gas (LPG). Population increase in the urban areas continues to increase the demand for fuelwood and related forest products.

The Burkina Faso project, which started in 1997, has several elements: It introduces efficient carbonization techniques for charcoal production, community-based forest management, kerosene cooking stoves and solar photovoltaic (PV) systems to rural communities. The project enhances community participation and removes market barriers for improved kerosene stoves and PV solar systems. The overall objective of this multifaceted project is to contribute to meeting the rapidly growing urban demand for household fuels without further loss of forest cover or loss of the ecosystem's C sequestration potential.

The Burkina Faso project is being implemented by the Regional Program for the Traditional Energy Sector (RPTES) within the Africa regional department of the World Bank. The AIJ investment is thus part of a larger energy project in Burkina Faso. The AIJ/RPTES collaboration leveraged funds necessary for the implementation of the overall project. The following table outlines the financing arrangements for the project as a whole.

TABLE 5. Funding arrangements for the Burkina Faso AIJ project.

Sources of funds	**US$ million**
Government of Burkina Faso	1.0
Danish International Development Agency (DANIDA)	14.0
International Development Association (World Bank - IDA)	3.0
Government of Norway (AIJ)	2.4

Through its additional funding of US$2.4 million, the AIJ component co-finances and supports the development of low and non-C energy sources. Thus, the AIJ component contributes to the prevention of the felling of 130,000 Mg of wood for charcoal burning and opens, expands and strengthens the market for non-C energy sources.

4.2 Benefits and Costs by Project Components

The Burkina Faso sustainable energy management project contributes to CO_2 abatement. But more importantly, it helps improve environmental management in the country. The project continues to assist with the enhancement of the government's efforts and ability to address local and global environmental problems. Benefits for Burkina Faso flowing from the project include:

- promotion and a rapid expansion of community-based forest management schemes,
- introduction of improved charcoal processing schemes,
- increased access to energy in rural areas,
- promotion of solar PV systems for clean energy services, and
- improved health in households previously inhaling smoke from burning fuelwood.

The costs and benefits of the Burkina Faso project were analyzed separately for the four project components. The analysis included an assessment of the environmental and social benefits realized by each project component.

4.2.1 Installation of PV systems

In June 1998, a survey on the possibility to introduce solar PV systems into rural areas in Burkina Faso was completed in 250 villages. 30% of all households were found able to afford solar energy. A village in Burkina Faso has on average 900 people with an average household size of 8.4. Applying the 30% figure, 8,036 homes in the project area could afford to participate in a program for rural solar PV systems.

These households obtain rechargeable batteries and electric lamps from a supplier in their village. The trained supplier operates the solar PV equipment and recharges the batteries. He also provides maintenance for the PV systems and repairs system components. The solar systems are purchased and installed with grant money and all 250 villages are supported to invest in the installation of a solar water pump instead of a diesel pump.

The installation of PV systems for lighting and water pumping will have global benefits through avoided GHG emissions. The GHG emission reduction is calculated as the emissions that would have occurred if kerosene lamps and diesel pumps were used instead of the PV systems. For the kerosene lamps it was assumed that each participating household consumes 0.4 liter of kerosene equivalent to 4 hours of lighting, which would result in 1.6 kg CO_2 emission per day. In the case of the diesel pumps it is known that a pump (type Lister ST-1 diesel), which supplies an average village of 900 persons with 36 m^3 water per day, i.e. 40 liters of water per person per day, consumes 3.84 liters of diesel. The diesel used would result in 15.5 kg CO_2 emission per day.

The installation of the solar PV systems for lighting and water pumping in the 250 villages would thus avoid about 18,000 Mg of CO_2 in a period of 6 years. Funding from the AIJ program for this project component is US$ 280,000.

4.2.2 Kerosene cooking stoves

The introduction of kerosene cooking stoves aims at mitigating the over-exploitation of Burkina Faso's forest resources and at reducing GHG emissions from the unsustainable harvesting of fuelwood. The kerosene cooking stoves are used for cooking and lighting as well. The use of kerosene in household cooking stoves reduces the pressure to cut down forests for fuelwood. The demand for fuelwood is high in urban areas and continues to grow.

The calculation of the CO_2 benefits from the introduction of kerosene stoves is based on the 13 urban areas in the country over a period of 30 years. The calculation accounts for population growth and the relative share of the various urban cooking fuels (fuelwood, charcoal, LPG). Charcoal consumption is assumed to increase at an annual rate of 10% and LPG use is assumed to remain more or less constant, since only a small percentage of the population can afford it.

The CO_2 benefit is calculated based on the urban population growth and a ratio of charcoal to fuelwood use of 1.2. Each type of fuel is converted into fuelwood equivalents taking a gradual increase in carbonization efficiency in the production of charcoal into account (1 kg of LPG equals 7 kg fuelwood equivalents; 1 kg of charcoal equals 8.3 kg fuelwood equivalents at 12% and 4 kg at 25% carbonization efficiency). The fuelwood supply is converted into CO_2 emissions considering that 270,000 ha of existing forests and later a projected 300,000 ha of forests are under community-based sustainable forest management. About 0.6 Mg of fuel wood is produced per ha of forest per year. The net CO_2 abated through the introduction of the kerosene stoves is the difference between the CO_2 emissions resulting from unsustainable wood harvest outside the managed areas and those resulting from the consumption of kerosene as a substitute for charcoal.

The introduction of kerosene cooking stoves is funded through the AIJ program with US$ 500,000, which includes investment and technical assistance throughout the life cycle of the project. This project component avoids about 271,000 Mg of CO_2 in a period of 6 years. The use of kerosene cooking stoves contributes to the preservation of C sinks and the biodiversity in both the managed and unmanaged forest areas.

4.2.3 Improved carbonization techniques

A controlled carbonization technique can improve the efficiency of converting wood into charcoal from 12% to 25%. The wood harvested for charcoal processing is concentrated outside of the community managed forest areas. Each ton of fuelwood that is saved from unsustainable harvest results in a net CO_2 abatement of 1.7 Mg. The introduction of improved carbonization techniques is funded through the AIJ program with US$ 650,000. This project component avoids 916,000 Mg of CO_2 in a period of 6 years.

The use of efficiently produced charcoal in stoves instead of cooking with fuelwood in open stoves has other benefits. Carbonization lowers the direct emissions from household cooking and hence reduces indoor air pollution and improves the quality of welfare in households. The use of charcoal can also decrease transportation costs as the local communities can obtain the same energy supply from the transport of a few bags of processed charcoal as compared to many logs of unprocessed wood.

4.2.4 Forestry management

The 570,000 ha of community-based managed forests include areas that border national parks and reserves. The managed areas protect the parks and reserves against encroachment for agricultural activities, illegal wood harvesting and unsustainable forest practices. Sustainable management of the forests is funded through the AIJ program with US$ 970,000. This program component avoids 245,000 Mg of CO_2 in a period of 6 years. The calculation of the CO_2 benefit is based on the assumption that the production and use of 1 kg of charcoal leads to net CO_2 emissions of 6.8 kg.

The community-managed forests provide other benefits to the local community. These include: providing cattle feeds, serving as wind-breaks and protecting the crops, binding the soil and preventing erosion, providing shelter for recreation, herbal medicine, rainfall catchment and improving micro-climatic conditions.

4.3 Lessons Learned to Date

As the first AIJ project in Africa, the Burkina Faso project has to overcome a number of problems, which partly stem from the complexity of the project. A variety of unforeseen risks and difficulties affect the implementation of the project including the following:

- Delays in decisions on co-financing. This resulted in a slow start of the project's implementation.
- Withdrawal of co-financing for covering administrative cost. This required unforeseen budget re-allocations and resulted in further delays in the project.
- Insufficient preparation of the bidding process and the procurement of equipment by local institutions. This led to a rejection of the inadequately prepared bidding documents.
- Institutional restructuring and a shifting of the competencies for the execution of the project between national ministries.
- Inflation and depreciation of the local currency. This inflated the price of the solar units.
- Conflicts over forest ownership between the local communities holding traditional forest rights and the government, as well as with fuelwood wholesalers and charcoal processors who were issued government permits to use the forests. Increasing urban demand for energy aggravates these conflicts, which create fuelwood shortages and put pressure on existing forests.
- Lack of skills for using the solar PV systems and the improved carbonization technique. This has made the implementation of the project

difficult and the monitoring and recording of abated CO_2 emissions almost impossible.

These complications have resulted in a reallocation of funds in the project budget and other adjustments with a view to secure the completion of the implementation phase. The unforeseen budget reallocations to cover administrative costs will have an impact on the CO_2 abatement that can be achieved by the project.

The project has shown that baselines and monitoring plans prepared for the project were insufficient given the complexity of the project and of the individual project components. Local experts find it difficult to interpret the suggested baselines for the project components and apply them to determine actual emissions reductions. This suggests that project developers must address the local institutional and human capacities much more vigorously in the project design and involve local participation in the project from design through to implementation.

The local experts, who work on the project with only minor supervision, have improvised data collection and recording instead of using a systematic monitoring plan. The project has thus reinforced the lesson that appropriate training components must be included in project activities and that monitoring and measurement must be meticulously planned as part of the project design in order to ensure the proper monitoring, measurements and calculations of CO_2 emissions and other indicators.

This element will receive particular attention during the remaining time for project implementation, during the project's operation and its evaluation. It may be possible to use the project for more work on baselines and additionality as well as for designing appropriate monitoring and verification plans. This effort may then lead to a validation exercise for the project or for parts of it.

Further, it is intended to subject the project to a comprehensive review and evaluation and to hold a project workshop. Both activities will provide opportunities to draw more lessons from implementing the project and to discuss what might attract CDM investment to Africa including the institutional requirements, capacity building and technology transfer. The project would thus assist the World Bank and African governments as well as other interested parties to make better plans for future activities in Africa in the context of AIJ or the CDM.

The Burkina Faso AIJ pilot project continues to generate practical experience necessary for African countries that are willing to participate in the emerging GHG market. AIJ support has leveraged funding from other donors, which made it possible to broaden and replicate project activities in other parts of the country. The project is the first, and still, the only AIJ project in Africa. It is a valuable experience for West Africa and the Africa continent as a whole, in particular in the face of the otherwise total lack of AIJ activities in the region. The experience of this AIJ project shows that financial resources and an enhancement of institutional and human capabilities are urgently needed in Africa if these countries want to play a successful role in the GHG market. Without new resources that address the specific problems of African countries it will be difficult to implement AIJ and CDM project and other climate change related activities in Africa.

5. THE ANDHRA PRADESH AGRICULTURAL DEMAND-SIDE MANAGEMENT PROJECT

5.1 Project Description

Consumption of electricity by the agricultural sector in Andhra Pradesh in India presents a significant challenge for the Andhra Pradesh State Electricity Board (APSEB). This sector consumes a large proportion of the total electricity production, and is growing rapidly. At the same time, agricultural tariffs do not provide sufficient revenue to cover the cost of production, let alone provide funds for the necessary expansion of the system. Further, a number of studies have indicated that there are significant cost effective opportunities to increase the efficiency of electricity usage both in terms of improving the efficiency of the distribution system and in terms of increasing end-use efficiency. The implementation and evaluation of the Andhra Pradesh Integrated Agricultural DSM project is intended to help the APSEB respond to this challenge.

The project combines the following four technical measures in an integrated package to generate electricity savings and GHG ERs:

- Improvements of the distribution system efficiency by converting from low voltage (LV) feeders to high voltage (HV) feeders.
- Reduction in system demand and line losses, as well as improved service to non-agricultural customers through the use of automated load control.
- The provision of customer meters to provide information on energy consumption.
- Improvements in end-use efficiency by replacement of pumpsets, and associated pipes and valves with efficient products.

Due to the substantial differential in tariffs between industrial and agricultural customers, it is possible for these technical measures to pay for themselves through the resale of saved energy to industrial customers who otherwise face power cuts.

5.2 Project Baseline

The APSEB placed agricultural DSM in the context of the rapidly growing electricity demand, a current peak power deficit of 28% and an energy demand shortage in the range of 18-20%. The growing share of coal thermal power in the APSEB's mix had also increased pooled costs of generation. The simultaneously increasing agricultural consumption, which was highly subsidized, is weakening the APSEB financially. In this context the APSEB was eager to implement measures to increase efficiency and contain demand through DSM measures.

The agricultural sector forms a major load of the Andhra Pradesh system. Out of the total sales of 23.55 GWh during 1995-96, agricultural consumption accounts for around 49%. The number of agricultural connections has increased ten-fold between 1970 and 1996 and strong growth in demand continues. In this context, the regulation of energy supplies to the agricultural sector has a significant bearing on the finances of APSEB, as this sector is heavily subsidized. A flat rate tariff of Rs. 75 per horse-power (HP) per annum is collected from the agricultural consumers, irrespective of the energy consumed by the pumpset.

Detailed baseline and metering studies for the purpose of assessing GHG emission reductions from the project are currently underway. It is also planned to develop a monitoring and verification protocol for the project and have the project design, the baseline, and the MVP validated by an independent third party before the physical implementation of the project, which is likely to start in late 1999.

5.3 Proposed alternative (AIJ Project)

Agricultural loads are believed to be deferrable loads and thus power can be provided at off-peak periods of the day without sacrificing the agricultural yield which will facilitate implementation of load management techniques to bring down the system demand. Great potential for energy savings exists through improving the pumpset efficiency on one hand and by decreasing the system demand by grouping the agricultural pumps and transformers and extending supply to each group at a time on the other hand. Providing a single-phase new pumpset improves the pumpset efficiency. Providing frictionless foot valves and piping alone is expected to result in energy savings of up to 30% which, given current consumption, would work out to 3.5 GWh.

The agricultural demand-side management is proposed to be implemented in two typical areas. The Low Voltage Distribution System

(LVDS) existing in these areas will be converted into a High Voltage Distribution System (HVDS) with small capacity, single phase distribution transformers. These transformers are grouped into two categories and extend supply to one group at a time by employing one way VHF radio communication (Krishna Rao, 1995). The end-use efficiency will be improved by replacing the existing three phase motors with single phase motors and the existing piping with lower friction piping. The project will encompass approximately 5,800 pumpsets on 8 feeders in two geographically separate areas of the state of Andhra Pradesh in southern India.

5.4 Project Costs and Benefits

The cost of this project is estimated at US$ 4.60 million. The bulk of the project cost is capital costs. In addition to these one time costs, implementation of this project will present an ongoing increase in annual operations and maintenance (O&M) costs for the distribution automation system, but will reduce costs for transformer maintenance and the requirement for power factor correction.

TABLE 6. Capital cost and changes in O&M costs attributed to the India DSM project.

Investment by item	Cost (in mil. Rs)
High voltage (11kV) distribution	55.459
Distribution automation	30.504
Meters	10.463
Pumpsets replacement and rectification	58.678
Evaluation	3.102
Total investment cost	158.206
Operations and maintenance by item	Cost (in mil. Rs per annum)
Additional O&M (3% of capital cost)	0.915
Reduced transformer maintenance	(0.342)
Reduced power factor correction	(0.113)
Total annual costs	0.455

There are substantial socioeconomic and development benefits resulting from the project, such as benefits of the use of conserved energy and employment benefits which are difficult to quantify even though they are known to exist. The quantifiable benefits accrue to three groups: customers, and especially farmers who participate in the project, the APSEB, and society as a whole.

The combined benefits of the project measures, which will be implemented, include:

- lower failure rates of distribution transformers;

- reduced maintenance costs of motors;
- increased system reliability and quality;
- release of power system capacity resulting in investment savings;
- increase in employment opportunities; and, finally
- Reduction in GHG emissions.

The benefits to customers include:

- new and more efficient pumpsets and piping,
- improved voltage control leading to reduction in the need to rewind motors, and
- improvement in quality and reliability of service.

The APSEB will have the following quantifiable benefits:

- freed system capacity and energy, including lower loading on distribution feeders (by as much as 50%);
- lower system reinforcement costs for new customers; and
- an estimated 20% reduction in pilferage of power with a corresponding increase in sales and revenue.

5.5 Economic and Financial Analysis

An economic analysis indicates that the project will provide significant benefits to both the farmers who participate in the project and to the SEB. The economic benefit / cost ratio exceeds 4:1 and the economic cost of the saved energy is less than 0.70 Rs/kWh (2 cents/kWh). By comparison, the equivalent cost of energy to the APSEB, as purchased from independent power producers, is approximately 2.25 Rs/kWh (over 6.4 cents/kWh).

A financial analysis of the project indicates that if the saved energy can be resold to the industrial and commercial customers in the higher tariff category, the additional revenue is sufficient to repay the cost of the project within 4 years. If the energy is resold at the average rate to all metered customers (excluding agricultural customers, this is estimated at Rs 1/kWh), the cost will be repaid over 7 years. The project's financial viability from the point of view of the utility depends critically on assumptions regarding the rate category to which the saved electricity is resold. The project is expected to result in a reduction of approximately 7,500 Mg C annually with approximately 150,000 Mg C abatement over the project lifetime.

The economic analysis reviews the impact of proceeding with the Andhra Pradesh Agricultural DSM project from the perspective of the overall economy. Not all benefits can be monetized and are not included in this analysis although they are expected. The benefits included in this analysis are:

- The freeing of electricity in the system due to the increased efficiency of the pumping systems and the reduction of line losses in the distribution system. The marginal value of this energy has been estimated at 2.25 Rs/kWh and is based on the cost paid for new energy from Independent Power Producers. A 20% line loss between the 11kV and 33kV substations are assumed.
- A reduction in the need for rewinding motors. Studies have estimated that farmers spend an average of Rs 1500-2000 annually on pumpset maintenance, including 800 -1000 Rs on motor rewinds. For the purpose of this analysis, it has been assumed that one rewind per year, with a value of 1000 Rs will be saved.
- A reduction in the need to purchase replacement pumpsets during the analysis period. The expected life of a pumpset is about 20 years. This project will provide farmers with new pumpsets, and hence allows them to defer this expense. It is assumed that the cost of the pumpset to the farmer is 11000 Rs, and that without the program, 5% of the pumpsets would be replaced each year.

The results, based on the above assumptions, are summarized in Table 7 below. The economic costs and benefits of the project are given in terms of their present value at 8% rate of discount.

TABLE 7. Comparison of costs and benefits from the India DSM project.

Economic cost of the project	155.5 million Rs
Economic benefit of the project	703.9 million Rs
Cost benefit ratio	4.50
Economic cost of energy from the project	0.69 Rs/kWh

These results show that implementation of the project will provide excellent economic benefits for the region. Approximately 12% of the economic benefits accrue directly to the farmers who participate in the program due to reduced maintenance and capital costs.

The financial analysis reviews this pilot project from the perspective of the APSEB and intends to determine whether this type of project is sustainable. The decisive question is whether the freed energy from the agricultural sector can be re-sold to the industrial or other sectors for sufficient funds to cover the costs of the project. Following are the mayor assumptions:

- Pilferage of energy is a major loss of revenue. For the area covered by the pilot project, it is estimated that pilferage accounts for 20% of the energy used. After the conversion to HV distribution, this pilferage will no longer be possible. It is assumed that the pilfered energy is used for agricultural pumpsets and that after the conversion the same energy will be used, but that these pumpsets will appear as additional agricultural customers.

- Currently all non-agricultural customers on these feeders receive limited hours of service. Once the distribution system has been converted, they will receive additional hours of service, which may result in additional consumption. For the purpose of this analysis, the additional consumption has been considered to be negligible, but will be verified as part of the project evaluation.
- The remaining energy is available for re-sale to other customers at differing rates as indicated in the Table below.

TABLE 8. Financial analysis for the India Agricultural DSM project.

	Million Rs	Rate at which saved electricity is sold		
		Rs 3/kWh	Rs 2/kWh	Rs 1/kWh
Financial cost	158.2			
Interest at 16%	25.3			
Increase in annual revenue				
- agricultural		2.81	2.81	2.81
- resold energy		539	360	288
Total annual revenue		542	362	291
Years to repay cost*		4	7	9
Years to repay cost**		5	12	∞

*Assuming a 10% annual tariff increase and 5% inflation rate. **Assuming constant dollars.

The financial analysis shows that:

- The viability of the project from the point of view of the utility depends on the tariff category to which the saved electricity is sold. If the saved electricity is utilized in the agricultural sector, the project is much less attractive than if it is sold to industrial and commercial consumers at a much higher price.
- When the tariff increases as projected, the project is sufficiently financially viable for the APSEB to conduct it on its own. If the tariff increases do not occur, the project is viable as long as the saved energy is resold for more than 2 Rs/kWh.

5.6 GHG Emissions Analysis

The analysis of GHG ERs takes the freed energy noted above and converts it into an equivalent reduction in GHG emission. The average specific coal and oil consumption at the thermal generating plants in the APSEB is 0.826 kg/kWh and 3 ml/kWh respectively for the year 1995-96.

This is a composite number derived from overall consumption of both fuels and total generation. Hence, this analysis assumes that each kWh saved results in a C reduction from both coal and oil. Assuming a 20-year life for this DSM project, the expected C reduction is estimated at 150,000 Mg as shown in the following Table 9.

TABLE 9. Expected GHG emission reductions from the India DSM project.

Fuel	Specific consumption	C(%)	Energy saved (mil. kWh)	C reduction (Mg)
Coal	0.826 kg/kWh	50.00	17.98	7,462
	APPROXIMATE REDUCTION IN 20 YEAR			150,000

5.7 Lessons so far

After long discussions with the relevant Indian authorities, the Andhra Pradesh Agricultural DSM project was approved as an AIJ project by the Government of India in 1998. It is in a very early stage of implementation and substantive experience from the project does, therefore, not yet exist. However, the project has already been instrumental in fertilizing the thinking about AIJ and similar project arrangements inside the Indian government and in demonstrating the potential benefits of such projects for India.

As the most recent project in the World Bank's AIJ pipeline, it will benefit from substantial knowledge transfer based on experience with earlier AIJ projects. In particular, the lessons learned from the ILUMEX project in Mexico, which is also a DSM project, are expected to allow a fruitful cross-fertilization. The India AIJ project is likely to be the first project which will host all elements of a full project cycle including validation and verification/certification components, for which funds have been set aside.

The project will predominantly be executed with the help of the Indian private sector and other local partners, which is expected to lead to further learning and capacity building in India. In particular, the DSM components of the project offer a wide potential for domestic replication in the supply shortage-ridden Indian power sector.

6. CONCLUSIONS

The four AIJ projects discussed above provide many experiences and lessons which contribute to the overall value of the AIJ pilot phase. The analysis of each project highlights these experiences. Beyond this, the World Bank's participation in the AIJ pilot phase permits also an interpretation of these experiences and lessons in the light of preparing for the operation of the future Kyoto mechanisms:

- Organizational learning: As a result of each of these projects, all parties involved, including the World Bank, have gained a good deal of valuable experience. For instance, at the beginning of the AIJ pilot program there was little neither conceptual nor operational knowledge in the World Bank on how the AIJ mechanisms might work, and the political context of AIJ and similar mechanisms was not well understood. In this regard, the Bank's involvement in the AIJ pilot phase has catalyzed considerable improvement and has allowed the Bank to contribute significantly to the discussion on JI and the CDM in the Kyoto Protocol. Similarly, on the side of AIJ host countries, a remarkable built-up of experience and human and institutional capacity has taken place, which has increased their interest in the Kyoto mechanisms and permits them to more fully participate in the international negotiations on the mechanisms. At the same time, it should be recognized that the AIJ experience has only laid the corner stone, which now allows many potential JI and CDM host countries to more clearly define their policies, internal procedures and authorities for project development. Finally, the AIJ experience will become extremely valuable in assisting these countries negotiate projects resulting in actual credits. The Andhra Pradesh project, for example, was the first AIJ project to pass fully through the cautious approval process of the Indian government. As such, the project provides insight into the process for those pursuing future projects and provides procedural precedence for those subsequent projects.
- Benefits beyond AIJ: All four AIJ projects studied above can potentially be replicated and provide other benefits to the host country, which may lead to more energy savings and further ERs. For example, once the India project is fully implemented, the monitoring and verification protocol established for the project will set an example for the monitoring, measurement and verification of energy savings, GHG emissions and other relevant indicators in future energy savings project. The enhanced confidence in the actual energy savings in combination with domestic capacity building through AIJ and similar projects will increase the incentive for energy service companies (ESCOs) to enter into performance contracts. Energy users or producers invest in projects that result in further energy savings and possibly GHG ERs. Considering the fact that profitable opportunities for energy savings exist in many countries, it is possible that the ancillary GHG reductions from pure energy savings operations, kicked off by experience from AIJ and similar projects, will outpace the reductions achieved within the boundaries of future JI or

CDM projects. These projects may eventually bend GHG emissions path downward more strongly than would be the case without today's AIJ and without future JI and CDM projects. Already, Mexico's plans to replicate the ILUMEX project demonstrate this effect.

- Project supply: Even with only four projects, the Bank's experience with the AIJ pilot phase suggests clearly that viable projects are quite available and possible in regions and/or countries which have been mostly ignored. The passage of the India project through the approval process is one example. The sustainable energy project in Burkina Faso is an even better showcase of a complex AIJ project being implemented in a difficult environment in the Africa region. However, the Bank's experience also suggests very clearly that a close collaboration with the host country's authorities and full respect for host country priorities and project criteria are crucial elements for AIJ projects to succeed. It is probably fair to expect that this will be even more true for JI and CDM projects, which will be negotiated in an emerging market environment and which, in the case of the CDM, must satisfy also the objective of assisting host countries in achieving sustainable development. What all this illustrates most plainly is the need for continued and increased activities in AIJ to build confidence and capacity and to improve the institutional environment in host countries, in particular in the Africa region. Whether or not this can be accomplished through a transitional process acknowledging both AIJ and CDM and includes the potential for crediting will be decided by the Parties to the Convention in future negotiations.
- Reporting: From the inception of the AIJ program, the complicated nature of AIJ projects with the requirements to establish baselines, satisfy additionality, calculate ERs, meet several other criteria and prepare detailed reports has created a need for capacity-building. Host countries have indicated the importance of the process of reporting and its initiation by the host country itself. During the last three years, various complications have arisen related to timely submissions to the FCCC illustrating continued need for institutional capacity-building, in particular with a view to the possible start of the CDM in the year 2000. However, marked improvement is also evident on the side of the host countries in terms of timeliness, quality and initiative in preparing AIJ reports.
- Transaction costs: Access to the Bank's regular portfolio of projects and to the Bank's extensive methodological and conceptual work and experience with the GEF over the years did result in a reduction of risks and transaction costs for AIJ projects. Still, however, the experience of the Bank suggested that implementation of bilateral deals could become excessively costly and increase the risks for individual countries and the private sector. Therefore, much thought has been put into the concept of a

portfolio approach whereby financing for projects can be pooled into a fund mechanism. In this case, the transaction costs and risks can be minimized and diversified through a portfolio that de-links potential investors from specific projects and their risks. In addition, a portfolio of projects offers learning opportunities, which may eventually allow the employment of more standardized approaches to baseline establishment and project preparation and drive down the related costs.

- Baselines: Discussions surrounding the AIJ pilot phase and the future of JI and the CDM frequently focus on the difficulties associated with the determination of the appropriate baseline. The experience of the World Bank AIJ Program has been that, on a project level, the determination of the technical reference scenario need not be particularly difficult as often only a few technical options with considerably different costs are available for meeting the identified demand. A cost-benefit or investment analysis can often be used to determine the baseline in large stand-alone power projects. However, the World Bank AIJ projects demonstrate unsurprisingly that it is much more difficult and demanding in terms of data availability to correct the relatively simple engineering approach to ERs for any indirect and behavioral effects, such as free riders and free drivers, that might have an impact on actual emissions. Again, as demonstrated by the World Bank AIJ pipeline, such effects are likely to play a significant role in DSM projects, which is why such projects require a particularly careful baseline assessment and planning as to the identification and the monitoring and measurement of relevant indicators including through consumer surveys and comparison group analysis. A careful assessment and planning at the project design stage will allow calculation and verification of emission reductions and will permit to establish the project's compliance with other relevant criteria. The difficulties encountered in the evaluation of emission reductions achieved by the AIJ projects and with the ILUMEX verification and certification exercise regarding the availability of useful data sets on customer behavior, testifies to the fact that this should be a predominant concern for any future investor interested in obtaining certifiable and valid GHG credits.
- Private sector: The objectives of the AIJ Program included improving the participation of the private sector in the AIJ pilot phase. The program was relatively unsuccessful in identifying a partnership with the private sector in its various projects. The necessary incentives for private sector participation, including the potential for crediting, were lacking in the pilot phase. However, the World Bank has experienced considerable interest from the private sector in its plans for its possible future contributions to

the GHG market, in particular through the PFC. This experience suggests that private sector entities are not only interested in obtaining or selling ready-made GHG credits, but with improved incentives, are also increasingly interested in the concrete knowledge generation and learning experience that such advanced mechanisms can provide. Knowledge dissemination to the private sector is also necessary on a wide basis, because the understanding of the key concept of additionality, the political nature of the GHG market and the likely market development appears to be still very immature in many private sector entities and potential investors in GHG reductions.

- The knowledge base: It is, in deed, quite essential that the requirements of transparency are fully met and the lessons and valuable knowledge gained from the AIJ experience are disseminated to all parties interested in AIJ and the Kyoto mechanisms. This should be effected not only through the evaluation and dissemination of past experiences, but also through the continuation of the projects that were begun under the AIJ pilot phase and through new activities that meet the changing needs of the host countries in their various stages of preparedness for and participation in the evolving GHG market. The World Bank, therefore, is continuing its efforts in the AIJ program and also in its other related initiatives. A great amount of attention will be given to contributing to and building upon the knowledge base gained through the AIJ pilot phase as well as to the generation of new knowledge through initiatives like the PCF. Moreover, the World Bank is keen to transfer this know-how to the participants in the evolving GHG market, and, most importantly, to its client developing countries.

The conclusion to be drawn from the above is value should be attributed to the experiences of the AIJ pilot phase. Capacity building in many organizations and individuals support the Kyoto Protocol discussions and provides data and experience for identification of principles, rules, guidelines and modalities for JI and the CDM. It appears that the AIJ pilot phase has fulfilled and continues to fulfill its purpose to provide practical hands-on learning for the FCCC Parties.

REFERENCES

Det Norske Veritas (1999a) Pilot Certification and Verification of the ILUMEX High Efficiency Lighting Project: Concept Paper and Work Program. Report No. 99-3021, The World Bank, Washington, DC.

Det Norske Veritas (1999b) ILUMEX: Verification Report. Report No. 99-3258, The World Bank, Washington, DC.

Det Norske Veritas (1999c) ILUMEX: Draft Model Monitoring and Verification Protocol. Report No. 99-3288, The World Bank, Washington, DC.

Det Norske Veritas (1999d) ILUMEX: Lessons Learned. Report No. 99-3287, The World Bank, Washington, DC.

Haugland, T. (1994) The Cost-Efficiency of Financial Investment Support in Energy Conservation Policy, The World Bank, Washington, DC.

Joskow, P.I., D.B. Marron (1992) What does a Megawatt Really Cost? Evidence from Utility Conservation Programmes. The Energy Journal 13: 4.

Khazzoom, J.D. (1986) An Econometric Model Integrating Conservation Measures in the Estimation of the Residential Demand for Electricity, JAI Press, Greenwich.

Krishna Rao, M.V., C.R. Krishna (1991) Development of Agricultural Demand Side Management Project. IEEE Trans. Power Systems. 6:1466-1472.

Krishna Rao, M.V., R. Peri, K.L. Clinard, C.R. Krishna (1995) Development and Evaluation of a Distribution Automation System for an Indian Utility. IEEE Trans. Power Delivery. 10:452-458.

Telnes, E. (1999) Verification and Certification of GHG Projects: Experience from the ILUMEX Verification Pilot. Background paper for the UNFCCC Technical Workshop on Kyoto Mechanisms, Det Norske Veritas, Oslo.

TERI (1996) Demand-side Management in the Agricultural Sector of Uttar Pradesh: Investment Strategies and Pilot Design, Tata Energy Research Institute, Report 95/EM/52, Delhi.

Quintanilla Martinez, J., R. Selrod (1999) Evaluation of the Mexico ILUMEX AIJ Demonstration Project, The World Bank, Washington, DC.

Selrod R., O. Veiby, E. Obryk (1999) Evaluation of the Poland AIJ Coal to Gas Boiler Conversion Project, The World Bank, Washington, DC.

Chapter 12

LEGAL DIMENSIONS OF AIJ PROJECT DEVELOPMENT ACTIVITIES

K DANISH[1], E BRENES[2], J ROTTER[3]
[1]*Hunton & Williams,* [2]*The World Bank,* [3]*The Nature Conservancy*

Key words: joint implementation, international relations, law contracts

Abstract: Participants in the UN Framework Convention on Climate Change (FCCC) Activities Implemented Jointly (AIJ) pilot phase have made effective use of both time-honored and innovative contractual techniques to create robust legal infrastructures for AIJ transactions. The requirement of host government approval effectively has made those government parties to AIJ transactions, giving them substantial leverage to demand that projects generate a variety of local co-benefits. Now, the Kyoto Protocol holds the promise of establishing the clear rules for joint implementation (JI) and the Clean Development Mechanism (CDM) the pilot has lacked and ensuring that important international interests are protected in individual transactions. However, a number of the proposed rules would be redundant or overly rigid, leading to high transaction costs and unnecessarily constraining the ability of project developers and host governments to develop mutually beneficial arrangements.

1. INTRODUCTION

Before the commencement of the AIJ pilot phase, not only opponents but also many proponents of JI insisted that JI would be untenable in the absence of a comprehensive international regulatory framework supported by a strong multilateral institution (Jepma 1995). Decisions at FCCC First Conference of the Parties (COP-1) produced the AIJ pilot phase but set forth only broad guidelines, and did not establish any kind of international JI institution empowered to review and approve projects, enforce compliance, and settle disputes (Carraro 1999). The AIJ national programs also have been minimalist, establishing few and imprecise guidelines and most were operated without organic authorizing legislation.

The laissez-faire quality of the AIJ pilot regime has had two consequences. On the one hand, participants have faced considerable uncertainty about criteria for approval of projects (e.g., what is necessary to demonstrate emissions additionality). The lack of transparency produced by ambiguous rules has led to high transaction costs (Jepma and van der Gaast 1999). Yet, the absence of an overarching, strictly enumerated regulatory framework has allowed participants to devise a variety of approaches for financing projects and for distributing the project's potential benefits and risks between the parties to the transaction. On a transaction-by-transaction, contract-by-contract basis, participants have developed robust and innovative legal infrastructures that have ensured that projects would achieve their greenhouse gas (GHG) mitigation aims. In addition, the requirement that the host government approve any AIJ project has given host governments considerable bargaining power to demand that they receive a share of any emission credits resulting from projects and/or that projects generate a variety of local co-benefits.

Therefore, the AIJ pilot phase has:

- revealed the diversity of participants and financing approaches possible for JI transactions;
- exposed the legal concerns that can arise in such transactions; and,
- helped clarify which of those concerns should be addressed through an international JI regime and which can be addressed more effectively through arrangements negotiated by the project developers and the participating governments.

International negotiations are underway to elaborate regulatory frameworks for Articles 6 and 12 of the FCCC Kyoto Protocol. As part of these negotiations, the Parties are considering a range of possible criteria and requirements for project eligibility. Given the lessons learned from the AIJ pilot, the first question with respect to each proposed rule should be whether the rule is necessary to protect an important interest of the international community that might not be adequately protected through contractual structures established by the project developers and participating governments. Assuming the rule is warranted, the second question is how the rule can be established to exclude flawed projects without also discouraging meritorious ones.

This chapter will explore the various legal issues implicated by JI transactions and the different contractual arrangements AIJ pilot participants have developed to resolve those issues. The chapter then will survey a number of the criteria and requirements that have been proposed in recent discussions regarding Articles 6 and 12 of the Kyoto Protocol and what effects the proposed requirements might have on the development of a market for JI projects.

2. LEGAL DIMENSIONS OF AIJ PROJECTS

2.1 A TYPOLOGY OF FINANCING ARRANGEMENTS AND PROJECT PARTICIPANTS

A lesson learned from the AIJ pilot phase is that JI transactions can involve a variety of different financing arrangements and project participants. Different transactions implicate different legal concerns and contractual solutions. FCCC COP-1 decisions and guidelines developed in concert with individual national AIJ programs have established, in effect, only three basic constraints on AIJ pilot projects. First, and most fundamentally, project developers have been obliged to demonstrate that their project would bring about environmental benefits related to the mitigation of climate change that would not have occurred in the absence of the project. This requirement is often referred to as the emissions additionality requirement. The second restriction is that, to the extent that a project investor uses public funds to finance the project, it must demonstrate that it did not divert those funds from the investor government's official development assistance or its required contributions to the Global Environment Facility (GEF). This is often referred to as the financial additionality requirement. The third restriction is that that no matter what entities are parties to the underlying transaction, the government of the country in which the project is physically situated, the host government, and the government of the country in which the investor is situated must approve the project.

These three constraints have left a great deal of room for experimentation with different project financing approaches and different kinds of participating entities. As a result, projects have been financed both on a project-by-project basis and according to a portfolio approach. Project developers have included not only governments but also a range of non-governmental entities playing a variety of roles. An AIJ project typology is considered in the following sections.

2.1.1 Project-by-Project Financing

The most common method of financing projects under the AIJ pilot has been on a project-by-project basis. Each project financed according to such an arrangement has involved an investor on one side of the transaction and a host or implementing entity on the other side of the transaction. A lesson learned from the AIJ pilot is that a range of entities can participate on either side of such a transaction.

- Government – Government. In the earlier years of the AIJ pilot, projects typically involved an Annex I government investor and a non-Annex I government host. Examples included transactions involving the government of Norway on the investor side and the host governments of Poland and Costa Rica.
- Non-governmental Entity – Government. A number of AIJ projects have involved a private investor from an Annex I country and a host government agency in a non-Annex I country. Most of these projects have involved environmental non-governmental organizations (NGOs) as investors and government agencies involved in land or energy administration on the host side. Examples of such projects can be found in Bolivia, Belize, and Russia.
- Non-governmental Entity – Non-governmental Entity. As the AIJ pilot has matured, an increasing number of projects involving private investors and private hosts or project implementers have developed. Examples can be found in Bolivia, Belize, and Russia. While the underlying transaction of such projects may involve private actors, the participants still have been required to obtain the approval of their governments.

2.1.2 Portfolio Financing

The second method of financing projects under the AIJ pilot has been the portfolio or bundled project approach. Under this approach, an investor buys a share in a number of projects assembled by a host or intermediary. The advantage of this approach is that it avoids many of the transaction costs involved in developing, evaluating, and marketing individual projects. On the other hand, a potential investor may feel that this approach does not afford it sufficient oversight over the implementation of a project or projects upon which it is relying for credit.

The most prominent example of the portfolio financing approach under the AIJ pilot has been the Costa Rican government's Certifiable Tradable Offset (CTO) program. (Castro and Tattenbach, 1997). Under this program, an investor may purchase a CTO, which represents a specific number of units of GHG emissions reduced or removed. The Costa Rican government has established a verification process to certify that its CTOs represent sufficiently real, measurable, and additional mitigation.

In 1999, the World Bank launched the Prototype Carbon Fund (PCF), an initiative unconnected to the AIJ pilot phase. (World Bank, 1999). The PCF is intended to serve as a pilot program for portfolio financing of JI projects. Under the PCF, the World Bank will serve as program Administrator, acting on behalf of contributors. The World Bank will enter into separate project agreements with host country governments and, where appropriate, with non-governmental project implementers. The Bank will direct PCF funding to the host entities to implement the projects. Each Participant then will receive a share of the carbon offsets generated by all of the PCF financed projects based on the amount of that Participant's contribution to the fund.

As with the project-by-project financing method, transactions financed according the portfolio method can involve a range of participating entities. Under Costa Rica's program, while the host government has played the exclusive role of assembling projects, investors have included governments and also one private business, Centre Financial Products, a commodities broker from the Chicago Board of Trade, Chicago, IL, USA. Under the PCF, the World Bank, an international financial institution, will play the role of an intermediary between hosts and Participant investors. So far, Participants have included several governments, utility companies, and private financial institutions.

One of the lessons learned during the AIJ pilot, therefore, is that a variety of financing methods and project participants is possible for JI-type transactions. The important common denominator among all such arrangements, however, is that the governments on both the investor and host sides have had to approve the underlying transaction.

2.2 LEGAL CONCERNS AND CONTRACTURAL SOLUTIONS FOR INDIVIDUAL TRANSACTIONS

Confronting minimal regulatory constraints, participants in the AIJ pilot have enjoyed substantial freedom to identify and resolve legal issues on a transaction-by-transaction basis. As a result, during the course of the AIJ pilot, project developers have developed a variety of means of establishing stable and effective legal structures to support projects.

Many observers believe that a JI transaction is an exotic in international finance, presenting a variety of unprecedented legal issues. In fact, the AIJ pilot has shown that such transactions have much in common with ordinary direct foreign investment transactions. As with any project-based direct foreign investment transaction, the parties to a JI transaction must determine in advance how they will arrange for feasibility studies, project governance, allocation of risks, currency exchange, resolution of disputes, insurance, and distribution of benefits.

Of course, the AIJ pilot also has demonstrated that JI transactions have some important distinguishing features for which unique contractual solutions have proved necessary. For example, while the primary benefits an ordinary project finance deal are the monetary revenues generated by the project, the primary benefits of the a JI transaction are rights to claim credits for emission reductions or removals. Obtaining the benefits of a JI transaction, moreover, necessitates making a technical determination to a third party empowered to certify the determination. That JI transactions involve an unusual commodity is not their only distinctive trait, however. Participants in the AIJ pilot have learned that other unique elements of the transactions have called for special terms in contracts. For example, unlike ordinary foreign investment transactions, AIJ pilot transactions have had to comply with specific monitoring and verification requirements. In addition, as discussed above, government approval has been mandatory for AIJ pilot transactions.

In its AIJ pilot land-use change and forestry (LUCF) projects in Belize (the Rio Bravo project) (Wisconsin Electric Power Company, et al., 1994) and Bolivia (the Noel Kempff project) (Fundacion Amigos de la Naturaleza, 1996), The Nature Conservancy (TNC) has confronted these various legal issues and developed a variety of contractual solutions to address them. In each project, TNC has served as an intermediary between Annex I investors, who have consisted of private businesses, and non-Annex I project implementers, which have consisted of local NGOs. TNC also has worked with investor and host government officials whose approval was required for project certification. The following sections examine some of the practical lessons that can be drawn from TNC's experience during the AIJ pilot. In developing each of its contracts, TNC has employed a combination of time-honored techniques used in other project finance transactions and some innovative responses to conditions unique to JI transactions.

2.2.1 Different Contracts for Different Project Stages

TNC has found that project development for both the Rio Bravo and Noel Kempff projects has required a series of increasingly detailed contracts leading up to the actual initiation of project activities. Such a phased approach can be found in a variety of project finance transactions including those involving acquisition of sizable areas of land or construction of large facilities. The first stage is the feasibility study stage. During this stage, having identified a site and a broad concept for the project, the participants enter into an agreement to undertake a project feasibility study. Such agreements are usually simple in form. The agreement outlines the basic structure of the project, anticipates roles the participants might play in implementation, and identifies conditions under which one or more of the participants agree to be involved in further stages.

The feasibility study should examine not only whether the project is technically workable but also whether the political and cultural environment is favorable. The study also should develop at least a rough estimate of the mitigation potential of the project using approved methods for baseline development and additionality calculations. But for the calculations of the project's mitigation potential, the feasibility stage is common to most ordinary project finance deals.

The second stage is the project proposal stage. This stage commences upon a determination that the project is feasible. At this point, the project participants' arrange for the preparation of a project proposal that is to be presented to whatever bodies must certify the project. For this stage, a more detailed agreement is ordinarily required. The agreement should outline the roles and responsibilities of the project participants in preparing the proposal. This agreement also can establish the consent of the parties to participate in the implementation of the project in the event that the project is approved. The project proposal agreement thus can anticipate and outline the parameters for the comprehensive agreement that follows project approval.

Once the project has been approved for implementation, the participants enter the comprehensive agreement stage, in which the participants negotiate the more detailed contract that will govern implementation of the project. The comprehensive agreement should establish the roles of the participants and the governance of the project. It summarizes the project activities, fixes the distribution of project benefits, divides any liabilities, and allocates any risks associated with the project. The comprehensive agreement also should describe dispute resolution procedures, financing, and monitoring and verification of the project. This more complex comprehensive agreement forms the basis for much of the discussion in the remaining parts of this section.

2.2.2 Delegation of Powers and Responsibilities

Some JI projects, particularly those involving LUCF activities, have very long lifetimes. TNC's Rio Bravo project contemplates a thirty-year implementation schedule and the Noel Kempff project will unfold over forty years. In establishing the legal infrastructure for these kind of multi-decade AIJ projects, pilot phase participants such as TNC have adopted techniques used to establish long-term contracts in other contexts.

TNC has found that a comprehensive agreement for a long-term project should provide for a management structure that is flexible and responsive to changing conditions. Such circumstances can range from changes in the international regulatory regime, transitions of governments, corporate succession, and alterations in the natural environment of the project site. The transaction costs involved in a negotiating an agreement comprehensive enough to provide for every such circumstance over multiple decades would be far too high. Therefore, TNC's agreements, like many other contracts intended to govern long-term relationships among the parties, establish management boards for each of the projects. With a board structure, it is possible for the parties to draft a less-detailed comprehensive agreement and empower the board to define or modify legal relations over time in response to new developments. The board is made up of representatives of the parties and meets on a regular basis to approve policies and budgets and to review the status of project implementation. The board for the Rio Bravo project consists of the funds manager, the site manager, and a representative of the investors.

In addition to establishing a board, a comprehensive agreement for a long-term project should establish in clear terms the roles and responsibilities of various other individuals and organizations involved in implementing the project. In TNC's projects, the site manager has some of the most important implementation responsibilities. The comprehensive agreement should list all of the tasks the site manager must accomplish to ensure the long-term success of the project. The site manager should be liable for the performance of these tasks. In addition, the agreement should establish, or empower the board to establish, where the site manager's authority ends and the board's authority begins. The site manager should have sufficient powers to implement the day-to-day, on-the-ground work of the project, so long as the manager's authority is subject to the parameters of the comprehensive agreement and the broader authority of the board.

TNC has found that its LUCF AIJ pilot projects require extensive local support, involvement, and knowledge to achieve long-term effectiveness. For this reason, local organizations have been parties to each of its projects during the AIJ pilot. In the case of the Rio Bravo project, the site manager is Fundacion Amigos de la Naturaleza, a conservation organization established under the laws of Bolivia, which has substantial knowledgeable about the local habitat. The site manager for the Noel Kempff project is Program for Belize, an in-country non-profit corporation dedicated to sustainable management of tropical hardwood forests.

In addition, each project provides for promotion of activities in the local community that will substitute for any income-generating activities displaced by the project. The Rio Bravo project, for example, provides for local employment in the form of new park guard positions and community development in the form of training in non-timber forest product utilization and commercialization. Similarly, the Noel Kempff project will finance employment of local individuals in conservation area stewardship. Such project activities not only serve the interests of the host country but also help ensure the projects long-term sustainability by developing local support for the project.

2.2.3 Financing

To sustain any kind of long-term project activity, the comprehensive agreement must include a durable financing mechanism. AIJ projects developed by TNC and others have provided for long-term endowments to ensure enduring financial support. Such endowments have become an increasingly common instrument for managing funds for environmental purposes in developing countries (Danish, 1996). Project developers have a number of options in establishing such mechanisms. Common law countries are likely to favor trusts while civil law countries will be more accustomed to foundations.

In both of TNC's projects, the investors, most of which are based in the USA, essentially have used TNC as the endowment instrument, empowering the TNC to invest their funds and disburse the investment income for project activities. The advantage of this approach is that money given to and disbursed by TNC is tax exempt under USA law because of TNC's status as a non-profit organization.

2.2.4 Risk Management and Division of Liabilities

In establishing an agreement for any kind of transaction, including a JI transaction, the parties must provide for the risk of project failure, either through default by a party or as a result of events not under the control of the parties. In almost all cases, means are available to reduce the risk of project failure resulting from default. Such risk management is a matter of creating incentives for full performance by the party or parties most able to ensure that performance. Accordingly, the investor parties should be liable for defaults in financing the project and the party responsible for on-the-ground implementation, such as the site manager, should be liable for failure to carry out tasks related to implementation. Investors usually will want to spread payments to the implementing party over time, so that the latter is encouraged to continue execution of the project.

Because JI projects are still new territory in many ways, pilot phase investors have found it to their benefit to finance capacity building for local implementers and for host government officials. As described above, investors in the Rio Bravo and Noel Kempff projects determined that money spent on capacity building would help ensure that the project is fully implemented. Indeed, in the case of the Noel Kempff project, the investors not only have funded local capacity building, but also have contributed to the development of a Bolivian AIJ office.

In addition, in the Noel Kempff project, the Bolivian government demanded a fifty- percent share of any credits generated by the project, presumably as a hedge against future obligations and/or to sell on a developing secondary market. Though the investing entities at first were reluctant to accede to this term, they ultimately determined that sharing credits with the government would be a valuable means of contributing to risk management. The Bolivian government now has a stake in the satisfactory performance of the project activities.

Credit seeking parties also have a risk-management incentive to provide for careful monitoring. If credit is available only after certain project-related activities have been completed, the parties to a JI type transaction who seek benefits in the form of credits will want to monitor the performance of the project so they can intervene if implementation is not proceeding as planned. Consequently, TNC's projects have included extensive monitoring provisions.

Some events resulting in incomplete project performance or failure cannot be prevented because they are not under the control of any of the parties to the contract. Such events include amendments to applicable international laws or a natural disaster at the project site. Over time, private insurance for JI transactions is likely to emerge. Until it does, project developers probably should be conservative in their estimates of the mitigation potential of projects. The risk of project failure also provides a rationale for portfolio financing because such risks can be spread among multiple projects.

2.2.5 Dispute Resolution

Clear dispute resolution provisions are a necessity for all types of transactions that envision a long-term relationship between the Parties. When such transactions involve Parties of different nationalities, such provisions require careful negotiation. Fortunately, many legal models exist. In its AIJ pilot projects, TNC has found that dispute resolution terms should encourage the parties to the agreement to first attempt to work out differences without outside intervention. If such efforts are not fruitful, the Parties should be required to use a mediator to attempt to craft a solution.

Finally, if mediation does not prove successful, the comprehensive agreement should provide for binding arbitration.

In crafting arbitration terms for the comprehensive agreement, the Parties must choose the law that will govern the arbitration, the procedures under which arbitration will take place, and the organization that will manage the arbitration. Parties to a JI transaction have numerous sets of arbitration rules and various organizations from which to choose. They should consult with local counsel in both the investor and host countries to determine which rules and organizations are recognized under the applicable national laws. Several regional and internationally recognized rules and institutions are available. In the Western Hemisphere, options include the Commercial Arbitration and Mediation Center for the Americas and the Inter-American Commercial Arbitration Commission. In addition, the UN Committee on International Trade Law has developed a set of arbitration rules that are widely recognized. The International Centre for Settlement of Investment Disputes and the International Chamber of Commerce also administer rules for dispute resolution.

A comprehensive agreement must provide for the possibility of litigation if these other dispute resolution options are unsuccessful. The agreement should establish what national law governs the agreement and in what forum disputes may be litigated. While investors may prefer that their national law and courts have jurisdiction over litigation, it may make sense in some instances to locate the forum in the host country in order to facilitate (1) access to witnesses and documents and (2) enforceability of any decisions. TNC's projects have located the forum in the investor's country in one case (Rio Bravo) and in the host's country in the other (Noel Kempff). The Noel Kempff agreement provides, however, that if both parties to the dispute are from the U.S., that country's law and courts will govern the litigation.

2.2.6 Preventing Leakage

A concern unique to JI transactions is that of GHG leakage. Leakage occurs when the project displaces GHG emissions-generating activities outside of the project boundaries. Leakage is a danger that is common to all types of JI projects and it is therefore likely that any future international regime will impose requirements on project developers to prevent leakage.

TNC has found that one method of leakage prevention is to effectively increase the boundaries of the project by means of the comprehensive agreement. The Bolivian project in which TNC is participating aims to prevent deforestation of a particular site through the purchase of a logging concession. In developing the project, TNC determined that there was a danger that the project would be undermined if the logging company from

which the concession was purchased simply used its payment to deforest another site. Accordingly, as part of the comprehensive agreement for the project, the other parties required the logging company to:
- use its payment only to pay certain taxes; and,
- notify the other parties of any future logging plans so that the Parties have an opportunity to intervene.

2.2.7 Contracting with Governments

Another issue that is unique to JI transactions is the requirement of government approval for the entire project. It is not unusual for a direct foreign investment transaction, such as a privately financed power plant, to require a variety of government-issued licenses or permits for certain aspects of its construction and operation. Under the rules of both the AIJ pilot and the Kyoto Protocol (Article 6.1a) and (Article. 12.5a), however, the investor and host governments must approve a JI project in its entirety.

As a result, even if the governments are not actual Parties to the contractual negotiations for the project, they are effectively Parties and hold substantial leverage over the development of terms and conditions. The host government, therefore, is in a position to demand not only that it receives a share of any credits resulting from the project, as the Bolivian government has done in TNC's Rio Bravo project, but also to demand that the project produces any of a variety of co-benefits. The AIJ pilot has demonstrated that JI projects can generate such co-benefits as technology transfer, capacity-building, improvement of local environmental quality, and poverty alleviation (World Bank, 1997). Because host governments have been empowered to terminate AIJ projects by walking away from the table, and investors have lacked any legal recourse for such termination, host governments have had significant power to require that project activities result in various desired benefits and avoid undesirable harms.

2.2.8 Conclusion

TNC's experience during the AIJ pilot, in addition to providing a series of specific practical lessons, demonstrates that, with clear rules, project developers have the ability to address a variety of legal issues in JI transactions using both traditional and innovative contracting methods. These contracts can protect and promote the various interests of the parties, the governments, and the international community.

3. THE EVOLVING ARTICLE 6 AND ARTICLE 12 REGIMES: PROPOSED REQUIREMENTS AND CRITERIA

Negotiations are currently underway to elaborate principles, modalities, rules, and guidelines for Kyoto Protocol Articles 6 and 12. These negotiations provide an opportunity to develop for each mechanism the clear regulatory framework that the AIJ pilot phase has lacked. Such a framework is necessary for two reasons:

- ensure that important interests of the international community are promoted and protected in Articles 6 and 12 transactions; and,
- encourage Articles 6 and 12 transactions by providing project participants certainty about the governing rules.

The challenge is to avoid developing redundant or poorly crafted rules that unnecessarily constrain the ability of project developers and host governments to arrive at mutually beneficial arrangements with the result that fewer worthy projects are implemented. The consequences of baselessly restricting the market for JI and CDM projects could have severe environmental consequences. Cost-effective GHG mitigation opportunities often have a short window. Failure to exploit those opportunities when presented may mean they are foreclosed later or available only at greatly increased cost. Non-Annex I countries, in particular, are in the process of making key economic development decisions with respect to electricity power generation and LUCF. Such decisions may lead to difficult-to-modify or even irreversible GHG emission paths.

The sections below survey a number of constraints, criteria and requirements proposed for the Articles 6 and 12 flexibility mechanisms. We have excluded discussion of proposed quotas on the use of JI and CDM and examine issues related to emissions additionality only briefly. With respect to each proposed rule discussed below, we examine its basis in the Kyoto Protocol and describe the policy concerns cited by its proponents. We then evaluate, based on the lessons learned during the AIJ pilot, and whether the rule is necessary to protect or promote an international interest not sufficiently protected or promoted in project-by-project contracts. To the extent that a rule is warranted, we further evaluate whether the proposed rule is the best means to achieve the desired end. In a number of cases, we find that another approach would accomplish the same goal without placing redundant constraints on the ability of project developers and governments to construct transactions that serve their interests.

3.1 EMISSIONS ADDITIONALITY REQUIREMENT

As discussed above, a fundamental criterion for project eligibility under the AIJ pilot and all national AIJ programs has been the emissions additionality demonstration. While both Articles 6 and 12 make explicit mention of emissions additionality as a requirement for obtaining credits, there is a question whether that requirement under Article 6 is binding or merely hortatory.

Article 6 provides that a project must provide a reduction in emissions by sources, or an enhancement of removals by sinks, that is additional to any that would otherwise occur. (Article. 6.1b). Presumably, if this was a binding criterion, the Kyoto Protocol would:

- provide for further rulemaking by the Parties to elaborate on the additionality requirement; and,
- as in the case of Article 12, establish that a third party designated by the COP or Meeting of the Parties (MOP) would be responsible for certifying additionality.

However, Article 6 provides only that the COP/MOP may further elaborate guidelines for implementation of Article 6 and Article 6 does not provide for operational entities or any other third-party bodies to certify the additionality of JI projects. (Art. 6.2).

This permissive, rather than mandatory, language in Article 6 regarding environmental additionality is sensible in light of Articles 3.10 and 3.11. According to those provisions, a Party may obtain emission reduction units (ERUs) and add them to its assigned amount only if they are subtracted from the host Party's assigned amount. (EDF 1998). This structure provides automatic, or built-in, environmental additionality:

> As in the case of trades of parts of Parties' assigned amounts under Article 17 and these same paragraphs of Article 3, this mandatory double-entry bookkeeping ensures the additionality of the transferred emission reduction units – guaranteeing . . . that trading under Article 6 can occur only in a zero-sum context. Because their transfer will result in a reduction of the transferring Party's assigned amount, and because that Party will be in compliance only if it achieves emissions reductions in addition to those transferred, the transferred reduction units will necessarily be surplus or additional with respect to the Party's assigned amount for each commitment period (EDF, 1998).

Thus, it makes sense to read the Kyoto Protocol as allowing each Annex I Party the discretion to develop its own guidelines for deciding whether to transfer part of its assigned amount, and how much it should transfer, on the basis of a JI project. It is very possible that individual Annex I countries will develop quite rigorous additionality tests. In any event, the requirement that Parties participating in Article 6 transactions be in compliance with national annual emissions inventory requirements provides extra assurances that the Article 6 mechanism will not be abused (Jepma and van der Gaast, 1999).

Unlike their Annex I counterparts, non-Annex I countries are not subject to assigned amounts, which makes it logical that the emissions additionality demonstration under Article 12 is a binding requirement. Because there is no guaranteed zero-sum context under Article 12, that provision mandates that project proponents demonstrate that their project will achieve emission reductions or removals that would not otherwise occur. Moreover, Article 12 provides that certified emission reductions (CERs) cannot be obtained without third-Party certification of this additionality demonstration. The CDM additionality requirement is an example of an area in which international interests are not adequately protected by deferring to the terms that project participants have bargained for in their contracts. The reason is that both parties to a CDM transaction have an incentive to inflate the baseline emissions.

Neither Articles 6 nor 12 offers clear guidance on methodologies for use in setting baselines, a necessary step in establishing emissions additionality. As noted above, a lesson learned from the AIJ pilot is that project developers need certainty about an acceptable baseline methodology or methodologies (OECD, 1999). Also, the availability of a simplified baseline methodology option or options as an alternative to setting baselines on a project-by-project basis could decrease transaction costs substantially (Hargrave, et al., 1998). These issues are the focus of ongoing international negotiations. The different alternatives for baselines are examined in greater detail in other chapters of this book. These options include a technology matrix approach, a benchmarking approach, a project-based approach, and a top-down or national baseline approach. Each alternative offers different advantages and disadvantages with respect to environmental integrity and ease of use.

Another approach to the additionality determination that is under discussion is a project profitability approach. Under this approach, a project that is found to be profitable to the investor or the host, either on the basis of a positive internal rate of return or some other metric, is presumed not to be additional. The rationale for this approach is that profitable mitigation projects should occur without the availability of credit and thus should not form the basis for a GHG offset (Jepma and van der Gaast 1999). Such a test is fundamentally flawed. First, that a proposed project has a positive internal rate of return does not mean it would have occurred without the availability of credit. Absent credit, the investors might have developed a non-mitigation project promising an even greater rate of economic return. Accordingly, accurate application of such a profitability test would require an international institution to determine whether the investing entity could have engaged in any more profitable projects anywhere in the world. Even if such a determination was feasible, it would require the institution to be empowered to probe deeply into the operations of the investor. It is unlikely many prospective investors would submit to such interference.

Another flaw with the profitability test for additionality is that it ignores host reasons why a profitable project might not occur without credit. In the last decade, analysts have identified numerous political, economic, and technical barriers to seemingly no-regrets mitigation projects (Trexler & Kosloff, 1998). For any number of reasons, many countries fail to undertake seemingly economically sound mitigation activities. In numerous cases, a profitable project will represent a change in the host country's business-as-usual emissions path, thereby meeting the fundamental emissions additionality test.

Accordingly, imposing the profitability test on top of the environmental additionality test adds no value and could discourage many otherwise meritorious projects. As the Environmental Defense Fund (1998) observes, by inhibiting efforts to harvest cost-effective and innovative emissions reductions in non-Annex I Parties, making such a test mandatory could be environmentally counterproductive. On the other hand, it might make sense to give project developers the option of meeting the additionality test by demonstrating that the project has zero or negative profitability. Project developers unwilling or unable to make such a demonstration then could rely on another approved mean of establishing additionality as an alternative, e.g., through a demonstration using one of the baseline methodologies.

3.2 FINANCIAL ADDITIONALITY REQUIREMENT

Both the AIJ pilot and national AIJ programs require project developers to demonstrate that the funds used to support a project have been not been diverted from required contributions to the Global Environment Facility (GEF) or from official development assistance. Neither Article 6 nor 12 explicitly includes this financial additionality requirement. The European Union (EU) and many developing countries support such a requirement under the CDM. The EU has proposed that if public funds are used, the project participants [must] provide information on the funding of the project activity proving that CDM investment will not result in a diversion of or competition with official development assistance [ODA] and GEF funding

The EU's formulation could place a potentially impossible burden of proof on the project developers. Trexler & Kosloff (1998) observe that ODA flows are notoriously difficult to monitor and compare. Moreover, it is not clear how developed country Parties will be able to establish, in the context of the declining overall flows of ODA, that investments are additional to resources that would or should have been committed to the GEF or to other sources of ODA. Given these concerns, the Parties should consider as an alternative a requirement that project developers using investor government funding provide a certification from the investor government that the funding has not been diverted from ODA or GEF support. Such a rule at least would place the burden of making the demonstration on the entity that controls the needed information.

3.3 ELIGIBILITY OF LUCF PROJECTS UNDER THE CDM

Some observers interpret the Kyoto Protocol such that LUCF mitigation projects are ineligible for credit under the CDM. They reason that the drafters of the Protocol must have intended such an exclusion because, while Article 6 explicitly provides for crediting both projects that result in a reduction in emissions by sources and projects that enhance removals by sinks, Article 12 refers only to projects that result in emission reductions (ER). Under this interpretation, Article 12 would exclude a broad range of LUCF-related projects; such projects have made up the majority of projects in the AIJ pilot. (Trexler and Kosloff, 1998).

This exclusionary interpretation is fundamentally flawed. First, it is evident that Article 12 does not contain any explicit preclusion of LUCF projects. Reading into Article 12 an implicit bar against LUCF projects would be inconsistent with the rest of Protocol, the preamble of which states that the Parties are to be guided by Article 3 of the FCCC. That article provides that measures . . . to anticipate, prevent or minimize the causes of climate change and mitigate its adverse effects . . . should . . . cover all relevant sources, sinks, and reservoirs of greenhouse gases. Given the presumption in favor of a comprehensive approach, it seems likely that, if the drafters intended to make LUCF projects ineligible under the CDM, they would have stated so explicitly.

In addition, other provisions in the Kyoto Protocol indicate that references to ER projects in Article 12 should be read as including both projects that reduce emissions from sources and projects that enhance removals by sinks. Throughout the Protocol, the drafters employ ER to describe both emission reduction and sink enhancement activities. For example, the Protocol calls the credits created through Article 6 projects emission reduction units (ERU) even though Article 6 explicitly states that sink enhancement projects are eligible for credits. In addition, the Protocol calls the Article 3 emission caps quantified emission limitation and reduction commitments, yet Article 3.3 explicitly establishes that, in determining their compliance with those commitments, Annex I Parties must account for changes in emissions resulting not only from ER activities but also from afforestation, reforestation, and deforestation. For these reasons, the argument in favor of an implicit bar against LUCF projects lacks merit.

Nevertheless, to the extent that Article 12 could be read to make CERs available for qualifying ER projects but unavailable for projects that enhance anthropogenic removals by sinks, this preclusion would apply only to a subset of LUCF projects. The LUCF category encompasses a number of different kinds of projects, including;

- deforestation reduction or prevention projects;
- biomass fuel-substitution projects; and,
- sink enhancement projects.

Under the exclusionary interpretation of Article 12, sink enhancement projects might be ineligible to receive CERs, but the other two types of projects would remain eligible. When forests are cut or cleared, they become a source of emissions. Thus, projects that reduce or prevent deforestation are ER projects. Likewise, biomass fuel substitution projects, by displacing more C intensive fuels, reduce emissions from sources. Accordingly, various LUCF projects still could go forward under an interpretation of Article 12 limiting eligibility to projects that reduce emissions from sources.

Other observers insist that, even if Article 12 does not exclude all LUCF projects, one must read into the Protocol a link between Article 12 and Articles 3.3 and 3.4. According to this interpretation, the scope of eligible LUCF projects under Article 12 is limited to the scope of activities recognized by Article 3.3, i.e., those involving afforestation, reforestation, and deforestation. This interpretation also is flawed. By its terms, Article 3.3 applies only to LUCF activities undertaken by Annex I Parties within their own countries, not to activities undertaken in non-Annex I countries. Article 3.3 establishes that the emissions impacts of LUCF activities should be measured with reference to changes in C stocks. Under the Protocol, only Annex I countries have obligations to inventory their C stocks. As a result, not only is there no explicit link between Article 12 and Article 3.3, it would be impossible to apply the formula established by Article 3.3 to LUCF projects under the CDM because such projects are sited in non-Annex I countries.

If an interpretation-linking Article 12 to Article 3.3 nevertheless prevails, how would this affect the scope of eligible CDM LUCF projects? First, this interpretation would limit projects in the category of sink enhancement to afforestation and reforestation (or other activities added under Article 3.4) however those terms are ultimately defined. With respect to emissions reduction projects, a range of such projects still fits squarely within the legal formula established by Article 3.3. Article 3.3 counts net changes in GHG emissions from sources and removals by sinks resulting from direct human-induced LUCF activities, limited to afforestation, reforestation, and deforestation. This language encompasses biomass fuel-switching projects, which reduce "emissions from sources" that were burning higher-GHG fuels. The language also encompasses deforestation reduction or prevention projects, which reduce emissions from sources [i.e., forests] resulting from human-induced LUCF activities . . . [such as] deforestation.

Advocates for these exclusionary interpretations of the Protocol cite a number of policy concerns about LUCF projects. They assert that LUCF projects present particularly difficult measurement problems and also argue that poorly designed LUCF projects will lead to environmental and socioeconomic harms. Forestry experts counter that both claims are overstated. Recent research shows that forestry projects do not present analytical issues that are significantly more difficult than those for other types of projects. Moreover, environmentally and ecologically appropriate projects in developing countries would yield sufficiently substantial ERs and removals that there is little reason to believe that the CDM would be flooded with inappropriate projects. (Trexler and Kosloff, 1998).

In any event, the constraints on project eligibility already established under the Protocol (that projects must generate real, measurable, and long-term mitigation benefits and that project must generate ERs additional to that would otherwise occur) are adequate to exclude flawed projects. A bar on LUCF projects, on the other hand, would needlessly exclude many high-quality projects that not only can contribute to GHG mitigation (deforestation resulted in 25% of global C emissions annually during the 1980s [Houghton, 1996]), but also lead to such co-benefits as rural poverty alleviation and biodiversity protection. In addition, such a bar likely would mean that only those non-Annex I countries with substantial industrial bases could host CDM projects. (International Institute for Sustainable Development, 1999). Accordingly, excluding all or most LUCF projects would have a dramatic impact on the reach of the Kyoto Protocol regime.

3.4 SUSTAINABLE DEVELOPMENT REQUIREMENTS UNDER THE CDM

A number of AIJ national programs include variously worded requirements suggesting that pilot phase projects should contribute to sustainable development in the host country. Generally, it has been unclear for project participants what they needed to do to demonstrate compliance with such requirements. Article 6 is silent on the subject of sustainable development. However, Article 12 states that the purpose of CDM shall be to assist FCCC Parties not included in Annex I in achieving sustainable development and in contributing to the ultimate objective of the FCCC. Article 12 also states CMD will assist Parties included in Annex I in achieving compliance with their quantified emission limitation and reduction commitments under Article 3. Given this language, contributing to sustainable development is ostensibly the CDM's primary purpose. The question is how to elaborate a clear guideline or guidelines to implement this provision.

The first issue is the degree to which international oversight is necessary to implement a sustainability provision. Host government approval is necessary for any CDM project. Presumably, a host government will not grant its approval to a project that would result in substantial harm to the country. Nonetheless, many observers have concluded that a sustainable development requirement cannot be entrusted to the sole discretion of the host government. Various reasons are cited for this conclusion, including:

- many non-Annex I governments lack the capacity to perform adequate environmental and socioeconomic impact assessments; and,
- even to the extent that non-Annex I governments have sufficient capacity for impact assessment, a CDM market will lead to a race to the bottom in which non-Annex I governments are forced to compete by lowering their protections against harmful projects.

Both claims are questionable as a basis for sweeping requirements for detailed impact assessments. That many non-Annex I governments do not have sufficient resources or expertise for impact assessments is undoubtedly the case. Nonetheless, the host government generally knows more about its country than other project participants. Before using a rule either to shift the full burden of an impact assessment onto an entity from another country, such as the investor, or require that all projects provide certain specific sustainable development benefits and avoid certain specific harms, the Parties should consider whether a better solution is to take steps to increase the assessment capacity of non-Annex I governments so that they are better positioned to make these demands themselves.

The second claim, that there will be a race to the bottom among non-Annex I countries, is subject to question. It presupposes that, in a future CDM marketplace, the supply of potential project opportunities will exceed demand, i.e., that prospective host countries will be vying for a limited number of prospective investors. Yet, there are reasons to believe it could be the quite opposite. Given the Protocol's stringent ER and limitation commitments, the demand for CDM project opportunities could be very substantial. In that case, host governments would hold substantial bargaining power and could afford to be selective. Yet, even if it turned out that investors had somewhat greater market power, it is not necessarily the case that the stringency of sustainable development requirements would be a decisive or even a substantial factor in investors' decisions among project opportunities in different non-Annex I countries. If investors are assured of a large number of credits, they might be willing to meet stringent requirements.

In any event, it still is sensible to impose on a CDM project's proponents an obligation to perform some form of project impact assessment and to establish guidelines for what such an assessment should cover. A number of internationally recognized models for environmental and socioeconomic impact assessments are available. Much would depend on the scope of the required assessment, however. A requirement that project developers undertake sweeping and detailed environmental and socioeconomic assessments, while appropriate perhaps for projects undertaken by multilateral development institutions, likely would result in the exclusion of a range of otherwise worthy privately funded projects.

The next issue then is what demonstration the project developers should be obliged to make to satisfy the requirement that host non-Annex I country will benefit from [the] project activity. The choice is between a flexible test and a specific test. The latter would require projects to generate certain mandatory co-benefits. Here, the justification for such international intervention is much weaker. Projects can produce a variety of co-benefits and there is no international consensus on the relative value of one kind versus another. Absent such a consensus, there is little rationale for overriding the ability of individual non-Annex I countries to bargain for the co-benefits that they value most. Moreover, requiring specific co-benefits could have the effect of excluding from eligibility certain types of projects. For example, a requirement that projects must result in technology transfer could preclude practically all LUCF projects. The best approach is to establish a requirement that places the onus on project developers to demonstrate that the project would generate benefits, but to allow project developers and host government's leeway to negotiate on a transaction-by-transaction basis what those benefits will be.

3.5 CDM CREDIT SHARING

According to another proposed CDM project requirement, Annex I project participants would be obliged to share a pre-set percentage of CERs generated by a project with the non-Annex I host government. Two different justifications are offered for such a credit-sharing requirement. The first is a desire that non-Annex I countries start accumulating credits to hedge against the future cost of binding commitments. According to this rationale, credit sharing is necessary so that non-Annex I countries have an obligatory bank account of CERs. (Hanafi 1998).

The other justification cited for a credit sharing requirement is that non-Annex I governments seeking credits will be undercut by Annex I participants who have superior bargaining capacities. Consistent with this concern, some advocates of credit sharing also propose a requirement of surplus sharing. According to this proposal an Annex I investor would be required to share with the non-Annex I host a percentage of the cost differential between the project's $MgC cost and the prevailing marginal mitigation cost in the Annex I investor's country (Hamwey 1998).

This credit-sharing requirement has a number of the same drawbacks as the proposed requirement that projects generate certain specific co-benefits. As discussed above, such requirements constrain the ability of project developers and host governments to arrive at mutually agreeable bargains. For example, a non-Annex I host government could calculate that it would demand a smaller number of CERs or no CERs at all in return for a project that upgraded pollution control equipment at a particular facility and thereby improved local air quality. Imposing on project developers and host governments a particular formula for distribution of benefits (and risks) will discourage many potential investors and hosts from undertaking projects. In particular, requiring not only credit sharing but also surplus sharing would almost certainly reduce the number and scope of CDM projects dramatically. In any event, as discussed above, many investors will find, as TNC has found, that it is in their best interest to insist that the host government take credits as part of their payment for a CDM transaction because it gives the host an stake in ensuring the long-term performance of the project. Here, as with other proposed project requirements, a flexible approach makes most sense.

3.6 TRANSPARENCY AND PUBLIC PARTICIPATION REQUIREMENTS

A number of observers have proposed that project developers be subject to certain obligations with respect to transparency and public participation. Neither Article 6 nor 12 contain explicit requirements with respect to public participation, although Article 12's reference to sustainable development as a purpose of the CDM has been cited as a basis for such requirements. Such requirements are discussed most frequently in the context of CDM projects. Proposals for extensive requirements appear to be motivated by two concerns:

- desire to provide various interested parties, particularly environmental NGOs, both access to project information and rights of appeal so that they can be in a position to enforce compliance with various project requirements (Hare & Stevens 1995); and,

- claim that host governments will not adequately protect the interests of local communities and, in particular, local indigenous communities. (Goldberg, et al. 1998).

Advocates for extensive transparency and public participation requirements argue that the substantial interests of the international community in many of the details of JI and CDM projects warrant such requirements. Reflecting on the AIJ pilot, the Environmental Law Institute (1996) has observed:

> Because JI must serve a multiplicity of objectives – including reductions in greenhouse gas emissions, compatibility with national environmental and development objectives, provision of technology and assistance, and additionality of financing – it is important that decision-making in the JI process include ample and sufficient involvement of those most likely to be affected by the practical consequences of implementation.

Accordingly, the USA Environmental Law Institute also proposes sweeping transparency and public participation requirements including disclosure and access to engineering and management documents and performance reports, employment of members of the local community, use of public or third-party auditors and complaint and dispute resolution forums. Goldberg, et al. (1998), as part of a review of potential LUCF projects under the CDM, have proposed a set of obligations that would go farther. Under their proposal, all potentially affected constituencies, which include local peoples, NGOs, businesses, and governments, would have a right to object to projects and the CDM would be required to give substantial weight to their objection in deciding whether to affirm the project.

Proposals for such sweeping requirements raise a number of issues. For example, obligations to make available cost and design data may have been warranted during the Pilot Phase, which was intended as an educational process for policy-makers, but would seem less appropriate in an active and fully-regulated JI and CDM market. (Trexler and Kosloff 1998). To the extent that any such requirements are imposed, they make more sense for projects developed by governments or international financial institutions than for those that are privately funded. Where private organizations are involved, by contrast, competing public concerns are at stake: on the one hand, an interest in verifying compliance with various criteria, and, on the other hand, an interest in protecting confidential commercial information and intellectual property rights. If the latter interests are substantially jeopardized by transparency and reporting requirements, it is likely that private organizations will avoid the Article 6 and 12 markets.

It may also make sense to vary the extent of public participation requirements with the scope and type of project. While some form of local notice and participation requirements may be appropriate for some of the large-scale LUCF projects, less extensive obligations likely would be sufficient for a range of other types of projects such as those involving fuel-switching or technology upgrades. Even with respect to LUCF projects, reasonable requirements for environmental and socioeconomic impact assessments may provide sufficient protection for local communities.

3.7 COMPLIANCE WITH REPORTING PROVISIONS AS AN ELIGIBILITY CRITERION

While Article 6 requires compliance with the annual reporting provisions of the Kyoto Protocol as a prerequisite to obtaining ERUs, Article 12 does not contain such a requirement. Some observers have suggested that a similar compliance prerequisite apply to Article 12 activities (Werksman 1998). Yet, the reasons why such a requirement is necessary under Article 6 do not apply under Article 12. Under Article 6, as discussed above, the environmental additionality test is not mandatory; a project earns ERUs if the host Annex I country agrees to transfer some of its Assigned Amount Units on the basis of the project. In the Article 6 context, a problem arises if the host is on track to substantially exceed its Assigned Amount. In that event, no reduction in global emissions might result from the project. Therefore, it makes sense to require the host to be in compliance with its obligation to submit annual emission inventory reports; it is the primary means of ensuring the host government is not giving away that which it does not have to give. The same concerns do not apply in the CDM context, since CERs are not generated unless the project results in real ERs. Article 12, with its requirement of third party certification, provides for built-in international oversight for each project.

3.8 ELIGIBILITY OF PROJECTS UNDER THE CDM EARLY START PROVISION

Article 12.10 provides that CERs obtained from the year 2000 up to the beginning of the first commitment period can be used for compliance with emission reduction commitments. The scope of projects eligible under this provision is subject to different interpretations. The question is whether only those CDM projects initiated after the trigger date could generate CERs or whether projects initiated before the trigger date could generate CERs for all of the emission reductions they obtain after the trigger date. Under the latter interpretation, projects initiated under the AIJ Pilot Phase that qualify under guidelines established for the CDM could generate CERs with respect to emission reductions they achieve after the trigger date.

Some observers have asserted that the crediting of projects undertaken in the AIJ pilot is forbidden under the COP guidelines. Decision 5 of COP-1 states that no credits shall accrue . . . during the Pilot Phase from activities implemented jointly. While this language does prohibits crediting during the pilot phase, it does not preclude credits for AIJ projects after the Pilot Phase ends (Gosseries, 1999). Accordingly, AIJ projects that meet the CDM standards could be eligible for CERs.

4. CONCLUSION

The AIJ pilot has provided many practical lessons about the variety of contractual approaches available for financing, risk management, allocation of responsibilities, and dispute resolution. Moreover, the AIJ pilot has demonstrated the broader lesson that, even with a minimal regulatory framework, project developers and participating governments are capable of assembling robust and effective legal infrastructures for JI projects. Given the basic requirement that projects produce additional and long-term mitigation benefits, project developers have had strong incentives to provide for effective project management, extensive monitoring, and local benefits. In addition, the requirement that each project have full government approval has given host governments substantial bargaining power, allowing them to demand that projects generate a variety of co-benefits and avoid environmental and socioeconomic harms.

It will be important to apply these lessons of the AIJ pilot in elaborating further rules to govern the Kyoto Protocol's JI and CDM. While certain basic requirements are appropriate in order to promote certain interests of the international community that cannot be entrusted to the discretion of the parties to an individual transaction, some of the proposed requirements would impose unfounded or unnecessary constraints on the ability of project developers and host governments to construct mutually beneficial transactions. Clear and flexible rules, on the other hand, will promote the development of JI and CDM projects that achieve a variety of benefits for all parties involved.

ACKNOWLEDGEMENTS

Annie Petsonk, Jeffrey Dunoff, Donald Goldberg, Pamela Wexler and Laura Kosloff provided comments and suggestions that improved the manuscript.

REFERENCES

Blaustein, R.J. (1996). Joint Implementation Essentials for Lawyers, Environmental Law Reporter 26: 10364.

Brenes, E., Chomitz, K., and Constantino, L. (1998) Financing Environmental Services: The Costa Rican Experience, Economic Notes, Central America Country Management Unit, The World Bank, Washington, DC.

Carraro, C. (Ed.) (1999) International Environmental Agreements on Climate Change, Kluwer Academic Publishers, Dordrecht, in press.

Castro, R. and Tattenbach, F. (1997) The Costa Rican Experience with Market Instruments to Mitigate Climate Change and Conserve Biodiversity, Presentation at the

Global Conference on Knowledge for Development in the Information Age, Toronto.

Danish, K. (1996) National Environmental Funds, in *Greening International Institutions*, J. Werksman, ed., Washington, DC.

Ellis, J. (1999) Experience with Emission Baselines Under the AIJ Pilot Phase, OECD, Paris.

Environmental Defense Fund (1998) Cooperative Mechanisms Under the Kyoto Protocol: The Path Forward, Washington, DC.

Environmental Law Institute (1996) Incorporating Public Participation in Joint Implementation of the Framework Convention on Climate Change, Washington, D.C.

Fundacion Amigos de la Naturaleza, The Nature Conservancy, and American Electric Power System (1996) Noel *Kempff Mercado Climate Action Project UISJI Pilot Project Proposal,* US IJI, Washington, DC.

Goldberg, D.M., et al. (1998) Carbon Conservation: Climate Change, Forests, and the Clean Development Mechanism, Center for International Environmental Law (CIEL) and Centro de Derecho Ambiental y de los Recursos Naturales (CEDARENA).

Gosseries, A.P. (1999) The Legal Architecture of Joint Implementation: What Do We Learn From the Pilot Phase? New York University Environmental Law Journal 7:49-118.

Hamwey, R. (1998) Can Attractive Models for Joint Implementation and the Clean Development Mechanism be Developed for Buenos Aires? Report of a Policy Dialogue, International Academy of the Environment, Geneva.

Houghton, R. (1996) Converting Terrestrial Ecosystems from Sources to Sinks of Carbon. Ambio 25: 267-272.

Hanafi, A.G. (1998) Joint Implementation: Legal and Institutional Issues for an Effective International Program to Combat Climate Change. *Harvard Environmental Law Review* 22: 441-508.

International Institute for Sustainable Development. (1999) Summary Report of the Alliance of Small Island States Workshop on the Clean Development Mechanism of the Kyoto Protocol, Boston.

Jepma, C. and van der Gaast (1999) On *the Compatibility of Flexible Instruments,* Kluwer Academic Publisher, Dosdrecht, in press.

Jepma, C.J. (Ed.) (1995) *The Feasibility of Joint Implementation*, Kluwer Academic Publishers, Dordrecht, 386p.

Kuik, O., (Ed.) (1994) Joint Implementation to Curb climate Change: Legal and Economic Aspects, Kluwer Academic Publisher, Dosdrecht, 212p.

Llobet, G. (1998) Trust But Verify: Verification in the Joint Implementation Regime, *George Washington Journal of International Law & Economics* 31: 233-269.

Trexler, M.C. and Kosloff, L.H. (1998) The 1997 Kyoto Protocol: What Does It Mean for Project-Based Climate Change Mitigation? *Mitigation and Adaptation Strategies for Global Change* 3: 1-58.

United Nations (1999) Mechanisms Pursuant to Articles 6, 12, and 17 of the Kyoto Protocol: Synthesis of Proposals by Parties on Principles, Modalities, Rules, and Guidelines. FCCC/SB/1999/INF.2/Add.1, Bonn.

Werksman, J. (1998) The Clean Development Mechanism: Unwrapping the Kyoto Surprise. *Review of European Community and International Environmental Law* 7: 147.

Wisconsin Electric Power Company, The Nature Conservancy, and Program for Belize, (1994). *Rio Bravo Conservation and Management Area USIJI Pilot Project Proposal*, USIJI, Washington, DC.

World Bank (1999) Prototype Carbon Fund, Global Environment Division, Global Climate Change Unit, The World Bank, Washington, DC.

World Bank, Government of Burkina Faso, and Government of Norway (1997) Report of the African Regional Workshop on Activities Implemented Jointly Under the U.N. Framework Convention on Climate Change, Ouagadougou.

Chapter 13

MONITORING, EVALUATION, REPORTING, VERIFICATION, AND CERTIFICATION OF CLIMATE CHANGE MITIGATION PROJECTS

EDWARD VINE, JAYANT SATHAYE, AND WILLY MAKUNDI, JED JONES
Lawrence Berkeley National Laboratory, Lloyd's Register

Key words: monitoring, evaluation, reporting, verification, certification, guidelines, methods, institutions, Clean Development Mechanism, joint implementation, emissions trading, Kyoto Protocol, Activities Implemented Jointly, UNFCCC, additionality, baseline, project leakage, free riders, market transformation, environmental impacts, socioeconomic impacts

Abstract: Monitoring and evaluation of climate change mitigation projects is needed to accurately determine the net greenhouse gas (GHG), and other, benefits and costs, and to ensure that the global climate is protected and that country obligations are met. One of the objectives of this chapter is to examine the Activities Implemented Jointly (AIJ) Pilot experience in monitoring and evaluation, in order to assist the development of modalities for monitoring, evaluation, reporting, verification, and certification (MERVC) for Clean Development Mechanisms (CDM) and joint implementation (JI) projects. Another objective of this chapter is to provide guidance on how one should go about monitoring, evaluating, reporting, verifying, and certifying climate change mitigation projects based on a review of the literature and our previous work on developing MERVC guidelines for energy-efficiency and land-use change and forestry (LUCF) projects.

The five-year AIJ Pilot is still in its early stages (most projects were started in 1997-98) and offers little insight into the experience of monitoring and evaluation so far. Evaluation of some of these projects is starting to improve our understanding of key MERVC issues, and we anticipate more evaluations of these projects in the near future. Nevertheless, we agree with the findings of the second synthesis study on the AIJ Pilot that concluded that work needed to be conducted on methodological, technical and institutional issues, including modalities for measurement, reporting and assessment. We also agree with the findings from a recent OECD study on emission baselines for AIJ projects which concluded that simple reporting measures were needed for improving the transparency and comparability of different projects (for AIJ, JI and CDM projects), including project-specific emission baselines.

Some MERVC issues have been examined in the AIJ pilot, but future investments are needed to refine methods and protocols in support of the Kyoto Protocol mechanisms. Some progress has been made in the development of guidance documents for the MERVC of energy-efficiency and forestry projects, but more work needs to be done for developing internationally agreed MERVC guidelines: e.g., evaluation of additionality, free riders, project leakage, positive project spillover, market transformation, environmental impacts, and socioeconomic impacts. A community of MERVC evaluators and verifiers has been developed in response to the UN Framework Convention on Climate Change (FCCC) AIJ pilot, and these individuals and organizations are involved in working on the details of implementing the modalities for measurement, reporting and assessment. Institutional and human capacity building is sorely needed to implement future CDM and JI projects, as well as emissions trading regimes.

1. INTRODUCTION

The key provisions of the Kyoto Protocol (eg, Article 6 on JI and Article 12 on CDM remain to be developed in more detail as negotiations clarify the existing text of the Protocol. As countries start to seriously respond to the mandate of the Kyoto Protocol, project developers of climate change mitigation projects will be asked to demonstrate how their project will reduce GHG emissions or sequester carbon (C). As a result, monitoring, evaluation, reporting, verification, and certification (MERVC) will be the key activities conducted at the project level. MERVC definitions are provided in Box 1. These activities will build upon the work conducted in previous projects, including those implemented in the AIJ pilot.

Under JI, the reduction in emissions by sources, or an enhancement of removals by sinks, must be additional to any that would otherwise occur, entailing project evaluation (Article 6). The emission reduction units (ERUs) from these projects can be used to meet Annex I Party's commitment under Article 3 of the Kyoto Protocol, necessitating all MERVC activities to be conducted. Similarly, under the CDM, emission reductions (ERs) must not only be additional, but be certified as real, measurable, and deriving from projects that contribute to sustainable development, again requiring the performance of all MERVC activities (Article 12).

There is little experience and insight from the AIJ pilot, due to the fact that most of the AIJ projects have not been fully implemented and the lack of project crediting has reduced the need for verification and certification.

Furthermore, internationally agreed MERVC guidelines have not been developed by AIJ sponsors for assisting project developers and evaluators. We expect such guidelines to be established for JI and CDM projects under the Kyoto Protocol, since they are needed to:

- increase the reliability of data for estimating GHG impacts;
- provide real-time data so programs and plans can be revised mid-course;
- introduce consistency and transparency across project types, sectors, and reporters;
- enhance the credibility of the projects with stakeholders;
- reduce costs by providing an international, industry consensus approach and methodologies; and,
- reduce financing costs, allowing project bundling and pooled project financing.

Box 1. Monitoring, Evaluation, Reporting, Verification and Certification Definitions

Estimation: refers to making a judgement on the likely or approximate C stock, GHG emissions, and socioeconomic and environmental benefits and costs in the with- and without-project (baseline) scenarios. Estimation can occur throughout the lifetime of the project, but plays a central role during the project design stage when the project proposal is being developed.

Monitoring: refers to the measurement of C stocks, GHG emissions, and socioeconomic and environmental benefits and costs that occur as a result of a project. Monitoring does *not* involve the calculation of GHG reductions nor does it involve comparisons with previous baseline measurements. For example, monitoring could involve the number of hectares (ha) preserved by a forestry project. The objectives of monitoring are to inform interested parties about the performance of a project, to adjust project development, to identify measures that can improve project quality, to make the project more cost-effective, to improve planning and measuring processes, and to be part of a learning process for all participants (De Jong et al., 1997). Monitoring is often conducted internally by the project developers.

Evaluation: refers to both impact and process evaluations of a particular project, typically entailing a more in-depth and rigorous analysis of a project compared to monitoring emissions. Project evaluation usually involves comparisons requiring information from outside the project in time, area, or population (De Jong et al., 1997). The calculation of GHG reductions is conducted at this stage. Project evaluation would include GHG impacts and non-GHG impacts (i.e., environmental, economic, and social impacts), and the re-estimation of the baseline, leakage, positive project spillover, etc., which were estimated during the project design stage. Evaluation organizes and analyzes the information collected by the monitoring procedures, compares this information with information collected in other ways, and presents the resulting analysis of the overall performance of a project. Project evaluations will be used to determine the official level of GHG emissions reductions that should be assigned to the project. The focus of evaluation is on projects that have been implemented for a period of time, not on proposals (i.e., project development and assessment). While it is true that similar activities may be conducted during the project design stage (eg, estimating a baseline, leakage, or spillover), this type of analysis is estimation and not the type of evaluation that is described in this chapter and which is based on the collection of data.

Reporting refers to *measured* GHG and non-GHG impacts of a project (in some cases, organizations may report on their *estimated* impacts, prior to project implementation, but this is not the focus of this chapter). Reporting occurs throughout the MERVC process (e.g., periodic reporting of monitored results and a final report once the project has ended).

Verification refers to establishing whether the measured GHG reductions actually occurred, similar to an accounting audit performed by an objective, accredited party not directly involved with the project. Verification can occur without certification.

Certification refers to certifying whether the measured GHG reductions actually occurred. Certification is expected to be the outcome of a verification process. The value-added function of certification is in the transfer of liability/responsibility to the certifier.

In the longer term, MERVC guidelines will be a necessary element of any international C trading system, as proposed in the Kyoto Protocol. A country could generate C credits by implementing projects that result in a net reduction in emissions. The validation of such projects will require MERVC guidelines that are acceptable to all parties. These guidelines will lead to verified findings, conducted on an *ex-post facto* basis (i.e., actual as opposed to predicted (*ex-ante*) project performance).

One of the objectives of this chapter is to briefly examine the AIJ experience in monitoring and evaluation, in order to assist the development of modalities for monitoring, evaluation, reporting, verification, and certification for CDM and JI projects. Another objective of this chapter is to provide guidance on how one should go about monitoring, evaluating, reporting, verifying, and certifying climate change mitigation projects based on a review of the literature and our previous work on developing MERVC guidelines for energy-efficiency and LUCF projects.

1.1 AIJ Experience in Monitoring and Evaluation

The FCCC First Conference of Parties (COP-1) in 1995 authorized a five-year pilot phase of AIJ to learn more about the operation of JI projects, to build confidence in the approach, and to more fully develop a framework for the international implementation of JI. Under the pilot phase, countries whose emitters undertake AIJ can identify emissions reductions (ERs) achieved but cannot claim pre-2000 ERs through JI as credit against current obligations to reduce emissions. This means that in practice, incentives for private investment in JI are limited to win-win projects that create net economic benefits independent of GHG reductions or projects where the investor is willing to speculate against the possibility that post-2000 emissions reductions might be creditable and thus gain market value. JI and AIJ projects must also be additional in the sense of not displacing any ongoing foreign assistance programs to recipient countries.

As an example of what is required, the U.S. Initiative on Joint Implementation (USIJI) prepared project proposal guidelines for organizations seeking funding from investors to reduce GHG emissions (USIJI 1996). The guidelines request information on the proposed project, including the identification of all GHG sources and sinks included in the emissions baseline as well as those affected by the proposed project, and net impacts. The guidelines also ask for information on the estimates of GHG emissions and C sequestration, including methodologies, type of data used, calculations, assumptions, references and key uncertainties affecting the emissions estimates. The estimates include the baseline estimate of emissions or sequestration of GHG without measures and the estimate of emissions or sequestration of GHG with measures.

The guidelines require applicants to describe the process used to monitor GHG reductions, including the Parties responsible for monitoring GHG emissions and reductions, the specific data that will be collected in monitoring GHG reductions, and data collection procedures (sampling methodologies, emissions monitoring equipment, and estimation methodologies). Furthermore, the guidelines ask the applicant to describe the provisions in the project for external verification of GHG emission reductions or sequestration. USIJI requires participants to allow external verification of GHG emissions reductions or sequestration by an evaluation panel, its designee, or a party(ies) named at a later date subject to approval by the evaluation panel. Such verification may include third-party inspection of documentation of emissions reductions, or site visits to the project.

The Subsidiary Body for Implementation (SBI) has the responsibility of assisting the Conference of Parties in reviewing the progress of the AIJ pilot based on the inputs from the Subsidiary Body for Scientific and Technological Advice (SBSTA). Two reviews have been conducted for the SBI, and the findings in this section are based on the second synthesis report on the AIJ pilot (UNFCCC, 1998a). Despite the clear differences between the project-based mechanisms of the Kyoto Protocol and the AIJ pilot, there are a number of areas where lessons learned during the AIJ pilot could be usefully employed in the design, development and operation of the project-based mechanisms, thus avoiding duplication of effort. In this section, we focus primarily on the monitoring and evaluation experience in the AIJ pilot.

Today, approximately 130 projects were reported and found to be in accordance with the criteria for reporting under the AIJ pilot (SBSTA, 1997). Most of the projects dealt with renewable energy (40) and energy efficiency (36), followed by forest preservation, reforestation or restoration projects (11). The quality of reporting varied in terms of structure, completeness and coverage (UNFCCC, 1998a):

- The basis for calculating costs and GHG mitigation effects is often insufficiently explained. Most of these data still remain at the level of estimates of varying accuracy, mainly because of uncertainty about appropriate procedures for establishing baselines and definitional and conceptual problems. Definitions of the costs of the AIJ component and other reporting items, such as the lifetime of the activity and technical data, are not consistent.
- Environmental, sociocultural and economic benefits and negative impacts are generally not described in a detailed manner.

TABLE 1. USIJI Projects with GHG Monitoring Plans.

Country	Proje˘ ct	Project Start Date	Date of Monitoring Plan
Belize	Rio Bravo Carbon Sequestration Pilot Project	Jan-95	Feb-96
Bolivia	Noel Kempff Mercado Climate Action Project	1997	Apr-97
Costa Rica	Dona Julia Hydroelectric Project	Oct-96	Early 1997
Costa Rica	ECOLAND: Piedras Blancas National Park	Jan-95	Nov-96
Costa Rica	Klinki Forestry Project	Jun-97	Jun-98
Costa Rica	Territorial and Financial Consolidation of Costa Rican National Parks and Biological Reserves	Jan-98	Nov-98
Czech	City of Decin: Fuel Switching for District Heating	Jan-95	Sep-95
Honduras	Solar-Based Rural Electrification in Honduras	Sep-97	Yes (date not known)
Mexico	APS/CFE Renewable Energy Mini Grid Project, San Juanico, Baja California Sur	May-98	Dec-98
Nicaragua	El Hoyo – Monte Galan Geothermal Project	Jan-99	Dec-01
Russian Federation	RUSAFOR: Saratov Afforestation Project	Oct-93	Oct-98
Russian Federation	RUSAGAS: Fugitive Gas Capture Project	Sep-98	Sep-98

Estimated emission baselines without the activity (project baseline), in most cases, were provided. Some baselines assumed no change in the pattern or level of energy consumption over the lifetime of the activity while others reported a continuation of present trends (e.g., assuming declining C stocks or unsustainable energy consumption patterns). When estimating the emissions with the activity, secondary effects of implementing an activity were indicated in some cases: (e.g., if in a cogeneration scheme renewables were to replace or coal, gas would still be needed for generating the heat component). However, in most cases, defining the proper baseline and the activity scenario did not sufficiently address such issues as monitoring domain and project leakage.

As implementation is proceeding, an increasing number of projects have developed monitoring plans (eg, for USA AIJ projects, see Table 1). Given the lack of centralized guidance and inherent project-specific variations, it is not surprising that the monitoring plans and methodologies used to calculate GHG emissions reduced or sequestered in AIJ projects are highly diverse (see OECD, 1999, and Puhl, 1998). The AIJ projects will be providing data on measured GHG emissions reduced or sequestered; however, most projects started in 1997-98 and had no emissions data at the time of the synthesis report. Coverage of gases other than CO_2 was available in only a few cases.

As a result of these findings, the synthesis study concluded that work needed to be conducted on the following methodological, technical and institutional issues:
- the determination of environmental benefits;
- the consideration of costs;
- transfer of environmentally sound technologies and know-how;
- modalities for measurement, reporting and assessment;
- endogenous capacity building; and,
- institutional arrangements.

The FCCC Secretariat will be participating in a series of workshops and seminars organized by other bodies on issues of monitoring, verification, and certification and on lessons learned from the AIJ Pilot.

A recent study on the experience with GHG emission baselines under the AIJ pilot came up with similar findings (OECD, 1999). They found that information on the methodology and assumptions used in calculating emission baselines for AIJ projects was neither complete nor transparent in many of the AIJ project reports submitted to the FCCC. In fact, many reports contained surprisingly little information on the emission baseline assumed in that project, and on the means by which the baseline was established. While some progress has been made, other methodologies (e.g., bench marking need to be investigated.

2. THE MERVC PROCESS

In the remainder of this chapter, we provide guidance on how one should go about monitoring, evaluating, reporting, verifying, and certifying climate change mitigation projects. Our suggestions are based on a review of the literature (Vine and Sathaye, 1997) and on our work on developing MERVC guidelines for energy-efficiency projects (Vine and Sathaye, 1999) and for LUCF projects (Vine, Sathaye, and Makundi, 1999). We are indebted to the efforts and guidance of many individuals and organizations (Box 2).

Climate change mitigation projects (not just for the CDMor JI) will likely involve several tasks (Figure 1.). We expect that there will be different types of arrangements for implementing these projects:
- project developer might implement the project with his/her own money;
- developer might borrow money from a financial institution to implement the project;

- developer might work with a third party who would be responsible for many project activities; etc.

While the flow of funds might change as a result of these different arrangements the guidelines should be relevant to all Parties, independent of the arrangement.

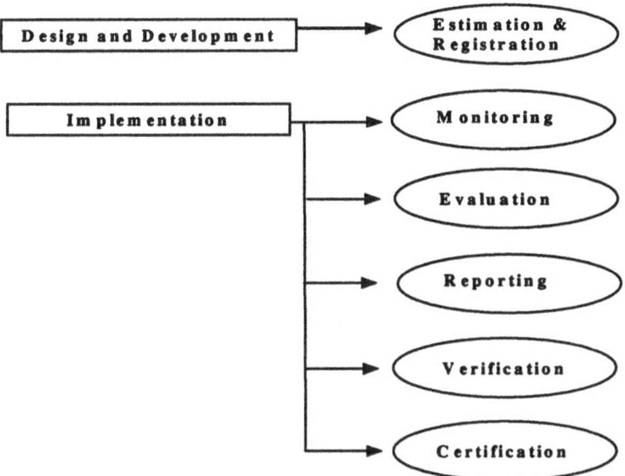

Figure 1. CDM and JI project tasks in monitoring, evaluation, reporting, verification, and certification.

Box 2

Organizations Involved in GHG Monitoring and Evaluation.

ASHRAE GPC 14P. The American Society of Heating, Refrigeration and Air-Conditioning Engineers (ASHRAE) GPC 14P Committee is currently writing guidelines for the measurement of energy and demand savings. When completed, these **guidelines** will be used to modify the IPMVP.

California's Measurement and Evaluation Protocols. Protocols and procedures for the measurement and evaluation of California's utility energy-efficiency programs were developed in response to the shareholder earnings mechanisms established for the four largest investor-owned utilities to acquire demand-side resources (CPUC, 1998). The protocols are targeted to the evaluation of programs, rather than an individual building, and have very detailed requirements.

Face Foundation. The Face Foundation in the Netherlands has worked on joint implementation projects in the forestry sector for many years and has used satellite imagery for evaluating these projects (Face Foundation, 1997).

Forest Stewardship Council's Principles and Criteria for Forest Management. The Forest Stewardship Council (FSC) is an international body that accredits certification organizations in order to guarantee the authenticity of their claims (Forest Stewardship Council, 1996). The FSC's Principles and Criteria for Forest Management apply to all tropical, temperate and boreal forests and more detailed standards may be prepared at national and local levels. The Principles and Criteria are to be incorporated into the evaluation systems and standards of all certification organizations seeking accreditation by the FSC.

Greenhouse Challenge Vegetation Sinks Workbook. The Greenhouse Challenge is a joint initiative of the Commonwealth Government and Australian industry, working in partnership to reduce Australia's GHG emissions. The methodologies in the sinks workbook are generic and can be applied to any type of C sequestration project based on afforestation, reforestation or forest management (Greenhouse Challenge, 1998).

SGS Forestry's C Offset Verification Service. SGS Forestry's C Offset Verification Service is the first international third-party verification service of forestry-based C offset projects (EcoSecurities, 1998). The service consists of a formal analysis of project concept and design, and an independent quantification and verification of projected and achieved savings in C stock derived from the project.

U.S. DOE's International Performance Measurement and Verification Protocol. The U.S. Department of Energy's International Performance Measurement and Verification Protocol (IPMVP) is a consensus document for measuring and verifying energy savings from energy-efficiency projects (Kats et al., 1996 and 1997; Kromer and Schiller, 1996; USDOE, 1997).

U.S. DOE's Voluntary Reporting of Greenhouse Gases. The U.S. Department of Energy (DOE) prepared guidelines and forms for the voluntary reporting of greenhouse gases (USDOE, 1994a and 1994b). The guidelines and forms can be **used** by corporations, government agencies, households and voluntary organizations to report on actions taken that have reduced or avoided emissions of GHGs. The documents offer guidance on recording historic and current GHG emissions, emissions reductions, and C sequestration.

Box 2

U.S. EPA Conservation Protocols. The U.S. Environmental Protection Agency's Conservation Verification Protocols are designed to verify electricity savings from utility demand-side management programs for the purpose of awarding sulfur dioxide (SO_2) allowances under EPA's Acid Rain Program (Meier and Solomon, 1995; USEPA, 1995 and 1996).

U.S. Federal Energy Management Program. The U.S. Department of Energy's Federal Energy Management Program (FEMP) was established, in part, to reduce energy costs to the U.S. Government from operating Federal facilities. FEMP assists Federal energy managers by identifying and procuring energy-efficiency projects. Part of this assistance included the development of an application of the IPMVP for the USA federal sector, which is called the FEMP Guidelines.

USIJI's Project Proposal Guidelines. The USIJI prepared project proposal guidelines for organizations seeking funding from investors to reduce GHG emissions (USIJI, 1996). The guidelines request information on the proposed project, including the identification of all GHG sources and sinks included in the emissions baseline as well as those affected by the proposed project, and net impacts. The guidelines also ask for additional information, such as the estimates of GHG emissions and sequestration, including methodologies, type of data used, calculations, assumptions, references and key uncertainties affecting the emissions estimates.

University of Edinburgh's provisional guidelines and standards. The University of Edinburgh's Institute of Ecology and Resource Management has developed provisional guidelines and standards for assessing C offset projects (University of Edinburgh, 1998). These guidelines are based on the experience of a community forestry and carbon sequestration project in Chiapas, Mexico, and overlap with the forestry standards of the Forest Stewardship Council.

Winrock's C Monitoring Guidelines. The Winrock International Institute for Agricultural Development published a guide to monitoring C sequestration in forestry and agroforestry projects (MacDicken, 1998). The guide describes a system of cost-effective methods for monitoring and verifying, on a commercial basis, the accumulation of C in forest plantations, managed natural forests and agroforestry land uses. This system is based on accepted principles and practices of forest inventory, soil science and ecological surveys.

World Bank's monitoring and evaluation guidelines. The World Bank prepared monitoring and evaluation guidelines for the Global Environment Facility (GEF), a multilateral funding program created to support projects that yield global environmental benefits but would not otherwise be implemented because of inadequate economic or financial returns to project investors (World Bank, 1994a).

In Figure 1, we differentiate registration from certification. Certification refers to certifying whether the measured GHG reductions actually occurred. This definition reflects the language in the Kyoto Protocol regarding the CDM and certified emission reductions (CERs). In contrast, when a host country approves a project for implementation, the project is registered

(UNFCCC, 1998b). For a project to be approved, each country will rely on project approval criteria that they developed:
- project funding sources must be additional to traditional project development funding sources;
- project must be consistent with the host country's national priorities (including sustainable development);
- confirmation of local stakeholder involvement;
- confirmation that adequate local capacity exists or will be developed;
- potential for long-term climate change mitigation;
- baseline and project scenarios; and,
- inclusion of a monitoring protocol (Watt et al., 1995).

A country may also use different administrative or legal requirements for registering projects. For example, the project proposal (containing construction and operation plans, proposed monitoring and evaluation of changes in C stock and energy use and emissions, and estimated changes in carbon stock, energy use and emissions) might have to be reviewed and assessed by independent reviewers. After this initial review, the project participants would have an opportunity to make adjustments to the project design and make appropriate adjustments to the expected changes in energy use, C stock, and emissions. The reviewers would then approve the project, and the project would be registered. Individuals or organizations voicing concerns about the project would have an opportunity to appeal the approval of the project, if desired.

2.1 Conceptual Framework

We use an example of a hypothetical climate change mitigation project to show the conceptual framework underlying the monitoring and evaluation that needs to be conducted. In this example, the project seeks to reduce GHG emissions rather than sequester carbon. The analysis of GHG emissions occurs when a project is being designed and during the implementation of the project. In the design stage, the first step is estimating the baseline (i.e., what would have happened to GHG emissions if the project had not been implemented) and the project impacts. Once these have been estimated, then the net GHG emissions are simply the difference between the estimated project impacts and the baseline (P-B, in Fig. 2). After a project has started to be implemented, the baseline can be re-estimated and the project impacts will be calculated based on monitoring and evaluation methods. The net savings will be the difference between the measured project impacts and the re-estimated baseline (P^-B^, in Fig. 2). The example in Fig. 2 illustrates a case where measured GHG emissions are lower than estimated as a result of a climate change mitigation project. On the other hand, GHG emissions in the re-estimated baseline are higher than what had been estimated at the project design stage. In this case, the calculated net GHG emissions (P^-B^) is larger than what was first estimated (P-B).

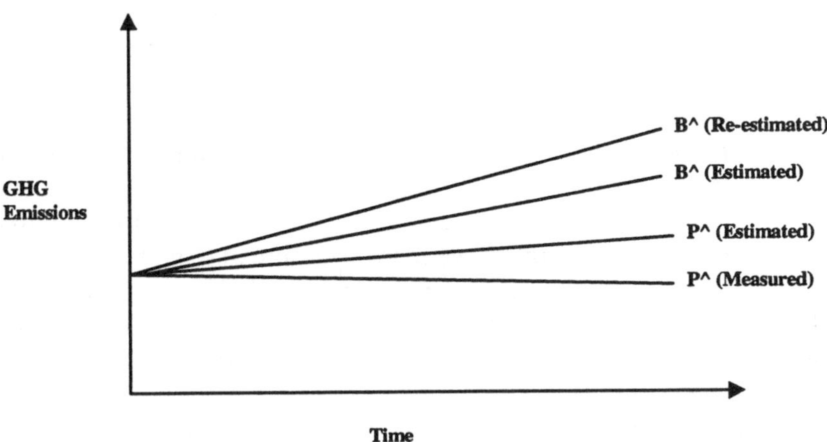

B: Estimated GHG emissions without project (baseline)
P: Estimated GHG emissions with project
P-B: Estimated net (additional) GHG emissions
B^: Re-estimated GHG emissions without project (baseline) (after monitoring and evaluation)
P^: Measured GHG emissions with project (after monitoring and evaluation)
P^-B^: Measured net (additional) GHG emissions (after monitoring and evaluation)

Figure 2. Diagram of shifting GHG emissions over time.

3. MONITORING AND EVALUATION

As an example of the type of monitoring and evaluation that is needed, we present in Figure 3 an overview of one approach used in evaluating gross and net changes in GHG emissions in a LUCF project (Vine, Sathaye and Makundi, 1999). A similar approach has been developed for energy-efficiency projects (Vine and Sathaye, 1999). In this section, we focus on some of the challenges involved in measuring the baseline and gross changes in GHG emissions, since the net change is simply the difference between the gross change and the baseline.

3.1 Establishing the Monitoring Domain

The domain that needs to be monitored is typically viewed as larger than the geographic and temporal boundaries of the project (Andrasko 1997 and MacDicken 1997). In order to compare GHG reductions across projects, a monitoring domain needs to be defined. Consideration of the domain needs to address the following issues: (1) the temporal and geographic extent of a project's direct impacts; and (2) coverage of project leakage, positive project spillover and market transformation.

The first monitoring domain issue concerns the appropriate geographic boundary for evaluating and reporting impacts. A LUCF project might have local (project-specific) impacts that are directly related to the project in question, or the project might have more widespread (e.g., regional) impacts. Thus, one must decide the appropriate geographic boundary for evaluating and reporting impacts. Similarly, the MERVC of changes in the C stock of LUCF projects can be conducted at the point of extraction (e.g., when trees are logged) or point of use (e.g., when trees are made into furniture), and when forests are later transformed to other uses (e.g., agriculture, grassland, or range). Thus, depending on the project developer's claims, one may decide to focus solely on the changes in the C stock from the logging of trees at the project site. This could entail monitoring the changes over time from the new land use type, or account for the wood products produced and traded outside project boundaries.

The same questions about the monitoring domain need to be asked for energy projects. For example, energy projects may impact energy supply and demand at the point of production, transmission, or end use. The MERVC of such impacts will become more complex and difficult as one attempts to monitor how GHG emission reductions are linked between energy end users and energy producers (e.g., tracking the emissions impact of 1,000 kWh saved by a household in a utility's generation system).

The second issue concerns coverage of project leakage (especially for C sequestration projects) and positive project spillover. It is important to note that not all secondary impacts can be predicted. In fact, many secondary impacts occur unexpectedly and cannot be foreseen. And when secondary impacts are recognized, a commitment needs to be made to ensure that resources are available to evaluate these impacts.

One could broaden the monitoring domain to include off-site baseline changes (which are normally perceived as occurring outside the monitoring domain). Widening the system boundary, however, will most likely entail greater MERVC costs and could bring in tertiary and even less direct effects that could overwhelm any attempt at project-specific calculations (Trexler and Kosloff, 1998).

In the beginning stages of a project, the secondary impacts of a project are likely to be modest as the project gets underway, so that the MERVC of such impacts may not be a priority. These effects are also likely to be insignificant or small for small projects. Under these circumstances, it may be justified to disregard these impacts and simply focus on energy savings or C sequestration from the project. This would help reduce MERVC costs. As the projects become larger or are more targeted to market transformation, these impacts should be evaluated.

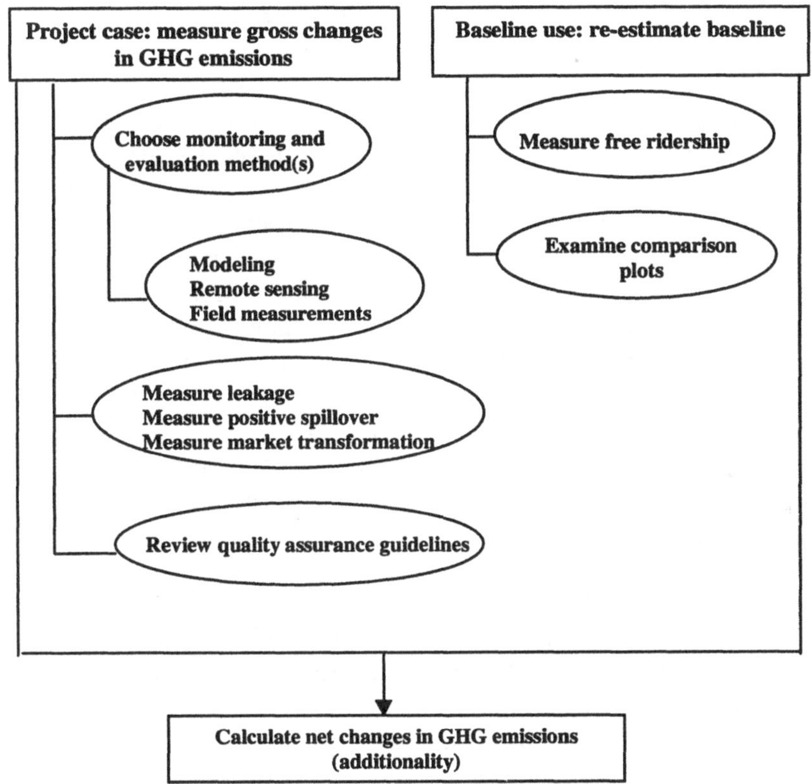

Figure 3. Evaluation of LUCF Project baselines and GHG emissions profile.

Currently, there are weak linkages in assessing multiple monitoring domains (e.g., local, regional and national). One potential solution to strengthening these linkages is the use of nested monitoring systems. This system can be employed where an individual project's monitoring domain is defined to capture the most significant energy savings and where provisions are made for monitoring energy use and GHG emissions outside of the project area by regional or national monitoring systems (Andrasko, 1997).

3.2 Monitoring and Evaluation Methods

3.2.1 Energy-efficiency Projects

For energy-efficiency projects, the first step in measuring emission reductions is the measurement of gross energy savings: comparing the observed energy use of project participants with pre-project energy consumption (Box 3). Several data collection and analysis methods are available which vary in cost, precision, and uncertainty. The data collection methods include engineering calculations, surveys, modeling, end-use metering, on-site audits and inspections, and collection of utility bill data. Most monitoring and evaluation activities focus on the collection of measured data. If measured data are not collected, then one may rely on engineering calculations and stipulated (or default) savings (as described in the Environmental Protection Agency's (EPA's) Conservation Verification Protocols (CVP) and in the U.S. Department of Energy's (DOE's) International Performance Measurement and Verification Protocol (IPMVP). Stipulated savings refer to two different types of stipulated savings methods:

- algorithms for calculating energy savings for specific measures; and,
- a set of criteria for using best engineering practices (USEPA 1995).

The rationale for the use of stipulated savings is that the performance of some energy-efficiency measures is well understood and may not be cost effective to monitor; stipulated savings should only be used for certain retrofits and conditions

Data analysis methods include engineering methods, basic statistical models, multivariate statistical models (including multiple regression models and conditional demand models), and integrative methods. If the focus of the monitoring and evaluation is an individual building, then some methods will not be utilized (e.g., basic statistical models, multivariate statistical models, and some integrative methods), since they are more appropriate for a group of buildings.

Box 3

Monitoring of an Energy AIJ Project:
City of Decin's Fuel Switching for District Heating

Natural gas consumption will be monitored annually. Using a fixed carbon content for natural gas, CO_2 emissions and emissions reductions will be calculated for each engine and boiler at the plant. In addition, data on total annual energy produced by the project will be collected, and the resulting GHG emissions reductions will be calculated. The CO_2 emissions will be monitored periodically throughout the year and will be certified by the Czech Ministry of Environment. The Czech Hydrometeorological Institution will develop and implement a monitoring and verification program and will review and assess: (1) the historic CO_2 emissions baseline for the plant, (2) the projected CO_2 emissions under the reference and project scenarios, (3) a report on potential leakage problems and shift of existing load to other sources of heat supply, (4) the Czech government policy on scoring CO_2 emission reductions from the plant relative to the Czech national plan, and the monitoring strategy and techniques proposed.
Source: USIJI (1998)

There is no one approach that is best in all circumstances (either for all project types, evaluation issues, or all stages of a particular project). The costs of alternative approaches will vary and the selection of evaluation methods should take into account project characteristics and the kind of load and schedule for the load before the retrofit. The load can be constant, variable, or variable but predictable, and the schedule can either be known (timed on/off schedule) or unknown/variable. The monitoring approach can be selected according to the type of load and schedule.

In addition to project characteristics, the appropriate approach depends on the type of information sought, the value of information, the cost of the approach, and the stage and circumstances of project implementation. The applications of these methods are not mutually exclusive; each approach has different advantages and disadvantages, and there are few instances where an evaluation method is not amenable to most energy-efficiency measures. Using more than one method can be informative. Employing multiple approaches, perhaps even conducting different analyses in parallel, and integrating the results will lead to a robust evaluation. Such an approach builds upon the strengths and overcomes the weaknesses of individual approaches. Also, each approach may be best used at different stages of the project life cycle and for different measures or projects. An evaluation plan should specify the use of various analytical methods throughout the life of the project and account for the financial constraints, staffing needs, and availability of data sources.

In an earlier report, we reviewed several protocols and guidelines that were developed for the MERVC of GHG emissions in the energy sectors by governments, nongovernmental organizations, and international agencies (Vine and Sathaye, 1997). Although not targeted to GHG emissions, we believe that the U.S. DOE IPMVP is the preferred approach for monitoring and evaluating energy-efficiency projects for individual buildings and for groups of buildings. IPMVP covers many of the issues discussed in these guidelines as well as offering several measurement and verification methods for user flexibility (Kats et al., 1996 and 1997; Kromer and Schiller 1996; USDOE 1997). North America's energy service companies have adopted the IPMVP as the industry standard approach to measurement and verification. States ranging from Texas to New York now require the use of the IPMVP for state-level energy efficiency retrofits. The USA, through DOE's Federal Energy Management Program (FEMP), uses the IPMVP approach for energy retrofits in federal buildings. Finally, countries ranging from Brazil to the Ukraine have adopted the IPMVP, and the Protocol is being translated into Bulgarian, Chinese, Czech, Hungarian, Polish, Portuguese, Russian, Spanish, Ukrainian and other languages.

3.2.2 LUCF Projects

Three general monitoring techniques can be used to monitor C fixed through LUCF projects (MacDicken, 1997):

- modeling;
- remote sensing; and,
- field/site measurements, including biomass surveys (which includes research studies; surveys; the monitoring of wood production and end products; and forest inventories) and destructive sampling.

Many of these techniques can be used together (Box 4).

Box 4
Monitoring of a Carbon Sequestration AIJ Project:
Noel Kempff Mercado Climate Action Project

The monitoring and verification plan includes specifications for: (1) monitoring forest biomass and C content of other forest components; (2) monitoring of secondary impact parameters; (3) establishing and maintaining monitoring plots; (4) conducting quality assurance tests and quality control procedures; and (5) developing a summary of the equations that will be used to convert raw data to CO_2 -equivalent units. Monitoring tasks will include: (1) routinely tracking the data elements for C contents, flux rates, and secondary impacts at three locations within the project area; (2) verifying the assumptions made to establish the reference case emissions and secondary impacts projections, and correcting or improving assumptions, as needed; and (3) comparing the documented changes in total C and secondary impacts at the three monitoring locations and making necessary adjustments to the reference case.
Source: USIJI (1998)

The unique features and diversity of LUCF projects, the monitoring domain and socioeconomic issues pertaining to LUCF projects, and the variety of C pools that might be impacted by LUCF projects makes the monitoring and evaluation of forestry projects very challenging. While LUCF projects offer the potential for significant C sequestration, the verification of C credit claims will necessitate significant technical and financial resources. A variety of monitoring techniques are available for LUCF projects for determining the amount of C sequestered by LUCF projects, each having its own advantages and disadvantages (Vine, Sathaye and Makundi, 1999). We expect the use of these techniques will vary by the size of the project area, region, type of forest, and the purpose of the project (e.g., to protect forests, supply energy, or provide wood products). One of the key decisions that will need to be made will be determining the optimal level of costs for implementing these techniques.

3.3 Re-estimating the Baseline

For JI (Article 6) and CDM (Article 12) projects implemented under the Kyoto Protocol, the emissions reductions from each project activity must be additional to any that would otherwise occur, also referred to as additionality criteria (Articles 6.1b and 12.5c). Determining additionality requires a baseline for the calculation of GHG emissions, i.e., a description of what would have happened to GHG emissions had the project not been implemented (Violette, Ragland and Stern, 1998). Additionality and baselines are inextricably linked and are a major source of debate (OECD, 1999; Trexler and Kosloff, 1998). Determining additionality is inherently problematic because it requires resolving a counter-factual question: What would have happened in the absence of the specific project?

Because investors and hosts of climate change mitigation projects have the same interest in an climate change mitigation project (i.e., they want to get maximum C savings from the project), they are likely to overstate and over-report the amount of C saved by the project (e.g., by overstating business-as-usual C emissions). Cheating may be widespread if there is no strong monitoring and verification of the projects. Even if projects are well monitored, it is still possible that the real amount of carbon saved is less than estimated values. Hence, there is a critical need for the establishment of realistic and credible baselines. As noted previously, it is difficult to determine the credibility of baselines developed in projects undertaken in the AIJ pilot because the available background data on the determination of baselines is lacking.

Future changes in GHG emissions may differ from past levels, even in the absence of the project, due to growth, technological changes, input and product prices, policy or regulatory shifts, social and population pressure, market barriers, and other exogenous factors. Consequently, the calculation of the baseline needs to account for likely changes in relevant regulations and laws, changes in key variables (e.g., population growth or decline, and economic growth or decline) (Andrasko, Carter and Gaast, 1996; Michaelowa, 1998).

Ideally, GHG emissions should be measured for at least a full year before the date of the initiation of the retrofit project and for each year after the initiation of the project during the lifetime of the project. However, some types of projects may not require a full year of monitoring prior to the retrofit: (e.g., in energy-efficiency projects) if the loads and operating conditions are constant over time, one-time spot measurement may be sufficient to estimate equipment performance and efficiency. The baseline will be re-estimated based on monitoring and evaluation data collected during project implementation. The re-estimated baseline should describe the existing technology or practices at the facility or site. Finally, in order to be credible, project-specific baselines need to account for free riders.

3.3.1 Free Riders

In climate change mitigation projects, it is possible that the reductions in GHG emissions are undertaken by participants who would have taken the same actions if there had been no project. These participants are called free riders. The carbon savings associated with free riders are not truly additional to what would occur otherwise (Vine, 1994). Hence, this is a test of both financial additionality as well as emissions additionality. Although free riders may be regarded as an unintended consequence of a climate change mitigation project, free ridership should still be estimated, if possible, during the estimation of the baseline. While free riders can also cause positive project spillover, this impact is typically considered to be insignificant compared to the impacts from other participants.

For energy-efficiency projects installing technologies in developing countries where the efficiency of these technologies would be regarded as conventional in developed countries, all project participants could be regarded as free riders. As a result, there would be few projects implemented. A possible solution to this problem would be the establishment of performance benchmarks (standards) that would indicate to project developers the type of energy-efficient equipment that would be allowed to be installed and that would pass the free rider test.

Free ridership can be evaluated either explicitly or implicitly (Goldberg and Schlegel, 1997; Saxonis, 1991). The most common method of developing explicit estimates of free ridership is to ask participants what they would have done in the absence of the project (also referred to as but for the project discussions). Based on answers to carefully designed survey questions, participants are classified as free riders (yes or no) or assigned a free ridership score. Project free ridership is then estimated as the proportion of participants who are classed as free riders. Two problems arise in using this approach:

- very inaccurate levels of free ridership may be estimated, due to questionnaire wording; and,
- there is no estimate of the level of inaccuracy, for adjusting confidence levels.

Another method of developing explicit estimates of free ridership is to use discrete choice models to estimate the effect of the program on customers' tendency to implement measures. The discrete choice is the customer's yes/no decision whether to implement a measure. The discrete choice model is estimated to determine the effect of various characteristics, including project participation, on the tendency to implement the measures.

For energy-efficiency projects, a method for calculating implicit estimates of free ridership is to develop an estimate of savings using billing analysis that may capture this effect, but does not isolate it from other impacts. Rather than taking simple differences between participants and a comparison group, however, regression models are used to control for factors that contribute to differences between the two groups (assuming that customers who choose to participate in projects are different from those who do not participate). The savings determined from the regression represent the savings associated with participation, over and above the change that would be expected for these customers due to other factors, including free ridership. This approach assumes:

- nonparticipants would naturally buy the energy-efficiency measure as much as participants would;
- savings from the measures have a significant impact on the bills of nonparticpants;and,
- sizeable proportion of nonparticipations buy/install the measure. These assumptions are not always valid.

The USA EPA's CVPs reward more rigorous methods of verifying free riders by allowing a higher share of the savings to qualify for tradable SO_2 allowances. Three options are available for verifying free riders:

- default net-to-gross factors for converting calculated gross energy savings to net energy savings; the net-to-gross factor is defined as net savings divided by gross savings. The gross savings are the savings directly attributed to the project and include the savings from all measures and from all participants; net savings are gross savings that are adjusted for free riders and positive project spillover. Multiplying the gross savings by the net-to-gross factor yields net savings;

- project-estimated net-to-gross factors, based on measurement and evaluation activities (e.g., market research, surveys, and inspections of nonparticipants); or
- if a developer does not do any monitoring nor provide documentation and the default net-to-gross factors are not used, then the net energy savings of a measure will be 50% of the first-year savings (Meier and Solomon, 1995; USEPA 1995 and 1996).

3.3.2 Performance Benchmarks

Concerned about an arduous project-by-project review that might impose prohibitive costs, some researchers have proposed an alternate approach, based on a combination of performance benchmarks and procedural guidelines that are tied to appropriate measures of output (e.g., Lashof, 1998; Michaelowa, 1998; Swisher, 1998; Trexler and Kosloff, 1998; Puhl, 1998). In all cases, measurement and verification of the actual performance of the project is required. The performance benchmarks for new projects could be chosen to represent the high performance end of the spectrum of current commercial practice (e.g., representing roughly the top 25th percentile of best performance). In this case, the benchmark serves as a goal to be achieved. In contrast, others might want to use benchmarks as a standard or default baseline which must be improved upon in order to generate valid emission reductions: an extension of existing technology, and not representing the best technology or process.

A panel of experts could determine a baseline for a number of project types, which could serve as a benchmark for the FCCC. This project categorization could be expanded to a categorization by regions or countries, resulting in a region-by-project matrix. Project developers could check the relevant element in the matrix to determine the baseline of their project. Most of the costs in this approach relate to the establishment of the matrix and its periodical update. Before moving forward with this approach, analysis is needed to consider the costs in developing the matrix and its update, the potential for projects to qualify, and the potential for free riders. The USA EPA is assessing the feasibility and desirability of implementing a benchmark approach for evaluating additionality (Hagler Bailly, 1998).

3.3.3 Comparison Groups

For many projects, comparison groups can be used for evaluating the impacts of climate change mitigation projects. Acting as a baseline, comparison groups can capture time trends that are unrelated to project participation. For example, if the comparison group shows an average reduction in GHG emissions of 5% between the pre- and post-periods, and the participants' bills show a reduction of 15%, then it may be reasonable to assume that the estimated project impacts will be 15% minus the 5% general trend for an estimated 10% reduction in use being attributed to the project.

3.4 Project Case: Monitoring and Evaluation

3.4.1 Project Leakage

Leakage occurs because the project boundary within which a project's benefits are calculated may not be able to encompass all potential indirect project effects. In this chapter, negative indirect effects are referred to as project leakage while positive indirect effects are referred to as positive project spillover. Leakage is likely more important for C sequestration projects than for energy projects. For example, projects affecting the supply of timber products can affect price signals for the rest of the market. Potentially this scenario could counteract a portion of the calculated benefits of the project: the establishment of forestry plantations could lead to a decrease in timber prices, leading to a higher incentive to convert forests to agricultural purposes. Another example of leakage occurs when a forest preservation project involves protecting land that was previously harvested by the local population for their personal consumption as fuel wood (MacDicken, 1998; Watt et al., 1995). Although this area is now protected from harvesting, people from the surrounding communities still require wood for fuel and construction. Preserving this forest area has shifted their demand for fuel wood to a nearby site, leading to increased deforestation. This off-site deforestation will at least partially offset the carbon sequestration at the project site. Furthermore, some projects may involve international leakage (e.g., in 1989, when all commercial logging in Thailand was banned, the logging shifted to neighboring countries such as Burma, Cambodia and Laos as well as to Brazil , Watt et al., 1995).

Leakage may occur not only after a project has been completed but also during project development. For example, in the Rio Bravo C Sequestration Pilot Project, a local timber company used the money from the sale of the land to project participants for upgrading their equipment, allowing for the possibility of an increase in output of plywood (Program for Belize, 1997). However, this increase in output did not occur. Similarly, the land purchases for the Rio Bravo project could also motivate competitors that had wanted to purchase that land to intensify clearance of the land already in their possession, or intensify production from the land, increasing emissions from agricultural inputs and machinery. However, this also has not occurred

(Programme for Belize, 1997).Leakage needs to be accounted for if off-site GHG emissions are to be accounted for, rather than those at a particular site. However, leakage can be difficult to identify and even more difficult to estimate and quantify.

3.4.2 Positive Project Spillover

For many programs, the number of eligible nonparticipants is far greater than the number of participants. For example, when measuring energy savings, it is possible that the actual reductions in GHG emissions are greater than measured because of changes in participant behavior not directly related to the project, as well as to changes in the behavior of other individuals not participating in the project (i.e., nonparticipants). These secondary impacts stemming from a climate change mitigation project are commonly referred to as positive project spillover. Positive project spillover may be regarded as an unintended consequence of a climate change mitigation project; however, as noted below, increasing positive project spillover may also be perceived as a strategic mechanism for reducing GHG emissions.

Spillover effects can occur through a variety of channels including:

- project participants that undertake additional, but unaided, actions based on positive experience with the project;
- manufacturers changing the efficiency of their products, or retailers and wholesalers changing the composition of their inventories to reflect the demand for more efficient goods or forestry products created through the project;
- governments adopting new building codes or appliance standards because of improvements to appliances resulting from one or more energy efficiency projects; or,
- technology transfer efforts by project participants which help reduce market barriers throughout a region or country.

The methods for estimating positive project spillover are similar to those used for free ridership (Goldberg and Schlegel, 1997; Weisbrod et al., 1994). Explicit estimates can be obtained by asking participants and nonparticipants survey questions, and discrete choice models can be used (e.g., the effect on implementation of program awareness, rather than program participation, is estimated). Participant and nonparticipant spillover effects can be included in savings estimates in billing analyses, similar to how gross savings are calculated.

3.4.3 Market Transformation

Project spillover is related to the more general concept of market transformation, defined as: the reduction in market barriers due to a market intervention, as evidenced by a set of market effects, that lasts after the intervention has been withdrawn, reduced or changed (Eto, Prahl and Schlegel, 1996). In contrast to project spillover, increasing market transformation is expected to be a strategic mechanism (i.e., an intended consequence) for reducing GHG emissions for the following reasons:

- increase the effectiveness of climate change mitigation projects: e.g., by examining market structures more closely, looking for ways to intervene in markets more broadly, and investigating alternative points of intervention.
- reduce reliance on incentive mechanisms: e.g., by strategic interventions in the market place with other market actors.
- take advantage of regional and national efforts and markets.
- increase focus on key market barriers other than cost.
- create permanent changes in the market.

Market transformation has emerged as a central policy objective for future publicly funded energy-efficiency projects in the USA, but the evaluation of such projects is still in its infancy. Furthermore, regulatory authorities have little experience in accepting savings from market transformation. Nevertheless, because of its importance, we encourage project developers to consider savings from market transformation, particularly since other countries are starting to implement market transformation programs (Box 5).

Market transformation is also relevant for C sequestration projects. As a hypothetical example, consider a bioenergy project that grows trees on a rotational basis and harvests the trees as an energy resource for a community hospital. The developer of the project needs to make sure there are no technical, financial, administrative, or policy barriers to the implementation of this project, and to determine if there are other large, energy-intensive end users who could take advantage of this resource (e.g., industrial customers?). The project developer could also examine what partnering opportunities exist for promoting the bioenergy project (e.g., developing a voluntary labeling program that labels customers as green energy users). Once the labeling program is in place, additional projects might emerge, creating an expanded market for bioenergy projects. Finally, the developer could try to extend the proposed labeling program to other regions, in order to enlarge the market for the project's trees.

Box 5

Market Transformation Programs Outside North America.

Market transformation programs are being implemented outside of North America, particularly in Sweden, Brazil, Thailand, India, Philippines, Sri Lanka, Poland, and China (Martinot, 1998; Meyers, 1998). We provide information on market transformation programs for the first three countries.

The ten-year old Swedish program for energy efficiency has produced 25 procurements within the residential, commercial and industrial sectors (Suvilehto and Öfverholm, 1998). Examples in the residential sector include refrigerators and freezers, washing machines and dryers; in the commercial sector, lighting and ventilation; and in the industrial sector: factory doors and fans. This program aims at establishing market transformation and consists of technology procurement and projects supporting market penetration. There is a wide variety of methods in use; each of them are designed according to the market barriers, its actors, decision makers, their interplay, and specific market needs, expectations and conditions.

Since 1995, Brazil's national electricity conservation program, PROCEL, has been involved in market transformation, including cooperative efforts with equipment manufacturers (Geller, 1997). PROCEL has had considerable success in transforming the efficiency of refrigerators and freezers, lighting, motors, and meters. PROCEL conducts or co-funds several other programs in the areas of research and development, consumer education, training, promotion and ESCO support. These programs are designed to introduce new technologies, increase awareness, change behavior, and stimulate investment in energy efficiency in Brazil.

The Thailand Promotion of Electricity Efficiency project is a comprehensive five-year utility DSM program that created a DSM office within the national electric utility (EGAT) (Martinot, 1998). The DSM office is developing and implementing a number of market intervention strategies in the residential, commercial and industrial sectors. The project provides for financing mechanisms, energy-efficiency codes and standards, appliance labeling, testing laboratories, monitoring and evaluation protocols and systems, development and training of energy service companies, integrated supply-side and demand-side planning, and load management programs. EGAT has tried to rely on voluntary agreements, market mechanisms, and intensive publicity and public education campaigns (including appliance energy labels).

Sources: (1) Suvilehto, H. and E. Öfverholm. 1998. "Swedish Procurement and Market Activities — Different Design Solutions on Different Markets," in the *Proceedings of the 1998 ACEEE Summer Study on Energy Efficiency in Buildings*. Vol. 7, pp. 311-322. Washington, D.C.: American Society for an Energy-Efficient Economy. (2) Geller, H. 1997. *Market Transformation through PROCEL: Brazil's National Electricity Conservation Program*. Washington, D.C.: American Council for an Energy-Efficient Economy. (3) Martinot, E. 1998. *Monitoring and Evaluation of Market Development in World Bank-GEF Climate Change Projects*. Washington,

Box 5

D.C.: The World Bank. (4) Meyers, S. 1998. *Improving Energy Efficiency: Strategies for Supporting Sustained Market Evolution in Developing and Transitioning Countries.* LBNL-41460. Berkeley, CA: Lawrence Berkeley National Laboratory.

Two examples in the LUCF sector show the beginnings of market transformation:

- the availability of improved biomass cook stoves, an important technology for reducing deforestation, has influenced many nonparticipants to purchase cook stoves as these programs develop (Bialy, 1991); and,
- a reduced impact logging (RIL) project in Malaysia is being replicated in Brazil and other parts of Indonesia (personal communication from Pedro Moura-Costa, EcoSecurities, Ltd., Sept. 15, 1998; Jepma, 1997).

Most evaluations of market transformation projects focus on market effects (Eto. Prahl and Schlegel, 1996; Schlegel, Prahl and Raab, 1997). For example, evaluating the effects of climate change mitigation projects on the structure of the market or the behavior of market actors that lead to increases in the adoption of products, services, or practices. In order to claim that a market has been transformed, project evaluators need to demonstrate the following (Schlegel, Prahl and Raab, 1997):

- There has been a change in the market that resulted in increases in the adoption and penetration of technologies or practices.
- That this change was due at least partially to a project (or program or initiative), based both on data and a logical explanation of the program's strategic intervention and influence.
- That this change is lasting, or at least that it will last after the project is scaled back or discontinued.

The first two conditions are needed to demonstrate market effects, while all three are needed to demonstrate market transformation. The third condition is related to persistence: if the changes are not lasting (i.e., they do not persist), then market transformation has not occurred. Because fundamental changes in the structure and functioning of markets may occur only slowly, evaluators should focus their efforts on the first two conditions, rather than waiting to prove that the effects will last.

To implement an evaluation system focused on market effects, one needs to carefully describe the scope of the market, the indicators of success, the intended indices of market effects and reductions in market barriers, and the methods used to evaluate market effects and reductions in market barriers (Schlegel, Prahl and Raab, 1997).

Evaluation activities will include one or more of the following:

- measuring the market baseline;
- tracking attitudes and values;
- tracking sales;
- modeling of market processes; and
- assessing the persistence of market changes (Prahl and Schlegel, 1993).

As one can see, these evaluation activities will rely on a large and diverse group of data collection and analysis methods, such as:

- surveys of customers, manufacturers, contractors, vendors, retailers, government organizations, energy providers, etc.;
- analytical and econometric studies of measure cost data, stocking patterns, sales data, and billing data; and,
- process evaluations.

4. ENVIRONMENTAL AND SOCIOECONOMIC IMPACTS

The Kyoto Protocol encourages developed countries, in fulfilling their obligations, to minimize negative social, environmental and economic impacts, particularly on developing countries (Articles 2.3 and 3.14). Furthermore, one of the primary goals of the CDM is sustainable development. At this time, it is unclear on what indicators of sustainable development need to be addressed in the evaluation of climate change mitigation projects. Once there is an understanding of this, then MERVC guidelines for those indicators will need to be designed. At a minimum, climate change mitigation projects should meet current country guidelines for non-CDM projects.

The persistence of GHG reductions and the sustainability of climate change mitigation projects depend on individuals and local organizations that help support a project during its lifetime. Both direct and indirect project benefits will influence the motivation and commitment of project participants. Hence, focusing only on GHG impacts would present a misleading picture of what is needed in making a project successful or making its GHG benefits sustainable. In addition, a diverse group of stakeholders (e.g., government officials, project managers, non-profit organizations, community groups, project participants, and international policymakers) are interested in, or involved in, climate change mitigation projects and are concerned about their multiple impacts. For example, in Lawrence Berkeley National Laboratory's (LBNL's) monitoring and verification forms, checklists are provided for developers, evaluators, and verifiers to qualitatively assess the impacts described in this section. These checklists are not exhaustive but are included to indicate areas that need to be assessed. Other existing guidelines are better suited for addressing these impacts, such as guidance documents developed for World Bank-supported projects (World Bank, 1989). LBNL's checklists should help to improve the credibility of the project (by showing stakeholders that these impacts have, at least, been considered) as well as to facilitate the review of climate change mitigation projects.

4.1 Environmental Impacts

Climate change mitigation projects have widespread and diverse environmental impacts that go beyond GHG impacts. The environmental benefits associated with climate change mitigation projects can be just as important as the global warming benefits. For example, potential environmental impacts that need to be considered for energy-efficiency projects are presented in Table 1. Direct and indirect project impacts need to be examined, as well as avoided negative environmental impacts (e.g., the deferral of the construction of a new power plant). Both gross and net impacts need to be evaluated.

TABLE 1. Evaluation of potential environmental impacts for energy-efficiency projects (World Bank 1989).

Impact Category	Comments
Dams and reservoirs	Implementation and operation
Effluents from power plants	Air, water and solid effluents from power plants (e.g., City of Decin's fuel switching for district heating project and Honduras' bio-gen biomass power generation project; USIJI 1998)
Hazardous and toxic materials	Manufacture, use, transport, storage and disposal
Indoor air quality	Measures to maintain and/or improve indoor air quality (Community of Guguletu et al., 1998; Chen and Vine, 1998)
Industrial hazards	Prevention and management
Insurance claims	Reduced losses in personal and commercial lines of coverage (Vine, Mills and Chen, 1998)
Occupational health and safety	Plans
Water quality	Protection and enhancement
Wildlife and habitat protection or enhancement	Protection and management

At a minimum, evaluators need to evaluate the environmental impacts associated with the project. Evaluators need to collect some minimal information on potential impacts via surveys or interviews with key stakeholders. The evaluator should also check to see:

– whether any existing laws require these impacts to be examined;

– if any proposed mitigation efforts were implemented; and,
– whether expected positive benefits ever materialized.

Evaluators may want to conduct some short-term monitoring to provide conservative estimates of environmental impacts. The extent and quality of available data, key data gaps, and uncertainties associated with estimates should be identified and estimated.

The information collected and analyzed by evaluators will be useful for better describing the stream of environmental services and benefits of a project, in order to attract additional investment and to characterize the project's chances of maintaining reduced GHG emissions over time. This information will, hopefully, also help in mitigating any potentially negative environmental impacts and encouraging positive environmental benefits.

4.2 Socioeconomic Impacts

In examining socioeconomic impacts, evaluators need to ask the following questions: who the key stakeholders are, what project impacts are likely and upon what groups, what key social issues are likely to affect project performance, what the relevant social boundaries and project delivery mechanisms are, and what social conflicts exist and how they can be resolved (World Bank, 1994b). To address these questions, evaluators could conduct informal sessions with representatives of affected groups and relevant non-governmental organizations.

After a project has been implemented, MERVC activities should assess whether the project led to any social and economic impacts and whether any mitigation was done (Table 2). Direct and indirect project impacts need to be examined, as well as avoided negative socioeconomic impacts (e.g., the preservation of an archaeological site as a result of the deferral of the construction of a new power plant).

TABLE 2. Evaluation of socioeconomic impacts for energy-efficiency projects (World Bank (1989) and EcoSecurities (1998).

Impacts
Cultural properties (archeological sites, historic monuments, and historic settlements)
Distribution of income and wealth
Employment rights
Gender equity
Induced development and other sociocultural aspects (secondary growth of settlements and infrastructure)
Long-term income opportunities for local populations plants (jobs)
Public participation and capacity building
Quality of life (local and regional)

Source: Adapted from World Bank (1989) and EcoSecurities (1998).

Evaluators need to review the checklist of socioeconomic impacts and should collect some minimal information on potential impacts via surveys or interviews with key stakeholders. The evaluator should also check to see if any proposed mitigation efforts were implemented and whether expected positive benefits ever materialized. The extent and quality of available data, key data gaps, and uncertainties associated with estimates may need to be identified and estimated.

5. INSTITUTIONAL ISSUES

It is unclear at this time which institutions have the authority and capability of conducting MERVC activities: government authorities, auditing companies, self-reporting by project developers or host countries, etc. We expect the roles and responsibilities will vary by MERVC activity, although some overlap is expected. We expect the division of labor to be a function of available resources and capabilities, the credibility of the person (or organization) in charge of the activity, and the cost of conducting the particular MERVC activity.

We believe that local institutions, in particular, should be evaluated during the evaluation of climate change mitigation projects. For example, if local community participants are not involved in the design or implementation of a project, then the sustainability of a project becomes problematic. Information on institutional capacity covers the credibility, experience and manpower situation in the executing agency, such as:

- size of staff (field operations, engineering support, planning, finance/administration, etc.) by function;
- academic qualifications, area of expertise, and years of experience of agency staff;
- supporting agencies (e.g., public sector agencies, private consultants, or international organizations); and,
- internal structure of the implementing agency.

Special attention needs to be paid to capacity issues as projects have to demonstrate:

- financial capacity (i.e., the organization must demonstrate that it has sufficient financial resources to implement the project throughout its time frame);
- management capacity (i.e., the organization must demonstrate its capacity to document and implement the project); and,
- infrastructure and technological capacity (i.e., the organization must demonstrate access to appropriate labor pools, technical skills,

technologies and techniques and general infrastructure necessary for the implementation and maintenance of the project throughout its time frame)

In sum, the MERVC guidelines should cover the administrative, institutional and political impacts of the climate change mitigation projects, such as: administrative burden (e.g., institutional capabilities), and, political impacts (e.g., sustained political support, consistent with other public policies).

5.1 Roles and responsibilities

Because of the diverse activities involved in the MERVC of GHG reductions, we expect that several organizations will be involved at different levels (local, state, regional, national, and international) (Table 3). It is imperative that the roles and responsibilities are clarified as early as possible, so that they are tailored to the appropriate organization; otherwise, delays in the designation will likely lead to delays and disputes later.

TABLE 3. Primary MERVC actors in prototype activities and projects.

	Monitoring	Evaluation	Reporting	Verification	Certification
Project developers	Π	Π	Π		
Consultant organizations [1]	Π	Π		Π	
Nongovernmental organizations [1]		Π		Π	
Governmental agencies			Π	Π	Π
International organizations			Π	Π	Π

[1] Consultants and nongovernmental organizations (NGOs) must first be accredited by a government organization or industry association to be able to verify a project or to issue a certificate.

One review of pilot AIJ projects suggests that project developers and project parties, who are most closely associated with the project and thus have access to the data and information, should play an instrumental role in the monitoring, evaluation, and reporting of climate change mitigation projects (Watt et al., 1995). These stakeholders would also rely on the assistance of technical consultants to conduct the monitoring and evaluation tasks; additional participants might include university staff, non-governmental organizations, and members of governmental agencies. If the evaluation of the project is to be more than calculation of GHG estimates (e.g., a process evaluation designed to improve project implementation) then outside consultants who are not involved in the project implementation should conduct the work due to their objective (independent) perspective. This recommendation is based on the assumption that distinctly different evaluation and implementation teams will enhance the credibility and integrity of evaluation. Because the separation of project evaluation and implementation functions is controversial, however, the costs and benefits of such a separation need to be examined in more depth. In addition to formal designations to ensure cooperation for conducting the MERVC activities, there will be a need for informal cooperation among all the parties involved: for example, through workshops and conferences at the regional, national, and international levels.

MERVC will entail significant resources, including the potential hiring and training of new staff (or contractors), equipment, and laboratory facilities. Because of the diverse individuals and organizations involved in the MERVC of energy savings and carbon sequestration with varying levels of technical expertise, qualification criteria are needed for allowing these people to report, monitor, evaluate and verify GHG reductions, so that the findings are perceived as objective and credible. Certification workshops may be needed to ensure that the activities are being conducted in a responsible and credible manner. Training and certification should be sector specific: e.g., a certified evaluator in LUCF (Watt et al., 1995).

5.2 Multiple Reporting

Several types of reporting might occur in climate change mitigation projects:

- impacts of a particular project could be reported at the project level and at the program level (where a program consists of two or more projects);
- impacts of a particular project could be reported at the project level and at the entity level (e.g., a utility company reports on the impacts of all of its projects); and,
- impacts of a particular project could be reported by two or more organizations as part of a joint venture (partnership) or two or more countries.

To mitigate the problem of multiple reporting, project-level reporters should indicate whether other entities might be reporting on the same activity and, if so, who. If there exists a clearinghouse with an inventory of stakeholders and projects, multiple reporting might not constitute a problem. For example, in their comments on an international emissions trading regime, Canada (on behalf of Australia, Iceland, Japan, New Zealand, Norway, Russian Federation, Ukraine and the United States) proposed a national recording system to record ownership and transfers of assigned amount units (i.e., C offsets) at the national level (UNFCCC, 1998b). A synthesis report could confirm, at an aggregate level, that bookkeeping was correct, reducing the possibility of discrepancies among Parties' reports on emissions trading activity.

5.3 Verification

If C credits become an internationally traded commodity, then verifying the amount of C reduced or fixed by projects will become a critical component of any trading system. Investors and host countries may have an incentive to overstate the GHG emissions reductions from a given project, because it will increase their earnings when excessive credits are granted. For example, these parties may overstate baseline emissions or understate the project's emissions. To resolve this problem, there is a need for external (third-party) verification.

Verifiers could be active from the beginning of the project's operations, but in our view, verification occurs after the project begins regular operations. After the project's first operational interval (e.g., one year), and periodically thereafter (e.g., annually), the verifier would verify the project's C sequestration in the preceding period.

Currently, no rules exist for what kinds of organizations will verify monitoring and evaluation results. Some possibilities include government agencies, private sector firms that specialize in verification, an intergovernmental body such as the FCCC subsidiary bodies, or groups of advisors recognized by the FCCC. The guidelines could also recommend that independent verification teams be established (Watt et al., 1995). The verification teams could either be composed of members from host and investor countries for joint implementation projects, or from an international agency, such as the United Nations (UN), for other projects.

Some resolution of disputes over verification results will also be needed:

> Because verification has the potential to be contentious, it should be possible for third parties, as well as the host and investor country parties, to challenge the verification results, in order to encourage watch-dogging between countries. Recourse in the event of disagreement about the results of a verification could include resolution by the initial verification team, introduction of a second verification team, development of new calculation methodologies, or recourse to a tribunal, depending on the project and the nature of the disagreement (Watt et al., 1995)

The tribunal might consist of people from the UN, or from a country. If the latter, someone may still be needed at the international level to monitor the activities of individual countries. The tribunal might also be responsible for developing a common set of standardized MERV guidelines. This is important not only for reporting GHG reductions internationally, but also for investment purposes: investors would probably welcome a standardized set rather than a diverse set of guidelines across different host countries.

5.4 Certification

Certification refers to certifying whether the measured GHG reductions actually occurred. This definition reflects the language in the Kyoto Protocol regarding the CDM and CERs. However, some argue that certification could be done ex-ante, to certify a proposed offset, assuming that it is carried out as planned. Similarly, some propose CDM projects to be certified when they are approved by a host country; however, in this situation, registered or validated appears to be a more accurate descriptor (UNFCCC, 1998b).

At this time, certification is expected to simply be the outcome of a verification process: i.e., no other measurement and evaluation activities are expected to be conducted. Each of the Kyoto Protocol's flexibility mechanisms (e.g., JI (Article 6), CDM (Article 12), and emissions trading (Article 17) requires some form of government approval either at the point of transfer, or under Article 3, at the point that the part of the assigned amount or ERU is added to or deducted from Annex I Parties' assigned amount. However, only Article 12 provides for a process of auditing and certification that would provide for an objective assessment of whether the transfer was likely to result in net ER. Hence, part of the discussions in implementing the Kyoto Protocol will focus on the establishment of certification procedures for ERUs generated and traded through these mechanisms.

Certification companies need to be accredited by some higher body (e.g., an international accreditation board, established under the auspices of the FCCC). This board would certify companies and make sure these companies are abiding by certain standards (e.g., via spot auditing). For instance, Societe Generale de Surveillance (SGS), Rainforest Alliance, and the Soil Association are certification companies that are accredited by the Forest Stewardship Council. They certify that forests meet the standards of the Forest Stewardship Council as set forth in their Principles and Criteria for Forest Management (personal communication from Pedro Moura-Costa, EcoSecurities, Ltd., Jan. 28, 1999).

Certification and verification of GHG emissions trading could be achieved by using a system of accreditation and certification, similar to that currently used for quality and environmental management systems certification, i.e. ISO 9000 and ISO 14000 respectively. In this section, accreditation is defined as the recognition, by a responsible authority, that an impartial body is competent to undertake defined activities (Jones, 1999).

Considerable experience exists throughout the world in operating these systems, and there are lessons that can be learned from that experience (Jones, 1999).

For management system certification, there are almost 50 accreditation bodies throughout the world. There are significant differences in the interpretation of international standards and accreditation criteria by the various bodies. With regard to certifying authorities, over 500 have been granted accreditation throughout the world, and this number is growing rapidly. Despite satisfying accreditation criteria, there can be a significant variation in the implementation of certification, not only between certifying authorities in different countries, but between certifying authorities within a country (Jones, 1999).

For emissions trading, the following are seen as the principle objectives of the accreditation and certification process, although this is by no means exhaustive (Jones, 1999):

- provide a service which instills, in all participants, confidence that the trading scheme regulations are being properly and consistently applied to maintain integrity world-wide;
- ensure that the trading scheme is regulated in such a way that it is equitable and free from anomalies, and that expanding the scheme does not disadvantage either the incumbents or the new entrants;
- be independent, auditable, rigorous and transparent.

For emissions trading, the certifying authority would:

- audit emission records of entities participating, or wishing to participate, in emission trading;
- validate, through the certification process, the permits participants wish to trade; and,
- provide other certification services as required by domestic arrangements.

Other tasks may be added to these for the certifying authority when the rules, modalities and guidelines for emissions trading have been agreed by the COP. Individual countries may also wish to extend these activities to suit national requirements. The structure of the accreditation and certification scheme is a very important factor in ensuring the integrity of emissions trading. It is in the interests of all parties to ensure that trading and compliance is properly applied in a consistent manner. The concept of an International Accreditation Body (IAB) has been proposed as one model for providing the best opportunity to ensure consistent certification standards throughout the world (Jones, 1999). However, this may be perceived by some governments as an infringement of their sovereignty. Even with an IAB, it is possible that some governments will appoint a national accreditation body, which will be responsible for accrediting certifying authorities within their country.

6. COSTS

Monitoring and evaluation costs will depend on what information is needed, what information and resources are already available, project type, the size of the project area, the monitoring methods to be used, and frequency of monitoring. Furthermore, some methods require high initial costs (e.g., in remote sensing for forestry projects, start-up costs in terms of equipment and personnel training may make a one-time digital image survey prohibitively expensive, while making multiple surveys exceedingly cost effective). The cost for monitoring a forestry project in India has been estimated at 8.5% of the total project cost, and it seems that monitoring similar projects would not exceed 10% of the total cost (Ravindranath and Bhat, 1997). In some cases, the monitoring and evaluation costs can be as high as 20% (personal communication from Margo Burnham, The Nature Conservancy, Jan. 28, 1999). Similarly, based on the experience of utilities and energy service companies, monitoring and evaluation activities can easily account for 5-10% of an energy-efficiency project's budget (Meier and Solomon, 1995; Raab and Violette, 1994; see also Kats et al., 1999). However, we expect monitoring and evaluation costs to decrease over time since the cost of measurement is coming down with the costs of communications, chips, computers, etc. (personal communication from Greg Kats, U.S. DOE, April 30, 1999).

Due to the availability of funding, we realize that some project developers and evaluators will not be able to conduct the most data intensive methods proposed in this chapter; however, we expect each project to undergo some evaluation and verification in order to receive C credits (especially, CERs). Moreover, we believe that monitored projects will save more C and offset the cost of the monitoring because installations following a monitoring and evaluation protocol should come in near or even above the projected level of C savings. Installations with some measurement of C emissions should tend to have higher levels of saved C initially and experience C savings that remain high during the lifetime of the measure (e.g., Kats et al., 1996). In the end, the cost of monitoring and evaluation will be partially determined by its value in reducing the uncertainty of C credits: e.g., will one be able to receive C credits with a value greater than 10% of project costs that are spent on monitoring and evaluation?

Because of concerns about high costs, MERVC activities cannot be too burdensome: in general, the higher the costs, the less likely organizations and countries will try to develop and implement climate change mitigation projects. However, in some cases, due to the enormous cost differential between the C reduction options of FCCC Parties, fairly high costs can be accommodated before these costs become prohibitive. Nevertheless,

MERVC costs should be as low as possible. In sum, actual (as well as perceived) MERVC costs may discourage some transactions from occurring. Tradeoffs are inevitable, and a balance needs to be made between project implementation and the level of detail (and costs) of MERVC reporting guidelines.

Project estimates of impacts could be adjusted, based on the amount of uncertainty associated with the estimates and potential leakage, without conducting project-specific analyses. Projects with less accurate or less precisely quantified benefit estimates would have their estimates adjusted and therefore have their benefits rendered policy-equivalent to credits from projects that can be more accurately quantified. The USA EPA's Conservation Verification Protocol reward more rigorous methods of verifying energy savings by allowing a higher share of the savings to qualify for tradable SO_2 allowances. Three options are available for verifying subsequent-year energy savings: monitoring, inspection and default option (Meier and Solomon, 1995). In the monitoring option, a utility can obtain credit for a greater fraction of the savings and for a longer period: biennial verification in subsequent years 1 and 3 (including inspection) is required, and savings for the remainder of physical lifetimes are the average of the last two measurements. The monitoring option requires a 75% confidence in subsequent-year savings (like in the first year). In contrast, the default option greatly restricts the allowable savings: 50% of first-year savings, and limited to one-half of the measure's lifetime. For the inspection option (confirming that the measures are both present and operating): a utility can obtain credit for 75% of first-year savings for units present and operating for half of physical lifetime (with biennial inspections), or 90% of first-year savings for physical lifetimes of measures that do not require active operation or maintenance (e.g., building shell insulation, pipe insulation and window improvements). Thus, utilities could use a simpler evaluation method at a lower cost and receive fewer credits, or they could use a more sophisticated method and receive more credits. A similar system could be applied to the crediting of LUCF projects.

7. CONCLUSIONS

Monitoring and evaluation of climate change mitigation projects is needed to accurately determine the net GHG, and other, benefits and costs, and to ensure that the global climate is protected and that country obligations are met. The five-year AIJ pilot phase is still in its early stages (most projects were started in 1997-98) and offers little insight into the experience of monitoring and evaluation so far. Evaluation of some of these projects is starting to improve our understanding of key MERVC issues (e.g., how slight changes in the estimates of deforestation can significantly affect the amount of C saved by a C offset project (Busch et al., 1999), and we anticipate more evaluations of these projects in the near future.

Nevertheless, we agree with the findings of the second synthesis study on the AIJ Pilot which concluded that work needed to be conducted on methodological, technical and institutional issues, including modalities for measurement, reporting and assessment (UNFCCC, 1998a). We also agree with the findings from the recent OECD study on emission baselines for AIJ projects which concluded that simple reporting measures were needed for improving the transparency and comparability of different projects (for AIJ, JI, and CDM projects), including project-specific emission baselines (OECD, 1999). Some MERVC issues have been examined in the AIJ Pilot, but future investments are needed to refine methods and protocols in support of the Kyoto Protocol flexibility mechanisms. Some progress has been made in the development of guidance documents for the MERVC of energy-efficiency and LUCF projects (Vine and Sathaye, 1999, and Vine et al., 1999). However, more work needs to be done for developing internationally agreed MERVC guidelines such as evaluation of additionality, free riders, project leakage, positive project spillover, market transformation, environmental impacts, and socioeconomic impacts. A community of MERVC evaluators and verifiers has been developed in response to the UNFCCC AIJ Pilot, and these individuals and organizations are involved in working on the details of implementing the modalities for measurement, reporting and assessment. Institutional and human capacity building is sorely needed to implement future CDM and JI projects, as well as emissions trading regimes.

ACKNOWLEDGMENTS

We would like to thank Maurice N. LeFranc, Jr. of the U.S. Environmental Protection Agency, Climate Policy and Program Division, Office of Economics and Environment, Office of Policy, Planning and Evaluation for his assistance. This work was supported by the U.S. Environmental Protection Agency through the U.S. Department of Energy under Contract No. DE-AC03-76SF00098.

REFERENCES

Andrasko, K. (1997) Forest Management for Greenhouse Gas Benefits: Resolving Monitoring Issues Across Project and National Boundaries, *Mitigation and Adaptation Strategies for Global Change*, 2: 117-132.

Andrasko, K., Carter, L., and van der Gaast, W. (1996) Technical Issues in JI/AIJ Projects: A Survey and Potential Responses, a background paper prepared for the Critical Issues Working Group, for the UNEP AIJ Conference. San Jose, Costa Rica.

Bialy, J. (1991) Improved Cookstove Programs in Sri Lanka: Perceptions of Success, Country Studies No. 3, in *Improved Biomass Cookstove Programs: A Global Evaluation*, East-West Center, Honolulu, Hawaii:.

Busch, C., Sathaye, J., and Sanchez-Azofeifa, A.. (1999) *Lessons for Greenhouse Gas Accounting: A Case Study of Costa Rica's Protected Areas Project*, LBNL-42289. : Lawrence Berkeley National Laboratory, Berkeley, CA.

California Public Utilities Commission (CPUC). (1998) *Protocols and Procedures for the Verification of Costs, Benefits, and Shareholder Earnings from Demand-Side Management Programs*. California Public Utilities Commission, San Francisco, CA.

Chen, A. and Vine, E. (1998). *A Scoping Study on the Costs of Indoor Air Quality Illnesses: An Insurance Loss Reduction Perspective*, LBNL-41919. Lawrence Berkeley National Laboratory, Berkeley, CA:. This report can be downloaded via the World Wide Web: http://eetd.lbl.gov/CBS/Insurance/CIpubs.html.

Community of Guguletu, PEER Consultants, P.C., and International Institute for Energy Conservation. (1998). Housing for a Sustainable South Africa: The Guguletu Eco-Homes Project, USIJI Project Proposal. International Institute for Energy Conservation, Washington, D.C.

De Jong, B., Tipper, R., and Taylor, J. (1997) A Framework for Monitoring and Evaluating Carbon Mitigation by Farm Forestry Projects: Example of a Demonstration Project in Chiapas, Mexico, *Mitigation and Adaptation Strategies for Global Change*, 2(2-3):231-246.

EcoSecurities, Ltd. (1998) *SGS Forestry Carbon Offset Verification Services*. Draft. SGS Forestry, Oxford, United Kingdom.

Eto, J., Prahl, R., and Schlegel, J. (1996) *A Scoping Study on Energy-Efficiency Market Transformation by California Utility DSM Programs*, LBNL-39058. Lawrence Berkeley National Laboratory, Berkeley, CA.

Face Foundation. (1997) *Annual Report 1996*. : Face Foundation, Arnheim, The Netherlands.

Forest Stewardship Council. (1996) *Principles and Criteria for Forest Management*. Forest Stewardship Council, U.S., Waterbury, VT.

Geller, H. (1997) *Market Transformation through PROCEL: Brazil's National Electricity Conservation Program*. American Council for an Energy-Efficient Economy. Washington, D.C.

Goldberg, M. and Schlegel, J. (1997) Technical and Statistical Evaluation Issues, Ch. 7 in J. Schlegel, M. Goldberg, J. Raab, R. Prahl, M. Keneipp, and D. Violette (eds.), *Evaluating Energy-Efficiency Programs in a Restructured Industry Environment: A Handbook for PUC Staff*, National Association of Regulatory Utility Commissioners, Washington, D.C.

Greenhouse Challenge. (1998) *Vegetation Sinks Workbook*. Australian Greenhouse Office, Canberra, Australia.

Hagler Bailly. (1998) *Evaluation of Using Benchmarks to Satisfy the Additionality Criterion for Joint Implementation Projects*, prepared for the U.S. Environmental Protection Agency. Hagler Bailly, Boulder, CO.

Jepma, C. (1997) Reduced-Impact Logging in Indonesia," *Joint Implementation Quarterly* 3:3(2).

Jones, J. (1999) Certification and Verification of Greenhouse Gas Emissions Trading, presented at Investing in the Clean Development Mechanism (CDM) & Emissions Trading Conference, The ANA Hotel, Sydney, Australia Marcy 24-25, 1999.

Kats, G., Rosenfeld, A., McIntosh, T., and McGaraghan, S. (1996) Energy Efficiency as a Commodity: The Emergence of an Efficiency Secondary Market for Savings in Commercial Buildings, in *Proceedings of the 1996 ACEEE Summer Study*, American Council for an Energy-Efficient Economy, Washington, D.C., Vol. 5, pp. 111-122.

Kats, G., Rosenfeld, A., and McGaraghan, S. (1997) Energy Efficiency as a Commodity: The Emergence of a Secondary Market for Efficiency Savings in Commercial Buildings, in *Proceedings of the 1997 ECEEE Summer Study*, European Council for an Energy-Efficient Economy, Paris, France, ID 176, pp. 1-14

Kats, G., Kumar, S., and Rosenfeld, A. (1999) The Role for an International Measurement & Verification Standard in Reducing Pollution, in *Proceedings of the 1999 ECEEE Summer Study*, European Council for an Energy-Efficient Economy Paris, France.

Kromer, J. and Schiller, S. (1996) National Measurement and Verification Protocols, in *Proceedings of the 1996 ACEEE Summer Study*, American Council for an Energy-Efficient Economy, Washington, D.C., Vol. 5, pp. 141-146.

Lashof, D. (1998) Additionality Under the Clean Development Mechanism, Natural Resources Defense Council, New York.

MacDicken, K. (1997) Project Specific Monitoring and Verification: State of the Art and Challenges, *Mitigation and Adaptation Strategies for Global Change*, 2(2-3): 191-202.

MacDicken, K. (1998) *A Guide to Monitoring Carbon Storage in Forestry and Agroforestry Projects*, Winrock International Institute for Agricultural Development, Arlington, VA.

Martinot, E. (1998) *Monitoring and Evaluation of Market Development in World Bank-GEF Climate Change Projects*. The World Bank, Washington, D.C.

Meier, A. and Solomon, B. (1995) The EPA's Protocols for Verifying Savings from Utility Energy-Conservation Programs, *Energy: The International Journal* 20(2): 105-115.

Meyers, S. (1998) *Improving Energy Efficiency: Strategies for Supporting Sustained Market Evolution in Developing and Transitioning Countries*, LBNL-41460. Lawrence Berkeley National Laboratory, Berkeley, CA.

Michaelowa, A. (1998) Joint Implementation – the Baseline Issue: Economic and Political Aspects, *Global Environmental Change* 8(1): 81-92.

Organization for Economic Co-operation and Development (OECD). (1999) Experience With Emission Baselines Under the AIJ Pilot Phase, Organization for Economic Co-operation and Development, Paris, France.

Prahl, R. and Schlegel, J. 1993. Evaluating Market Transformation, in the *Proceedings of the 1993 International Energy Program Evaluation Conference*. National Energy Program Evaluation Conference, Chicago, IL, pp. 469-477.

Programme for Belize. (1997) Rio Bravo Carbon Sequestration Pilot Project, Operating Protocols, Introduction, Programme for Belize.

Puhl, I. (1998) *Status of Research on Project Baselines Under the UNFCCC and the Kyoto Protocol*, OECD/IEA Information Paper, Organization for Economic Co-operation and Development, Paris, France.

Raab, J. and Violette, D. (1994) *Regulating DSM Program Evaluation: Policy and Administrative Issues for Public Utility Commissioners*. National Association of Regulatory Utility Commissioners, Washington, DC.

Ravindranath, N. H. and Bhat, P. R. (1997) Monitoring of Carbon Abatement in Forestry Projects—Case Study of Western Ghat Project, *Mitigation and Adaptation Strategies for Global Change*, 2(2-3):217-230.

Saxonis, W. (1991) Free Riders and Other Factors that Affect Net Program Impacts, in E. Hirst and J. Reed (eds.), *Handbook of Evaluation of Utility DSM Programs*, Oak Ridge National Laboratory, Oak Ridge, TN.

Schlegel, J., Prahl, R. and Raab, J. (1997) Nest Steps for Evaluation of Market Transformation Initiatives: An Update to the NARUC Guidebook, Ch. 4 in J. Schlegel, M. Goldberg, J. Raab, R. Prahl, M. Keneipp, and D. Violette (eds.), *Evaluating Energy-Efficiency Programs in a Restructured Industry Environment: A Handbook for PUC Staff*, National Association of Regulatory Utility Commissioners, Washington, D.C.

Subsidiary Body for Scientific and Technological Advice (SBSTA). (1997) Report of the Subsidiary Body for Scientific and Technological Advice on the Work of Its Fifth Session, Bonn, 25-28 February 1997. Annex III. Uniform Reporting Format: Activities Implemented Jointly Under the Pilot Phase. Framework Convention on Climate Change, United Nations.

Suvilehto, H. and Öfverholm, E. (1998) Swedish Procurement and Market Activities — Different Design Solutions on Different Markets, in *Proceedings of the 1998 ACEEE Summer Study on Energy Efficiency in Buildings*. American Society for an Energy-Efficient Economy, Washington, D.C., Vol. 7, pp. 311-322.

Swisher, J. (1998) Project Baselines and Additionality in the Clean Development Mechanism, presented at The Aspen Global Forum, Aspen, CO.

Trexler, M. and Kosloff, L. (1998) The 1997 Kyoto Protocol: What Does It Mean for Project-Based Climate Change Mitigation, *Mitigation and Adaptation Strategies for Global Change*, 3:1-58.

UNFCCC. (1998a) Second Synthesis Report on Activities Implemented Jointly, FCCC/CP/1998/2, at UNFCCC Web site (under CC:INFO Products): http://www.unfccc.de/ccinfo.

UNFCCC. (1998b) Non-paper on Principles, Modalities, Rules and Guidelines for an International Emissions Trading Regime, FCCC/SB/1998/MISC.1/Add.1/Rev.1, at UNFCCC Web site (under CC:INFO Products): http://www.unfccc.de/ccinfo.

U.S. Department of Energy (USDOE). (1997) *International Performance Measurement and Verification Protocol*. U.S. Department of Energy, Washington, D.C.

U.S. Department of Energy (USDOE). (1994a) General Guidelines for the Voluntary Reporting of Greenhouse Gases Under Section 1605(b) of the Energy Policy Act of 1992, DOE/PO-0028, U.S. Department of Energy, Washington, D.C., Vol. 1.

U.S. Department of Energy (USDOE). (1994b) Sector-Specific Issues and Reporting Methodologies Supporting the General Guidelines for the Voluntary Reporting of Greenhouse Gases Under Section 1605(b) of the Energy Policy Act of 1992, DOE/PO-0028, U.S. Department of Energy, Washington, D.C., Vols. 2 and 3.

U.S. Environmental Protection Agency (USEPA). (1995) *Conservation and Verification Protocols, Version 2.0*. EPA 430/B-95-012, U.S. Environmental Protection Agency, Washington, D.C.

U.S. Environmental Protection Agency (USEPA). (1996) *The User's Guide to the Conservation and Verification Protocols, Version 2.0.*, EPA 430/B-96-002, U.S. Environmental Protection Agency, Washington, D.C.

U.S. Initiative on Joint Implementation (USIJI). (1996) *Guidelines for a USIJI Project Proposal*. U.S. Initiative on Joint Implementation, Washington, D.C.

U.S. Initiative on Joint Implementation (USIJI). (1998) *Activities Implemented Jointly: Second Report to the Secretariat of the United Nations Framework Convention on Climate Change*. U.S. Initiative on Joint Implementation, Washington, D.C.

University of Edinburgh. (1998) Provisional Guidelines and Standards. See Web site: http://www.ed.ac.uk/~ebfr11/ecor.

Vine, E. (1994) The Human Dimension of Program Evaluation, *Energy - The International Journal*, 19(2): 165-178.

Vine, E. and Sathaye, J. (1997) *The Monitoring, Evaluation, Reporting, and Verification of Climate Change Mitigation Projects: Discussion of Issues and Methodologies and Review of Existing Protocols and Guidelines*, LBNL-40316. Lawrence Berkeley National Laboratory, Berkeley, CA: This report can be downloaded from the World Wide Web: http://eetd.lbl.gov/EA/ccm/MonitoringMitigation.pdf.

Vine, E. and Sathaye, J. (1999) *Guidelines for the Monitoring, Evaluation, Reporting, Verification, and Certification of Energy-Efficiency Projects for Climate Change Mitigation*, LBNL-41543. : Lawrence Berkeley National Laboratory, Berkeley, CA. This report can be downloaded from the World Wide Web: http://eetd.lbl.gov/ea/ccm/ccPubs.html.

Vine, E., Sathaye, J., and Makundi, W. (1999) *Guidelines for the Monitoring, Evaluation, Reporting, Verification, and Certification of Forestry Projects for Climate Change Mitigation*, LBNL-41877. Lawrence Berkeley National Laboratory, Berkeley, CA. This report can be downloaded from the World Wide Web: http://eetd.lbl.gov/ea/ccm/ccPubs.html.

Vine, E., Mills, E., and Chen, A. (1998) *Energy-Efficiency and Renewable Energy Options for Risk Management and Insurance Loss Reduction: An Inventory of Technologies, Research Capabilities, and Research Facilities at the U.S. Department of Energy's National Laboratories*, LBNL-41432, Lawrence Berkeley National Laboratory, Berkeley, CA. This report can be downloaded via the World Wide Web: http://eetd.lbl.gov/CBS/Insurance/CIpubs.html.

Violette, D., Ragland, S., and Stern, F. (1998) *Evaluating Greenhouse Gas Mitigation through DSM Projects: Lessons Learned from DSM Evaluation in the United States*, Hagler Bailly Consulting, Inc., Boulder, CO.

Watt, E., Sathaye, J., de Buen, O., Masera, O., Gelil, I., Ravindranath, N., Zhou, D., Li, J., and Intaraprvich, D. (1995) *The Institutional Needs of Joint Implementation Projects*, LBL-36453, Lawrence Berkeley National Laboratory, Berkeley, CA.

Weisbrod, G., Train, K., Hub, A., and Benenson, P. (1994) *DSM Program Spillover Effects*. Cambridge Systematics, Inc., Cambridge, MA:

World Bank. (1989). Checklist of Potential Issues for an Environmental Assessment, Operating Manual: Section OD 4.00 - Annex A2. World Bank, Washington, D.C.

World Bank (1994a) *Greenhouse Gas Abatement Investment Project Monitoring & Evaluation Guidelines*. World Bank, Washington, D.C.

World Bank. (1994b) *Incorporating Social Assessment and Participation into Biodiversity Conservation Projects*. World Bank, Washington, D.C.

Chapter 14

MARKET BASED FRAMEWORK FOR CDM TRANSACTIONS

PAUL HASSING[1] and MATTHEW S. MENDIS[2]
[1]Ministry of Foreign Affairs, Government of the Netherlands; [2]Alternative Energy Development, Inc.

Key words: Clean Development Mechanism, additionality, baselines, CDM validation, CDM project cycle, emission reduction units, certified emission reductions, market mechanisms, project finance.

Abstract: Based on the past experience of various Activities Implemented Jointly (AIJ), this chapter presents a step-by-step process by which greenhouse gas (GHG) mitigation projects can be identified, developed and validated as Clean Development Mechanism (CDM) projects and, based on their validation, reach financial closure and implementation. The chapter illustrates how the resulting emission reductions (ERs) must be monitored, reported, verified and ultimately certified to reach the status of certified emission reductions (CERs). The chapter identifies the roles of the key players in each of these steps and discusses some of the more relevant issues associated with the process of creating CERs. The paper concludes with a brief discussion of the international market opportunities for trading of CERs and the need for a national and international tracking system to account for the CERs used to meet the compliance requirements of the Annex I countries listed in Annex B of the Kyoto Protocol.

1. INTRODUCTION

1.1 Objectives

The principal objective of this chapter is to present the basic elements of an international market framework that are necessary to encourage and facilitate the development and implementation of CDM projects and the production of

CERs. The term CER is derived from Article 12 of the Kyoto Protocol. It refers to GHG ER that has received the certification of the CDM and can, therefore, be used to assist in achieving compliance in the first commitment period by the Parties listen in Annex B of the Protocol. The chapter discusses the process and issues associated with the identification, development, investment, implementation, monitoring, verification and certification cycle associated with a CDM project. It also briefly outlines the baseline definition and project validation process that is unique and crucial for the environmental integrity of CDM projects. The international market framework provides the basis for partnership between project developers, investors, national authorities in host and recipient countries and the international agencies that are responsible for the implementation of the Kyoto Protocol.

A successful international market framework for CDM transactions must be driven by a number of fundamental principles. The framework must:

- Result in agreed sustainable development that meets national objectives for the host country and not just CERs for the recipient country;
- Help maximize the generation or supply of cost-effective CERs;
- Provide reliable information and secure access for the buyers of CERs;
- Provide legal recourse for both buyers and sellers of CER's;
- Meet the needs of a wide spectrum of potentially diverse project types and proponents;
- Provide a real incentive for a broad base of investors to invest in CDM projects and not just attract a limited band of green investors; and
- Result in CDM projects that are additional to defined baselines.

1.2 Background

Significant time and resources have been spent to date to study, debate and develop a framework for the CDM (Jepma and van der Gaast 1999). Most of the discussions and efforts have focused on the political dimensions, equity issues, baseline definitions and regulatory aspects of the CDM while downplaying the framework for investments and market transactions that is necessary for the CDM to be effective. A review of past literature on the CDM indicates a number of areas that would benefit from further elaboration. These include:

- Market forces and mechanisms that must be employed if volume and efficiency are to be achieved in the production and sale of CERs. Haites (1998) and other estimate the potential annual market for CDM at between 5.3 to 25.9 billion $US.

- Investor's attitudes, behavior and requirements which must be clearly reflected in the emerging regulatory governance of the CDM;
- Clear and transparent rules and guidelines for the definition of baselines and additionality associated with CDM projects;
- The responsibility for and integration of CDM monitoring, verification and certification processes within the overall investment and implementation project cycle; and
- Legal recourse for failed or delayed transactions undertaken within the CDM framework.

1.3 Scope of the Paper

This chapter provides a blueprint for the main building blocks of a practical international market framework for investments in CDM projects. It builds on an earlier report by Hassing and Mendis (1998) which presents a preliminary structure for operating and managing transactions in the CDM. This chapter seeks to provide an integrated overview of the decision making process, key issues and relevant actors involved in the CDM project cycle. The blueprint takes into account the regulatory hurdles, approval process, permit and license requirements and legal and contractual agreements that are necessary for the implementation, operation and production of ERs in a CDM project. The chapter seeks to provide some guidance to host-country governments considering participation in CDM activities as well as investors seeking to understand the CDM business. It addresses the key steps in the creation of CERs.

Figure 1 diagrams the key steps in the decision-making process of a CDM project. In addition, the diagram identifies the corresponding sections in the chapter that identifies and elaborates the roles and responsibilities of the principal stakeholders and discusses the relevant issues associated with each step of the CDM project cycle. Particular attention is given to the process of baseline definition, CDM project validation and ER monitoring, verification and certification. The Kyoto Protocol refers to certified project activities (Article 12.6) and verification of project activities (Article 12.7). This paper makes a clear distinction between validation, verification and certification. The chapter shows how the successful completion of all the CDM project steps can lead to certified emission reductions (CERs) that meet the requirements of Article 12 of the Kyoto Protocol.

2. CDM GOVERNANCE

Establishing clear rules, regulations and guidelines for the CDM will help stimulate the identification and development of projects that are eligible for the CDM. An established regulatory framework will distinguish the CDM from the unregulated character of the AIJ pilot phase. In the AIJ pilot phase, project identification and implementation occurred in the absence of a uniform and formal regulatory infrastructure. Guidelines for AIJ projects were established in bilateral discussions and varied considerably from one Party to another.

National and international rules and regulations, that meet the requirements of the Kyoto Protocol, will provide the governance for the contribution of the CDM instrument to sustainable development, project financing and the global GHG mitigation (Carraro 1999). All non-Annex I countries, that ratify the Kyoto Protocol, are in principal eligible to participate in the CDM. However, a host country's willingness to participate in the CDM must be clearly established by the country and is voluntary in accordance with Article 12.5 (a). Additionally, the eventual acceptance of a project as a CDM certified project activity will be dependent on the international and national rules that govern the CDM. Thus, identification and development of projects for the CDM should not be undertaken independent of the governance of the CDM but in association with it.

2.1 International CDM Executive Board

Article 12 of the Kyoto Protocol indicates that the responsibility for the design of the international CDM regulatory framework falls under the authority of the Conference of the Parties (COP) U.N. Framework Convention on Climate Change (FCCC) to the Kyoto Protocol. It clearly indicates that the CDM will be subject to the authority and guidance of the COP and/or Meeting of the Parties (MOP) and will be supervised by an Executive Board of the CDM that is appointed by the COP/MOP. Article 12 endows the Executive Board of the CDM with a number of responsibilities. These include:

- Determining the participation in the CDM including which countries and non-Party entities (i.e. private sector, international organizations) are eligible;

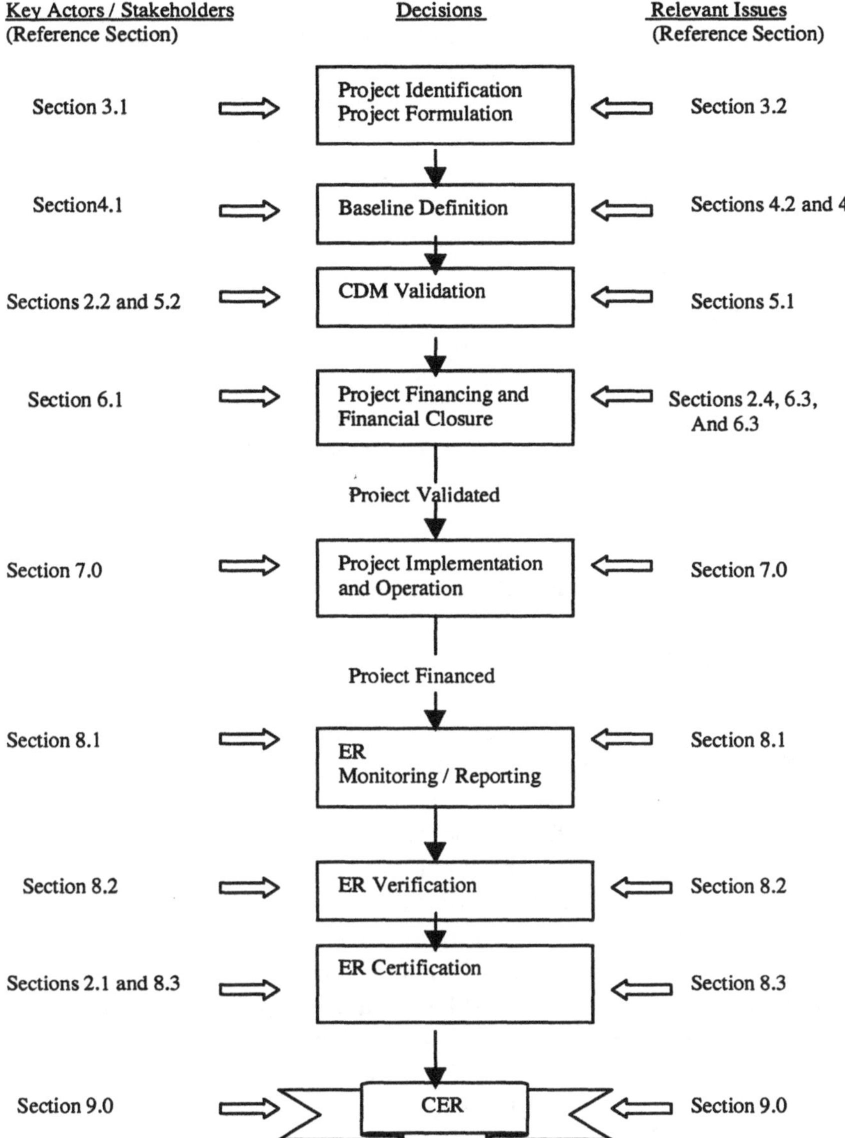

Figure 1. Key steps, actors, and issues in the decision making process of a CDM project.

- Setting the guidelines and rules for project eligibility and the determination of project baselines;
- Creating and overseeing a mechanism for the independent verification and certification of CDM project activities;

- Ensuring that CDM project emission reductions are additional to any that would occur in the absence of the certified project activity;
- Ensuring that certified or validated CDM projects have real, measurable and long-term benefits related to the mitigation of climate change;
- Assisting in arranging funding of certified project activities as necessary; and
- Ensuring that a share of the proceeds from certified project activities is used to cover CDM administrative expenses as well as to assist developing country Parties, that are particularly vulnerable to the adverse effects of climate change, meet the costs of adaptation.

The last item clearly anticipates the CDM Executive Board creating an administrative operational entity or body to implement the objectives, rules and guidelines of the CDM. This administrative body will serve to implement the international regulatory framework for the CDM and will interact with national entities undertaking CDM activities. The international CDM administrative body could also serve to establish an international certification and tracking system for CERs generated by CDM activities. This would help provide the necessary database for assessment of the compliance reports of Annex I countries that include CERs.

2.2 National CDM Board

The Kyoto Protocol does not require the establishment of a national CDM board or regulatory framework. However, to facilitate the generation of CDM activities at the national level and to ensure that proposed CDM activities are consistent with national and sustainable development objectives, it is anticipated that national agencies will be designated to review and approve proposed CDM activities. It is not immediately evident or necessary that the national CDM regulatory framework be consistent across recipient or host countries. What is important is that it is not contradictory to the international CDM regulatory framework. Therefore, while the process for recognition and approval of proposed CDM activities is likely to vary from country to country the resulting eligibility of projects for the CDM should have a consistent base across countries.

The responsibilities of the national CDM Board could include:

- Validating eligible CDM project activities that (a) meet national priorities; (b) contribute to sustainable development; and (c) result in real, measurable and long-term benefits related to mitigation of climate change;
- Validating the baselines associated with CDM project activities on the basis of criteria that are established by international rules and national development priorities;

- Facilitating investments in approved national CDM project activities;
- Establishing the rules and guidelines for monitoring of CDM project activities to ensure the availability of the data needed for independent verification of the resulting ERs; and
- Tracking and registering the production and transfer of ERs from approved CDM project activities.

2.3 Operational Elements of a CER Market Framework

In addition to the international and national regulatory framework outlined above, a number of operational elements need to emerge to support the development of a CER market. These include:

- Institutions, specialists, or consultants that provide technical inputs for identification, formulation and development of CDM projects and project baselines;
- Institutions, banks, or development agencies that provide or secure financial resources for developing and financing of CDM projects;
- Approved agents or agencies that are capable of providing monitoring, verification and certification services for CDM projects;
- Markets or information sites where potential sellers and buyers can obtain price and other relevant information relating to the supply and demand for CERs; and
- Brokers that bring potential buyers and sellers together to assist in the selling and purchasing of CERs and the recording of binding transactions.

Many of these elements are already evolving in anticipation of the CDM and the market for CERs. Some of these elements have been active in AIJ transactions while others have operated in other related emissions trading markets. The rules governing the operation of these elements will derive from the rules established by the CDM Executive Board and the national CDM authorities.

2.4 Legal Environment

A necessary but not sufficient pre-condition for attracting investments in CDM projects is a legal environment that fosters general project investments. An environment in which rules and regulations are not in place to protect and foster general project investments will certainly not be encouraging for CDM project investments as well. Additionally, environments in which internal and/or external investors have no legal recourse or have difficulty in gaining fair legal recourse will also inhibit CDM project investments. Thus, countries without favorable legal and regulatory investment environments will have difficulty in attracting CDM project investments as well.

3. CDM PROJECT IDENTIFICATION AND FORMULATION

The first and most important step in the CDM project cycle is the identification and formulation of potential CDM projects. A step in the wrong direction may lead to the expenditure of considerable time and resources which may result in a project that is not eligible or acceptable for the CDM. The lessons learned from the AIJ Pilot Program regarding the eligibility of projects and the definition of baselines need to be reviewed in order to provide clearer guidance to determine, at an early stage, the eligibility of proposed projects for the CDM. This would help conserve and channel scarce resources to project activities that will ultimately be certified by the CDM. To help conserve resources, many AIJ project developers resorted to forwarding projects that had already undergone considerable preparation and repackaged them for AIJ consideration. This process will be most likely repeated at the outset of the CDM to rapidly bring projects forward for consideration. Figure 2 illustrates the key actors, stakeholders, and relevant issues associated with the project identification and formulation step.

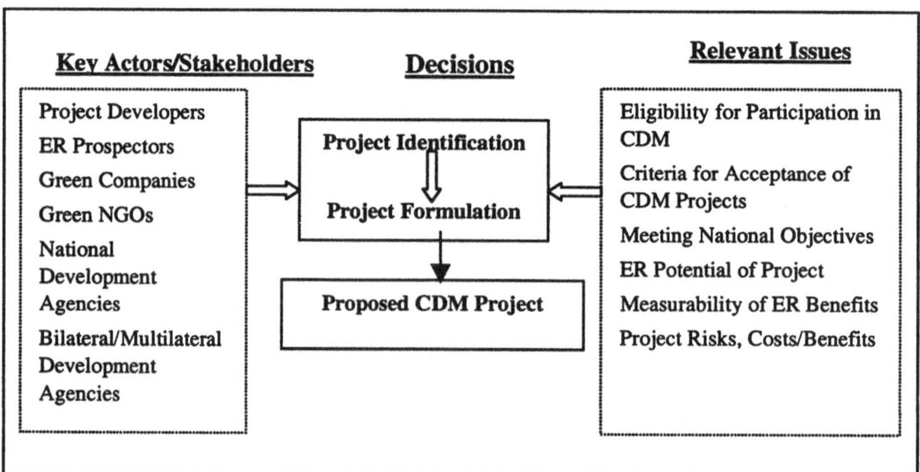

Figure 2: Key actors, stakeholders, decision and issues in CDM project identification and formulation.

3.1 Project Proponents

There is a broad spectrum of project proponents and parties involved in the project identification and formulation stages as identified in Figure 2. The ongoing pilot AIJ pilot has resulted in the emergence of those players interested in the development of ER credits. However, many of the key players have, to date, been reluctant to invest a great deal of time and resources as the rules of the AIJ pilot and the proposed CDM are not clearly defined or well established. Most players have ventured into the AIJ arena with support from Annex I development or environmental agencies using public funds. The mobilization and use of private funds has been limited in some countries (e.g., Japan), but not others (USA).

For the market in CDM projects to flourish, the rules of the CDM must be clear and transparent. The key players involved in project identification and formulation, recognizing the potential value of this market, will independently invest and undertake the front-end work needed to successfully develop CDM projects. By fostering the involvement of independent players, the CDM can avoid the need for expending public funds for project development while relying on the marketplace to mobilize and apply funds for development of projects for the CDM.

3.2 Project Eligibility Criteria

The Kyoto Protocol in Article 12 stipulates four principal eligibility criteria for CDM projects. They are:

- Non-Annex I Parties will benefit from project activities resulting in CERs;
- Projects must assist Non-Annex I Parties in achieving sustainable development and contributing to the ultimate objective of the Convention.
- Projects must result in real, measurable and long-term benefits related to the mitigation of climate change; and
- Projects must result in reductions in emissions that are additional to any that would occur in the absence of the certified project activity.

3.2.1 National Benefits from Project Activities

The first two criteria are clearly established to assist developing countries in achieving economic, social, environmental and sustainable development objectives while reducing GHG emissions. It clearly prohibits projects that do not have any direct benefits or may have negative benefits for the host country but still result in GHG emission reductions. This criterion is particularly important for projects that may produce GHG emission reductions or carbon dioxide (CO_2) sinks but which may not have any additional benefits for the host country. An example may be a project to reduce methane emissions from rice agriculture that does not reduce the cost of rice production or increase rice yields.

The determination of whether projects help host countries achieve sustainable development must lie with the host country but can be governed by a broad set of guidelines established by the CDM Executive Board. There is no operational or objective method to determine if a project contributes to a country's sustainable development. Attempts are underway to find indicators for sustainable development, but general acceptance of any resulting indicators will have political ramifications and will need to be validated through a political process. Therefore, host countries must decide for themselves if a proposed CDM project is likely to contribute to their sustainable development (Chatterjee 1997). However, in the absence of a common standard or a common reward system for a project's contribution to sustainable development, a race to the bottom might occur where competition between projects results in the negligence of sustainable development in favor of maximizing climate mitigation benefits. Hassing and Mendis (1998) offer a solution to this problem by limiting eligibility for CDM participation to project activities with clearly proven sustainable development impacts.

3.2.2 Measurability and Long-Term Benefits Related to Climate Change

The criteria of real, measurable and long-term benefits related to the mitigation of climate change requires that the ERs associated with projects for the CDM must posses some specific characteristics.

- The ERs must be based on real reductions of GHG emissions that are directly associated with the CDM project activity.
- The ERs that are produced must be measurable or quantifiable using reliable measuring, sampling or mass balance techniques. This essentially requires that projects that produce ERs that are not directly measurable should not be eligible for validation as CDM projects.
- The ERs must be permanent. Risks associated with the permanence of GHG reductions are directly related to whether reductions are reversible at a future point in time. The issue of permanence is of particular importance in the context of CO_2 sequestration. It also relates to the possibilities of delayed or displaced leakage in which the reductions

achieved by a CDM project are eventually offset by emissions from other related activities. This could occur due to an increase in a GHG emitting activities elsewhere, in the host country of possibly even outside the host country, that is a direct result of the CDM project activity. The classic example is for the case where a preservation of a GHG sink in one part of the globe leads directly to the accelerated destruction of another sink elsewhere in the globe.

3.2.3 Additionality of Reductions

GHG emissions from CDM projects must be lower than those that would have occurred in the absence of the CDM activity. To accurately estimate the additionality of reductions of a CDM project, it is important to have an accurate portrayal of the baseline. A discussion of various methods for defining baselines and their implications on the additionality of CDM projects is presented in section 4.

An important financial consideration emerges in the process of defining baselines that must also be considered within the context of the additionality requirement of the CDM. Specifically, a CDM project should also have financial additionality in comparison to the baseline option. If a proposed CDM project is financially more attractive than the project that would occur in the baseline, then the argument can be made that the CDM project belongs in the baseline and should replace the assumed baseline project. The issue of financial additionality is the subject of considerable debate. However, without the financial additionality criteria, there is no basis to determine if baselines are an accurate reflection of expected profit maximizing behavior. The possibilities will arise for defining baselines that maximize the eligibility of projects for the CDM and thereby result in ERs that are not additional to what would have occurred in the absence of the CDM.

4. BASELINE DEFINITIONS

The definition of the baseline, against which the ERs of a proposed CDM project are assessed, is a very important step in the CDM project cycle. A project baseline defines a level of expected emissions that is used to assess the mitigation performance of an alternative project. It is the basis from which the ERs for a CDM project activity must be measured. The quantity of ERs that a potential CDM project activity can generate provides the basis for attracting the additional investments that may be needed to support the CDM project activity. Therefore, the development of a baseline for a CDM project lies at the heart of the validation process for CDM projects.

To date most of the experience with project baselines for estimating ERs has been gained in the context of the AIJ pilot phase and the Global Environmental Facility (GEF). This experience has shown that current approaches for baseline setting do provide some guidance for the CDM. Key issues such as defining the expected baseline, setting the period for which baselines should be valid and defining system boundaries for the baseline activity should be clearly addressed. Additionally, the issue of macro-economic policies and regulations that inhibit the adoption of CDM type activities should also be carefully evaluated in order minimize the potential for rewarding bad policies with CDM projects. Figure 3 identifies the key actors/stakeholders and the relevant issues associated with the process of baseline definition for CDM projects.

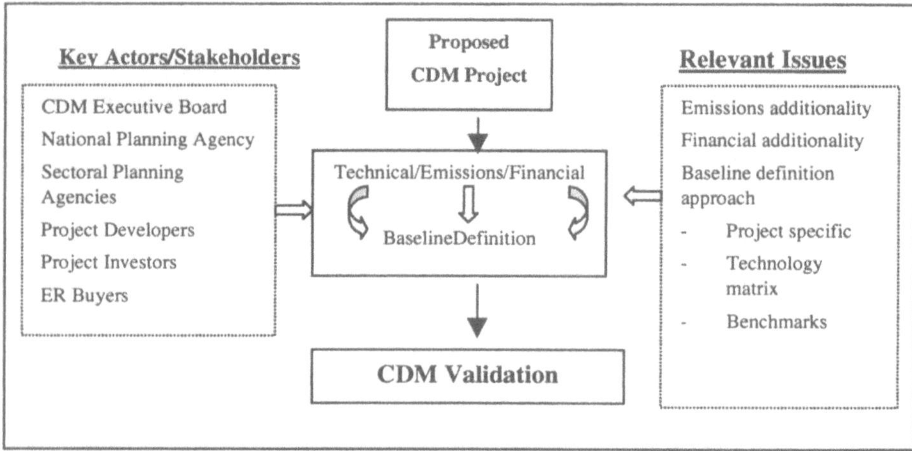

Figure 3. Key actors, stakeholders, decisions and issues in CDM baseline definition.

4.1 Responsibility

The principal responsibility for defining the baseline associated with a specific project will lie with the project developer or investor. However, the underlying assumptions and data that support a baseline definition must be derived from national or international authorities. For example, the sectoral growth rates, performance of baseline technologies, cost of baseline technologies and emission rates of baseline technologies will need to be derived from national data and ultimately validated at the national level. To the extent possible, the national authority for the CDM may establish the baseline parameters that are ultimately needed to help define project specific baselines. However, the guidelines for establishing these baseline parameters must derive from and be approved by the CDM Executive Board in order to ensure a degree of consistency across host countries.

4.2 Approaches

There are several approaches that can be used to define a baseline. These vary in the degree of aggregation and level of accuracy. There is at present an ongoing debate regarding the scope of baseline definition methodologies, the accuracy of various methodologies, the pros and cons of each methodology and the methodology that is best suited for the needs of the CDM. What is important is that if the market for ERs from the CDM is to grow, it will require that baseline definitions are consistent, transparent and relatively easy (i.e., not costly) to apply. Three methodologies for defining baselines that meet the above criteria and are presently gaining recognition in the literature. They are:

- project specific;
- technology matrix; and
- benchmarking.

These methodologies are briefly discussed below.

Project Specific: The project specific baseline approach has been established and used extensively by the Global Environment Facility (GEF). It requires that for each proposed CDM project activity, a specific baseline project must be defined which provides the equivalent normal economic benefits as does the proposed CDM project activity. The GHG emissions of the baseline project are then estimated and compared against that of the proposed CDM project activity. The difference in the GHG emissions of the CDM project activity from the baseline project is the resulting ERs of the CDM project. The principal advantage of the project specific approach is that it can provide a more reliable estimate of the ERs for a CDM project. However, the effort and costs required for undertaking this process may be considerable depending on the size and complexity of the proposed CDM project activity. Additionally, it is usually the responsibility of the project developer/investor to produce the data for the associated project baseline. The developer will have an incentive to skew the outcome of the baseline definition. Thus, the resulting baseline definition must be carefully and independently evaluated. Baseline definitions are very much like environmental impact assessments (EIAs). The EIA is usually the responsibility of the project developer but is subject to institutional review and approval by independent authorities. An additional advantage of the project specific approach is that it not as politically sensitive and can be applied without considerable up-front costs for host countries. This factor alone provides a strong argument for the use of the project specific approach in the initial stages of the CDM. Additionally, the GEF and a number of AIJ pilot programs have extensively applied the project specific baseline so there is a track record to build on.

Technology Matrix: The technology matrix approach requires that a number of pre-defined default technologies are identified as the baseline technologies for a country or region and for a specified time (Hargrave, 1998). The baseline emissions for a proposed project would be equivalent to the emissions from the pre-defined default technology that is applicable to the project. The emissions from the proposed CDM technology would be compared against this baseline and the ERs credited to the CDM project would be the difference between the resulting baseline and the CDM project. The technology matrix could be updated periodically to reflect changes in the technology base of a country or region. Thus, as new technologies are integrated into to the country or region, they would be added to the baseline technology matrix and would no longer be considered to be additional. The advantages of the technology matrix approach are that it would be simple to apply and that it would reduce opportunities for gaming by project proponents. The drawbacks of this approach are that it requires considerable up-front work to establish, it may not be accurate in estimating the expected baseline for specific projects and it requires host countries to identify default technologies which may have political implications and may not be easily adopted.

Benchmarking: In this approach, the emissions of a proposed CDM project would be compared to a benchmark emissions rate for the associated sector or sub-sector activity. For example, the baseline emissions rate for a new power plant would be set equivalent to a weighted average of the emissions rate of new power plants that are expected to be built in the absence of the CDM. This would be a forward-looking or projected benchmark. Alternately, historic or backward-looking benchmarks could also be established on the existing and past practice. Thus, a benchmark emissions rate for a power sector project activity could be defined in terms of tons of carbon (C) equivalent per unit of output (tons C_e/kWh). The degree of accuracy of benchmarks will be dependent on the degree of aggregation applied in defining benchmarks. The greater the level of aggregation, the lesser the level of accuracy for specific projects. Benchmarks could also be updated periodically to reflect changes in the baseline conditions. These are referred to as dynamic benchmarks. The advantages of the benchmark approach are similar to that of the technology matrix. However, the benchmark approach has the added advantage of weighting the basket of baseline technologies represented in the technology matrix to more accurately reflect the expected baseline emissions from a sector or sub-sector activity. The drawbacks of the benchmarking approach are also similar to that of the technology matrix approach. In addition, the benchmarking approach requires significantly more data if it is to increase the accuracy of the estimates of baseline emissions. Benchmarks could vary significantly within a country or region due to variations in fuel use or other factors. Additionally, benchmarks for sectors and sub-sectors may be viewed by the Non-Annex I countries as limits on emission per unit of activity which entail significant political ramifications. Therefore, it may be difficult to reach agreement on acceptable benchmarks.

4.3 Validity Period

The period for which a baseline is valid should be equivalent to the period for which the underlying baseline project is in fact replaced by the CDM project. However, this is not an operational definition as the underlying baseline project is counterfactual. The actual baseline conditions are dynamic and must be updated periodically (e.g., annually or every five years). It is quite conceivable that a technology used in a CDM project could become the baseline technology shortly after the initiation (and possibly because) of the CDM project.

Figure 4 illustrates a case where the cost per unit of output (e.g., \$/kWh) from the CDM project declines over time. This could result from advances in the technology or removal of market barriers up to a point in time (T) where the cost of the CDM technology is lower than the original baseline technology. Beyond time T, the CDM technology can no longer be considered financially additional to the baseline as it costs less and in fact becomes the new baseline. However, the more perplexing question is can the emissions from a CDM project initiated prior to time T continue to be considered additional to the baseline in the period after time T? The answer to this question is critical to the financial viability of CDM projects, which in principal will be dependent on the revenues from the sale of their ERs. If the financing of the CDM project is based on a revenue stream from the sale of ERs that goes beyond time T, a reclassification of the CDM project to be inclusive in the baseline would result in the loss of its ER revenue stream. This would have serious financial implications for the CDM project. Alternately, it is possible that a CDM project implemented prior to time T will continue to displace GHG emissions beyond the period time T if, in the absence of this CDM project, the baseline project would have been built and continues to operate beyond the period time T. The issue in fact requires one to answer the hypothetical question of the expected economic life of the baseline option that was replaced by the CDM option. The CDM option should receive credits for as long as the baseline option would have operated if the CDM option had not in fact displaced the baseline option.

An alternative approach is to define a fixed period (T_{b1}) for which the initial baseline definition is valid. Upon the completion of this period, the baseline would be reevaluated for a subsequent period (T_{b2}). If, in the new period, the CDM project has a lower cost than the baseline, then it would no longer be considered to be financially additional to the baseline and it would not be validated for additional ERs (even if its ERs continued to be additional to the baseline).

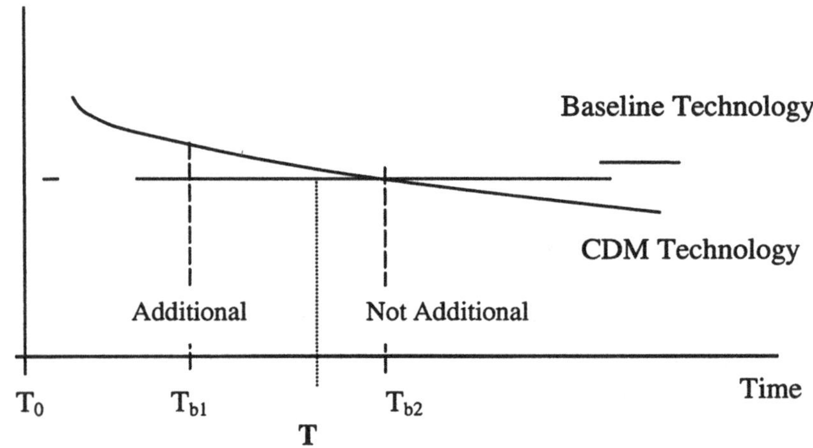

Figure 4: Validity period for CDM project financial additionality, based on case where cost per unit output decline over time.

Given these assumptions, the issue of the validity period for a baseline definition is crucial to the financing strategy of a CDM project. The shorter the validity period (T_b), the higher the risks associated with the acceptance of CDM project ERs. The net result is that the costs for the ERs will be higher. Therefore, the setting of the validity period for baselines is a critical factor that will influence the market and decision to invest in CDM projects.

The magnitude of the emissions reduction credited to a CDM project will vary depending on the validity period for the baseline as well as the definition of the "expected emissions" from the baseline. Figure 5 illustrates the impact, on the magnitude of the emissions reduction credited to a CDM project, when assuming different estimates of the emissions from the baseline for a given validity period. The emissions from the proposed CDM project are represented by "A" Mg C yr^{-1}. The emissions from the baseline project in time "T_0" is represented by "B" Mg C yr^{-1}. The baseline validity periods are represented by periods T_1, T_2 up to T_n, the expected life of the CDM project.

Three options for defining the emissions from the baseline are illustrated in Figure 5:

- step baseline;
- descending baseline; and
- static baseline.

In the case of the step baseline, the emissions from the baseline are adjusted periodically taking into account technological and financial changes that would result in lower emissions from the expected baseline option. The resulting emissions reduction credited to the CDM project can be estimated by the area of the triangle ABEA. In the case of the descending baseline, the emissions from the baseline are projected to decline annually up to a point in time, T_2, when the emissions from the baseline and CDM options are equal. The area of the triangle ABEA represents the resulting emissions reduction credited to the CDM project. Assuming that the emissions for the baseline continue to decline beyond time T_2, the CDM project would not be eligible for additional credits, as its emissions would exceed that of the expected baseline by the area EDGE.

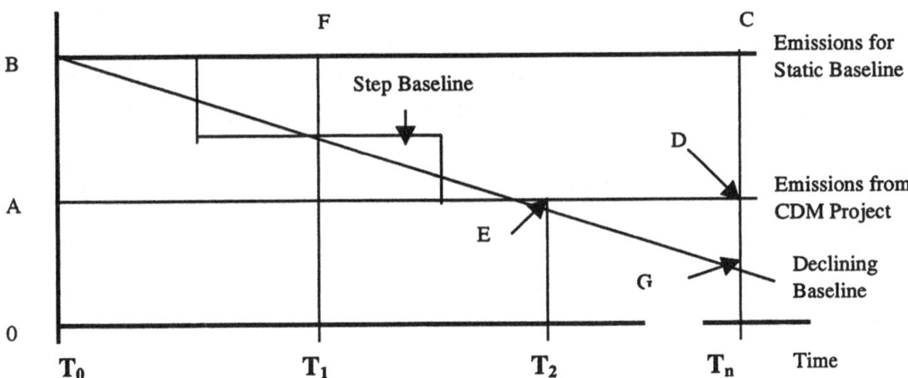

Figure 5: CDM project validity period for emissions additionality illustrating step baseline, descending baseline, and stated baseline.

B represents the emissions from the static baseline in Mg C yr^{-1} for the life of the CDM project. The area ABCD represents the resulting emissions reduction credited to the CDM project over its life. The impact of an adjustable baseline on the emissions reduction credited to a CDM project is significant. If the baseline project is built instead of the CDM project and assuming the baseline project has the same life as the CDM project, the resulting emissions from the baseline project is represented by the area $OBDT_n$. The difference between this area and the area of the emissions from the CDM project is the area ABCD. Thus, the net loss in credits from using an adjustable baseline is represented by the area "BCDE".

Clearly, there are significant implications for the expected emissions reduction of a CDM project that relate to the expected emissions and validity period of baselines. The CDM Executive Board will need to establish guidelines/criteria for determining the expected emissions and validity period of baselines. These guidelines or criteria must take into account the need to encourage the market for CDM projects while limiting the potential for rents on the resulting ERs.

5. VALIDATION OF CDM PROJECTS

Article 12.6 of the Kyoto Protocol states that CDM shall assist in arranging funding of certified project activities as necessary. The certification of CDM project activities is referred to as the validation of CDM project activities. This is to avoid confusion from the distinct and different action of the certification of ERs. Validation of a CDM project activity does not guarantee that the project's ERs will be certified. Certification of a project's ER can be done only after the project implementation and after the actual ERs have been generated. Therefore, for the purpose of this paper, validation is not the same as certification. Validation can be viewed as licensing a project to proceed to implementation.

Projects that are proposed for the CDM will need to be certified or validated as CDM eligible projects before they are able to secure all their financing (financial closure). This process is necessary in order to capitalize on the financial value of the project's ERs. Validation as a CDM project activity is tantamount to receiving a license to produce and sell ERs. Without the CDM validation, a project developer would not be able to negotiate an ER purchase agreement (ERPA) with potential buyers. This is similar to a power production license, which must precede the negotiation of a power purchase agreement (PPA). Figure 6 illustrates the key actors or stakeholders and relevant issues that are associated with the validation of a CDM project.

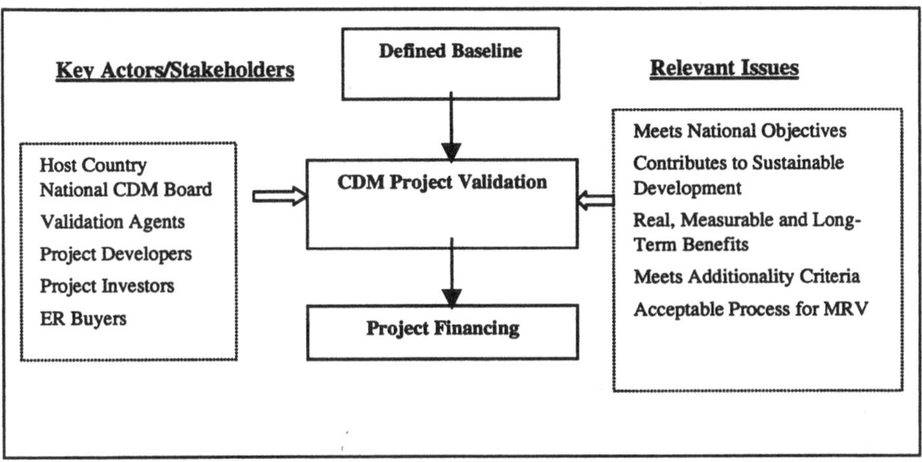

Figure 6. Key actors or stakeholders and issues associated with validation of CDM projects.

5.1 Conditions

The conditions for CDM validation are, to a large extent, dictated by Article 12 of the Kyoto Protocol and by the national objectives of host countries. Criteria for validation of CDM projects should, at a minimum, include the project eligibility criteria stated in Article 12 of the Kyoto Protocol and discussed earlier in section 3.2. In addition, validated projects should have an acceptable process for monitoring, reporting and verifying (MRV) its associated ERs.

5.2 Responsibility

The responsibility for CDM validation must lie with the host country and can be executed by the National CDM Board or some other designated national authority. The national designated authority must be capable of establishing (possibly in association with the CDM Executive Board) and applying the validation criteria during the project review and appraisal process. A CDM validation assessment (CVA) report, submitted by the project developer, can facilitate this process. The CVA must provide the documentary evidence to demonstrate that the proposed project does in fact meet the established validation criteria. The review and approval process for project validation can be similar to the process that is presently used to grant national environmental clearances. In the case of the later, the documentation provided is the projects environmental impact assessment (EIA).

A critical element of the CDM validation process is the assessment of the assumed baseline and the estimated production of ERs by the proposed project. The process must validate the estimated ER production of the proposed CDM project against the assumed baseline conditions. The validation, once issued, would serve as a license to produce and sell ERs that can be fully certified. As discussed in Section 4, the validity period for the baseline defines the period for which the ERs from a CDM project are considered to be additional. Thus the period of validation for a CDM project will be tied to the validity period for the associated baseline.

6. FINANCING CDM PROJECTS

Reaching financial closure for CDM projects is very similar to reaching financial closure for a project in general with one exception. The potential financial value of the resulting ERs from the CDM project must also be secured. The additionality requirement of CDM projects will translate to additional costs for CDM projects in comparison to the baseline options. Therefore, recovering these additional costs through the sale of the project's ERs is critical for financial closure of CDM projects. Structuring and having an ER purchase agreement (ERPA) will be necessary to reach financial closure for CDM projects. Figure 7 highlights the key actors or stakeholders and relevant issues associated with the financing of CDM projects.

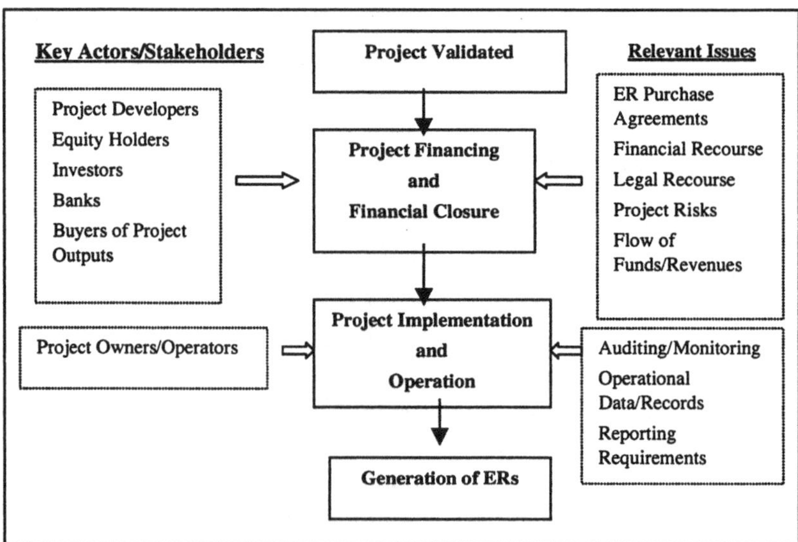

Figure 7: Key actors or stakeholders sand issues in financing and implementation of CDM projects.

This section explores a number of financial instruments and opportunities for securing an ER purchase agreement. If a sound ERPA can not be secured for a CDM project, it is unlikely to reach financial closure. This is the fate of many circulating AIJ pilot project proposals. Under current rules, C reduction units from AIJ pilot projects may not be used for compliance purposes under the Kyoto Protocol. In addition, there is a lack of financial instruments to capitalize the little value these C reduction units have and project proposals often do not fit into the portfolio of investors and/or are not well prepared.

6.1 Securing an ERPA

A number of common project elements need to be addressed in advance of securing an ERPA. These project elements do not have to be designed from scratch. Useful analogies can be found in the context of energy performance contracting as well as power purchasing agreements.

6.2 Establishing the MRV Process

At the heart of any financial instrument based on the sale of ERs are the rules that establish how the ERs are measured. This is because these rules have a fundamental impact on the capacity of a CDM project to generate ERs. ERs are not physical products that can be stored, counted and transferred. Therefore, the agreement on the monitoring, reporting and verification (MRV) process needs to be made between two contracting parties, one of them the performance contractor/project owner or seller and the other the investor/recipient or buyer. Both contracting parties have an incentive to set up the process that would deliver over-stated amounts of ERs at no additional costs to either party. However, this would work to the detriment of the CDM and the objectives of the FCCC. The independent certification of the resulting ERs must, by necessity, check against these potential abuses.

The MRV process for the project requires a detailed ex-ante description of the applicable process that is consistent with CDM requirements and will require approval by the national authority responsible for the CDM. The proposed MRV plan should be part of the information provided and approved during the validation process. The energy efficiency industry has created similar protocols for its own purposes and work to adopt these examples for the purposes of ER performance monitoring is underway.

6.3 Financial Models for Capitalization

A number of common financial models that can be applied to capitalize the potential ER revenue stream from CDM projects. ER benefits could be capitalized by equity inputs. Where debt or third party equity financing is sought in return for the sale of a project's ERs, CDM project developers, prior to financial closure, must negotiate an ER purchase agreement (ERPA) with potential buyers.

6.4 Achieving Financial Closure

Financial closure is achieved when all contractual arrangements related to project financing, construction, fuel supply, operation and maintenance, performance monitoring and product sales are finalized. During this process, performance risks, responsibilities and liabilities are allocated and key licenses and contracts for the import of equipment, plant construction and plant operation are secured. From the perspective of a project's ERs, it is important that the ER purchase agreement cover all relevant aspects that relate to its financial value. These include:

- Protocols for measurement, reporting and verification of resulting ERs;
- Quantity, price and delivery date of ERs;
- Responsibilities and liabilities in case of non-performance with regards to ERs;
- Required approvals from the host government and the CDM as applicable;
- The implications of a change in the status of the project's CDM validation and baseline reference; and
- Procedures to resolve impacts of future changes (in the regulatory environment or baseline definition) on a project's capacity to generate ERs.

7. PROJECT IMPLEMENTATION AND OPERATION

Upon achieving financial closure, a project moves quickly to the project implementation and operation phase. The effort and time required during this phase of the project cycle is highly dependent on the specifics of each project. For CDM projects, an important element is ensuring that monitoring and reporting procedures are implemented according to any agreed or required protocols. This is particularly important if the resulting project

ERs are to be verified and certified at a later date. Failure to do so could lead to the voiding of potential ERs.

Thus the principal objective during the project implementation and operation phase is to ensure the production of ERs and to establish a clear and practical audit trail for the monitoring and verification of the ERs. Figure 7 illustrates the key actors or stakeholders and the relevant issues associated with the project implementation and operation phase.

8. ER MONITORING, VERIFICATION AND CERTIFICATION

The ERs of a validated CDM project must be carefully monitored, verified and certified prior to being eligible for transfer to an Annex I country (that is listed in Annex B of the KP). The requirements for monitoring, verification and certification are interrelated while the entities responsible for monitoring, verification and certification are distinct. The key actors or stakeholders and the relevant issues associated with these last three steps in the CDM project cycle are presented in Figure 8. A brief discussion of the important processes and issues related to CDM project monitoring, verification and certification is presented below.

8.1 ER Monitoring

The CDM project owner or operator has the primary responsibility for the monitoring of ERs. The owner/operator is the initial seller of the ERs and therefore must have the principal responsibility for putting in place the required procedures and measures for monitoring of the project's resulting ERs. The owner or operator may contract or may have agreed to contract a third party to carry out the ER monitoring functions. However, because of concerns for other proprietary information, most project owner/operators may not be willing to allow ER buyers or other external parties to monitor daily project operations. Additionally, this may be impractical and costly and is best done by the owner or operator as long as the data produced can be audited and verified by the ER buyer or agent and the national and international CDM authorities.

The monitoring of CDM project ERs is analogous to the monitoring and accounting of other product streams associated with the project. However, as ERs are not easily measured physically, it is important that corresponding measurable physical outputs of the project be identified that can serve as surrogate indicators for ER production. For example, a wind power project that produces electricity to displace electricity generated from oil can easily be monitored for its ERs by monitoring the actual electricity production of the wind project and using the baseline definition of the GHG emissions associated with the generation of electricity from oil. However, not all CDM projects can be monitored as easily as the example given for the wind power project. Depending on its complexity, energy, industry, forestry and agriculture CDM projects will require the development of approved monitoring protocols in advance of project implementation and ER certification.

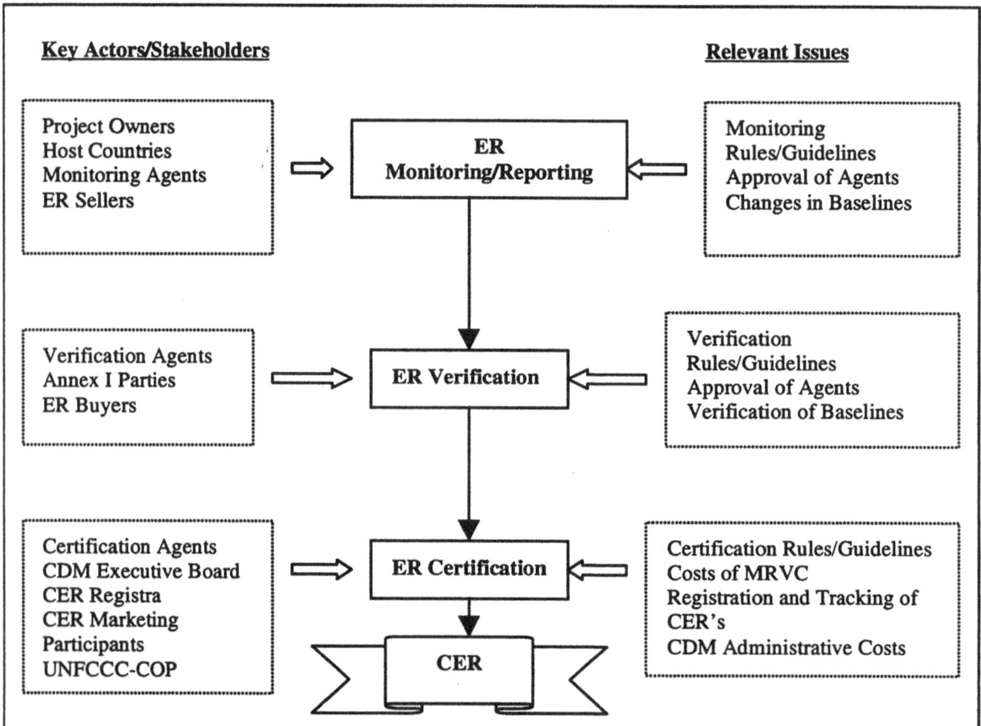

Figure 8: Key actors or stakeholders and issues associated with monitoring, verification and certification of emission reductions.

The end result of the ER monitoring process should be an ER monitoring report (EMR) that is subject to auditing and verification by CDM authorities and ER buyers. The EMR would serve much like the accounting books of a firm that are subject to auditing and verification by a public accountant or potential corporate buyer. The EMR would provide the documentary basis for ER verification and ultimate certification.

8.2 ER Verification

The ERs claimed by the owner or operators of CDM projects must be subject to verification by independent verifiers licensed by the Executive Board of the CDM. This is particularly important because there is no exchange of a physical commodity that transpires that would allow the CDM authority or buyer to unilaterally verify the production of valid ERs. Given the nature of ERs, the CDM Executive Board (or an authorized representative) and ER buyer would have to inspect the operating records of the CDM project. The auditing of the EMR can facilitate this process.

The process of ER verification is very similar to the process of independent product inspection and testing prior to payment by a buyer. The buyer must be assured of the receipt of what was contracted for prior to payment. Similarly, the independent verifier must also be assured that the ERs have been produced according to its guidelines and conditions as agreed to in the initial validation of the CDM project. Additionally, as the resulting ERs will be ultimately certified and transferred (exported), the national authority will need to record the transaction and in many instances want to extract administrative and royalty fees. This is similar to administrative and royalty fees (taxes) that are charged for mineral and other commodity exports.

Independent certified ER verification agents could undertake the actual verification process (Sathaye et al, 1997). The CDM Executive Board, in accordance with guidelines provided by the COP or MOP, would certify these agents. The agents would act much like certified public accountants and could in fact be drawn from national and international firms that presently provide such services. As the ER buyer might also require verification, the ER verification agents would need to be acceptable and accountable to the ER buyers as well. The ER verification agents would work with the project's ER monitoring report to carry out their audit and verification duties. It is anticipated that this would be an annual process. However, the frequency could be more or less depending on the nature of the project and the conditions agreed to in the ER purchase agreements.

8.3 ER Certification

Upon the completion of the verification process, the resulting ERs are ready for certification by the operational entity entitled by the CDM Executive Board. Certification will ensure that only ERs that meet the criteria of the CDM are ultimately certified emission reductions (CERs). If all of the earlier steps for CDM project validation and ER monitoring and verification have been accomplished successfully and in accordance with approved guidelines and regulations, then the certification of the resulting ERs would be procedural.

As part of the certification process, the CDM may levy a fee on certified ERs to cover administrative expenses as well as to assist developing country Parties that are particularly vulnerable to the adverse effects of climate change to meet the costs of adaptation (Kyoto Protocol Article 12.8). The issue of a certification fee is certain to receive significant attention and debate. It is tantamount to a tax on ERs and, if levied, should be designed to minimize the impact on the supply and demand of ERs.

The process of ER certification will also provide the data necessary to track CERs as they are transferred and traded. A mature market for CERs will result in multiple transfers and trades and will need to operate much like a commodity market. In fact, CERs could be traded in existing commodity markets as is currently being done with a limited number of C offset certificates from Costa Rica.

Certification signifies that a reported ER represents a real and measurable emission reduction according to approved protocols and that the data used to calculate the ER is a true representation of the project's performance. Upon obtaining certification, the ER is registered with the national and international CDM authorities and receives CER status. Considering the approval and acceptance process in the recipient country, the CDM registry would need to link all CERs to their originating project (including country and time-stamp). Additionally, the CDM registry would need to keep track of the transfer and ownership of all CERs to ensure that CERs are not double counted by Annex I countries in meeting their compliance targets.

9. CONCLUSIONS

A composite of all the key actors, stakeholders, decisions and relevant issues associated with the CDM project development cycle and market framework is presented in Figure 9. The steps in the CDM project development cycle are straightforward and logical. However, the smooth operation of the CDM will require the resolution of a number of critical and politically sensitive issues. The process for defining the baseline against which a CDM project will be assessed must be clear and transparent (Ellis 1999). The rules and guidelines for defining baselines and determining the additionality of CDM projects must be consistent. The process by which the ERs of CDM projects are monitored, verified and certified must be accepted by all Parties and must be administratively manageable and financially affordable. Finally, an acceptable formula must be established to ensure that a share of the proceeds from certified project activities is used to cover CDM administrative expenses as well as to assist developing country Parties that are particularly vulnerable to the adverse effects of climate change to meet the costs of adaptation.

A vibrant and active market for ERs from CDM projects will only evolve when the key issues associated with the CDM project development cycle are resolved and accepted by all Parties (Jepma and van der Gaast 1999). However, the best approach for resolving many of these crucial issues is to build on the results of the AIJ pilot and initiate a demonstration CDM program in a number of developing countries. These demonstration programs could be undertaken on a bilateral basis with the specific intention of engaging interested developing country Parties in a collaborative process of testing and resolving a workable process for the CDM. The process of implementing a demonstration CDM program will force the identification and resolution of many of the key issues surrounding the CDM. It can provide the UNFCCC's Subsidiary Body for Scientific and Technical Advice (SBSTA) the critical inputs that it will need to help define a workable process for the CDM.

The CDM has the potential to help reduce global GHG emissions while assisting developing countries achieve sustainable development (Hassing and Mendis 1998). A market framework for CDM transactions will permit the participation of a broad cross-section of critical players while attracting the financial resources that is necessary for the success of the CDM. This paper has presented a road map to jump-start of the CDM. It is time to start down the CDM road. It is inevitable that corrections may need to be made to fine-tune the CDM. These corrections can be made more effectively from the experiences gained in operating the CDM. They will not be resolved by protracted political debates.

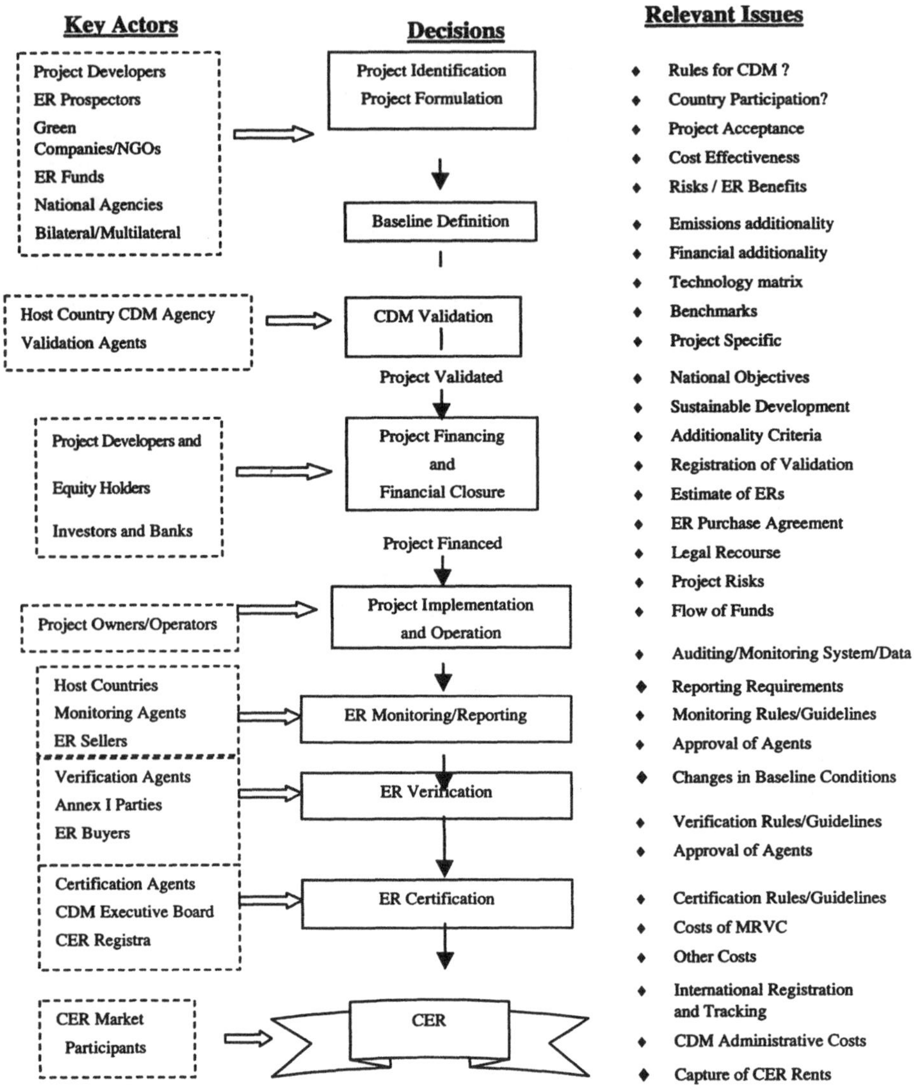

Figure 9: Composite diagram of key actors or stakeholders, issues, and decision framework for CDM project development cycle.

REFERENCES

Chatterjee, K. (Ed) (1997) Activities Implemented Jointly to Mitigate Climate Change: Developing Country Perspective, Development Alternatives, New Delhi.

Ellerman, A.D., et. al. (1998) The Effects on Developing Countries of the Kyoto Protocol and Carbon Dioxide Emissions Trading, Policy Working Paper 2019, The World Bank, Washington, DC.

Ellis, J. (1999) Emission Baselines for Clean Development Mechanism Projects: Lessons from the AIJ Pilot Phase, CDM Baselines Workshop, NEDO, Tokyo.

Friedman, S. (1999) The Use of Benchmarks to Determine Emissions Additionality in the Clean Development Mechanism, CDM Baselines Workshop, NEDO, Tokyo.

Haites, E. and Yamin, F. (1998) The Clean Development Mechanism: Proposal for its Operation and Governance, Margaree Consultants, Toronto.

Hargrave, T. et.al. (1999) Options for Simplifying Baseline Setting for Joint Implementation and Clean Development Mechanism Projects, Center for Clean Air Policy, Washington.

Hassing, P. and Mendis, M. S. (1998) Sustainable Development and GHG Reduction, *Issues and Options; The Clean Development Mechanism*, UNDP, New York.

Jepma, C and van der Gaast, W. (Eds.) (1999) *On the Compatibility of Flexible Instruments*, Kluwer Academic Publishers, Dosdrecht, in press.

Ċararo, C. (1999) *International Environmental Agreements on Climate Change*, Kluwer Academic Publishers, Dordrecht, in press.

Matsuo, N., et. al. (1998) Issues and Options in the Design of the Clean Development Mechanism, The Institute for Global Environmental Strategies, Hayama.

Puhl, I., (1998) Status of Research on Project Baselines under the UNFCCC and the Kyoto Protocol, OECD and IEA Information Paper, Paris.

Sathaye, J., et.al. (1997) The Monitoring, Reporting and Verification of Climate Change Mitigation Projects: Discussion of Issues and Methodologies and Review of Existing Protocols and Guidelines, Lawrence Berkeley Laboratories, Berkeley.

Werksman, J., (1998) Responding to Non-Compliance under the Climate Change Regime, OECD Information Paper, OECD, Paris.

Chapter 15

TOWARD AN INTEGRATED EMISSIONS MARKET KYOTO MECHANISMS

Perspectives for Future Development

NAOKI MATSUO

Institute for Global Environmental Strategies

Key words: Kyoto mechanisms, emissions market, emissions trading, joint implementation, clean development mechanism, allowance trading, credit trading, confinement effect, incompleteness of the market, market integration, emissions exchange, contract-based trading, derivatives, incentives for private sectors, domestic allocation

Abstract: The emissions market can supply Annex I countries cost-efficient options and also strengthen the compliance framework of the Kyoto Protocol. The market mechanism tries to confine total emissions below the cap sum of the stringent Kyoto Protocol's quantified commitments. In order to fully utilize this effect, the market must be integrated, fluid and open to invite many participants to meet demand and supply. In this regard, the role of project-based mechanisms, joint implementation (JI) and the Clean Development Mechanism (CDM) is clarified to improve the incompleteness (providing opportunities, technology transfer,etc.) of the emissions trading market. AIJ pilot provides useful information to address some of the designing issues. Each Kyoto mechanism is expected to play an important role in the integrated market. In the autonomous developing stage of the market, many kinds of contract-based derivative trading are expected. Incentives for the private sector in Annex I is also discussed focusing on domestic allocation.

1. OUTLINE OF THE THREE KYOTO MECHANISMS

1.1 Emissions Transfer Mechanisms in the Kyoto Protocol

Kyoto Protocol, adopted in December 1997, incorporates the epoch-making schemes of emissions transfer mechanisms between Parties (Jepma and Van der Gaast 1999). These transfer schemes have attracted considerable attention for they are innovative and are being tried internationally for the first time. These transfer instruments have not yet been implemented internationally or over sectors, except in the case of some emissions trading schemes for air pollutants within the USA and domestic greenhouse gas (GHG) credit acquisition scheme in the USA under the US Initiative on Joint Implementation (USIJI).

Flexibility mechanisms adopted in the Kyoto Protocol are called Kyoto Mechanisms, those include:

a) Emissions Trading (Article 17),
b) Joint Implementation (Article 6), and
c) Clean Development Mechanism (Article 12).

The bubble concept (article 4) is also recognized as flexibility clause. However, the bubble is also the reassignment of commitments in the countries concerned. After reassignment, each country in the Bubble can participate in the Kyoto mechanisms framework. Of these, (1) and (2) are emissions transfer mechanisms between Annex I Parties with emission commitments in Annex B of the Kyoto Protocol. A part of the assigned amount of each Annex I Party can be transferred under these mechanisms. A big difference between (1) and (2) is that the international emissions trading can start as early as the certificates/permits (allowances) are issued, while the JI credits would only be traded once they have occurred. Of course, each certificate/permit should correspond one-to-one to each tonne of actual volume of GHG emission (CO_2-equivalent). However, these activities will not increase/decrease the cap of total Annex I emissions.

The CDM (3) provides a credit-type transfer similar to that of JI (2) based on project activities. These project activities take place in a non-Annex I country with no emission limit under the Kyoto Protocol, and the certified emission reductions (CERs) generated by the project are to be added to the cap (total assigned amounts) of Annex I emissions. However, global emissions will not increase as a consequence of CDM activities. In addition, CDM has another important role, as emphasised in the Kyoto Protocol, of assisting host developing countries' sustainable development and, as being a potential financial source for supporting adaptation measures to climate change.

There are other differences between JI and CDM. For example, CDM can start generating credits in 2000, requires international certification, and some of the proceeds are devoted for adaptation assistance for vulnerable developing countries. On the other hand, JI cannot create credits until 2008, does not need to contribute to the adaptation assistance and may not need international certification process. Another difference is, for example, JI can include (some kind of) carbon (C) sequestration projects, while it is not certain for CDM.

As mentioned above, (2) and (3) are mechanisms developed from the AIJ pilot phase. Technically speaking, there is no direct linkage between AIJ pilot under the UN Framework Convention on Climate Change (FCCC) and CDM/JI under the Kyoto Protocol in spite of similarities. In terms of their functions as emission transfer mechanisms, these mechanisms (2) and (3) are expected to promote development of the entire climate change mitigation regime by integrating with the international emissions trading market under the emissions trading mechanism, and thereby vitalizing the private sector activity. The development of the emissions market has greater significance on the global GHG limitation. Therefore, in this chapter, I will discuss the differences between emissions trading (1) and project-based mechanisms (2) and (3), as well as the potential integration of these Kyoto mechanisms.

1.2 Relation to Domestic Policies and Measures

Regulated Party with quantified commitment under the Kyoto Protocol is each Annex B country, so each government is responsible for complying with the Protocol. Annex I country governments could be buyers and sellers of mechanisms (1) and (2) and buyers of (3), while the non-Annex I governments could be sellers of CDM credits (CERs). On the other hand, economic entities of both Annex I and non-Annex I countries could be the participants of the market. However, details of participation are to be settled at the COP-6.

In an Annex I country, the government may choose its domestic policies and measures to comply with the Protocol. These could be command-and-control type measures, economic instruments such as tax and subsidies, and other voluntary approaches and educational programs. Among them, some countries may install domestic emissions trading scheme linking it to the international framework, which allocates a part of its assigned amount to its domestic entities. All of the policies and measures are used to comply with the Protocol. If all of the GHG emissions are under the domestic trading scheme, the compliance of the country is theoretically automatic (Matsuo,

1998a). This does not need domestic allocation, governmental trading can meet this condition. The implicit condition of this automatic compliance is the existence of the market which functions properly; i.e., supply can meet demand always adjusted by market price.

It should be noted that emissions trading, especially those implemented domestically, could promote GHG reducing activities not covered by other (command-and-control type) policies and measures through market mechanism, even in OECD countries. In other words, integrated (international and domestic) emissions trading scheme is not merely a permit transfer scheme from countries with economies in transition (EITs) to OECD countries. It conceivably covers all activities in a cost-effective manner in Annex I regions.

2. DIFFERENCE BETWEEN EMISSIONS TRADING AND PROJECT-BASED MECHANISMS

2.1 Allowance Trading and Credit Trading

Transferred amount through the emissions trading (sometimes called allowance) and project-based mechanisms (called credit) have the same unit (Mg of C or CO_2-equivalent) and are roughly equivalent in value as well. Here, value implies both environmental value of a unit utilised to comply with the Protocol and also monetary value ($) in the market, once approved as a credit. There are, however, considerable differences between them in their implementation. The transferred emission rights are sometimes called allowances in the case of (cap-and-trade type) emissions trading, and credits in the case of project-based mechanisms. CDM credits are sometimes called CERs, and Article 6 JI credits are called emission reduction units (ERUs).

The allowance-type emissions trading system is a pre-verification type trading scheme in which assigned amount to be emitted in the future (emission allowance certificates) are open to trade. However, project-based credit trading mechanisms are based on generation/transference of certified emission reductions (or sink increases) generated and transferred after the official verification. However, contract-based trading can make participants pre-verification type trading for JI case also.

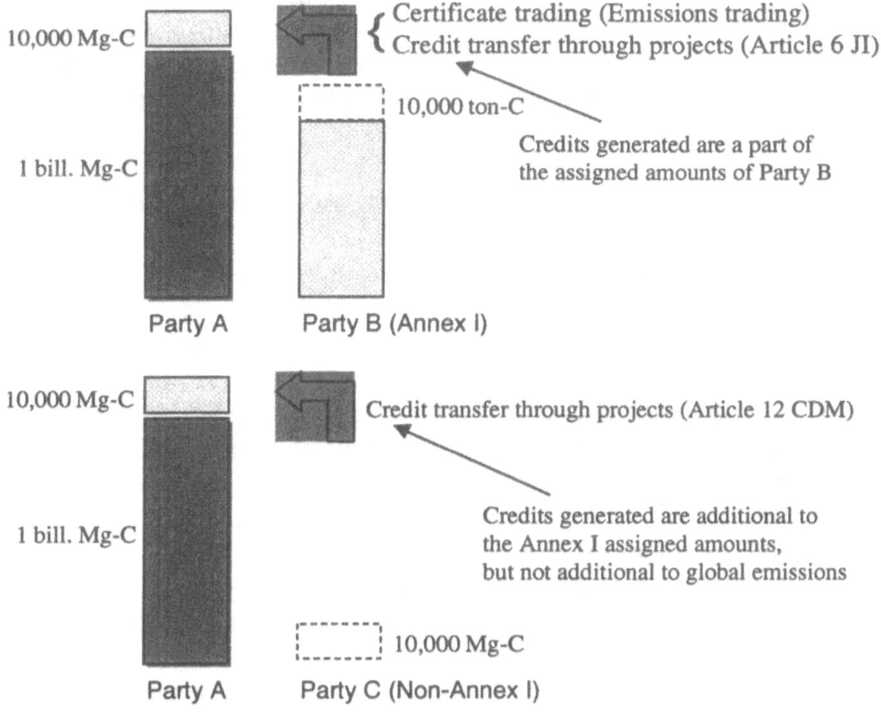

Figure 1 Illustration of differences between Kyoto flexibility mechanisms.

For example, suppose the assigned amount for an Annex I Party country (country-A) is one Pg C for the first commitment period (5-years). Emissions trading means the trading of the unused (or non-consumed) portion of this assigned amount. In contrast, JI under Article 6 allows the transfer of emissions reduction amount (say 10,000 Mg-C (eq)) resulting from JI project activity that Country A implemented in another Annex I Party country (country-B) after the actual reductions are verified annually. In this case, the assigned amount for country-A increases by 10,000 Mg-C (eq), while that of country-B decreases by the same amount. In the case of CDM, some portion (e.g., 10,000 Mg-C (eq)) of the reductions earned through the project activity that country-A implemented in a non-Annex I Party country (country-C) is transferred to the investor country-A after certification of the reduction amount. In this case, the assigned amount for the country-A increases by 10,000 Mg-C (eq), while country-C, with no quantified limit, suffers no demerits under the Protocol, although it can choose to receive the remaining portion of the certified reductions and sell it in the emissions market.

In general, project-based mechanisms are more complex and require higher transaction costs than a certificate (allowance)-based emissions trading scheme. This is especially true in the case of CDM, which calls for higher administrative expenses and further funding (or crediting) for adaptation measures.

2.2 Relation as Options for Policies and Measures

As a definite environmental impact, cap-and-trade type emissions trading ensures control of the overall emissions of Annex I Parties. The whole amount of emissions (cap) is allocated to each regulated entity in the case of cap-and-trade type emissions trading scheme. Each entity can trade a portion of the allocated amount once issued by the regulatory body. CDM credits might be added to the assigned amount of Annex I. However, the overall emissions in the world do not increase in general because CDM credits are reductions in the non-Annex I region. In contrast, (credit-type) project-based mechanisms alone cannot secure the control of the total emissions under a certain ceiling as it does not cover all emission sources.

However, project-based mechanisms can reduce emissions explicitly, offer ancillary benefits other than those of climate change mitigation and complement the deficiencies of the cap-and-trade emissions trading system within Annex I. These deficiencies are, for example, market imperfections especially at the initial stage of implementation, lack of potential for non-Annex I Parties' participation in the market, issues related to the trading of GHGs with uncertainties in monitoring, apparent technology transfer effect, etc. If the market works perfect, all of the potential low-cost measures below permit price in Annex I are implemented. Sometimes, potential host country cannot invest its low-cost options because of various reasons, e.g., lack of information, lack of technology, difficulty in raising funds, etc. Consequently, cap-and-trade and credit-type project-based mechanisms supplement each other.

2.3 Economic Interpretation

Here I explain the differences between these Kyoto mechanisms from the viewpoint of an economics textbook. If *L1* in Figure 2 is the marginal cost-curve for Annex I as a whole, then the market price (a rough average) of the allowance will equal the marginal cost of abatement at the level of (regulated) emission reduction target. Here the horizontal length of each step represents the technical potential of each measure times annual economical penetration rate in a country. It must be noted that the step function of this marginal cost curve is a function of time, i.e., both the economical penetration rate and cost for each measure vary over time. The negative-cost here implies the presence of policy/measure options with net benefit (in the case of implementing energy-saving projects, the benefits of energy cost reduction will exceed initial capital costs). *L2*, the cost-curve of investment without payback, shows the potential of low cost options for Article 6 JI projects. *L3*, as the cost-curve for developing countries (like *L2*), indicates the potential of CDM options.

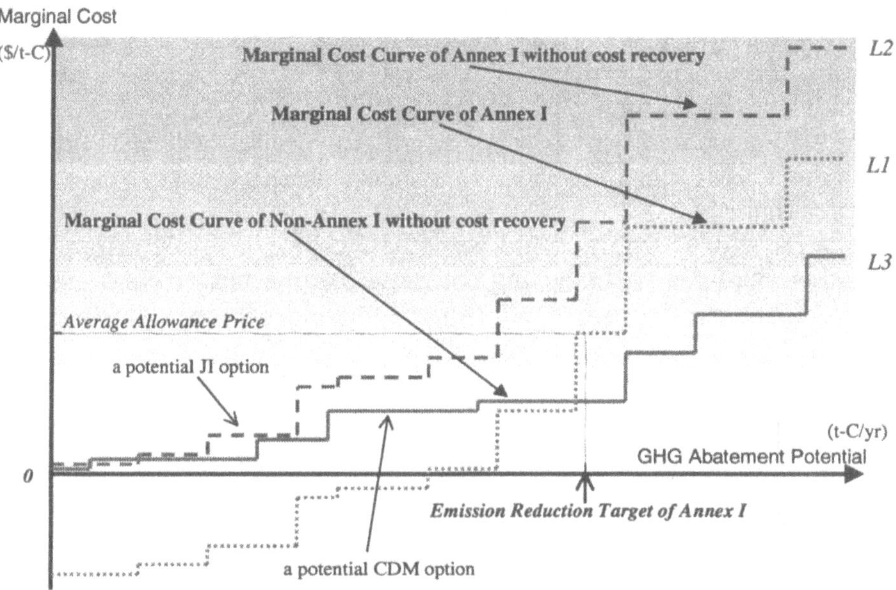

Figure 2. Cost differences between Kyoto flexibility mechanisms (JI and CDM)

Project based mechanisms utilise low-cost options roughly below the level of allowance price minus transaction costs. Investors in the real world will consider conditions like risks for project implementation and other factors in addition to the above cost condition.

For emissions trading, if the market functions properly, it can fully utilise or even enlarge (increase the market penetration ratio) the market potential of reductions (horizontal axis) in the (dynamically varying) cost-curve *L1*. Although, it is unclear whether the market can bring out the potential effectively, however the JI regime of Article 6 may complement any possibility of a market failure. On the other hand, CDM is not only a mechanism for providing low-cost options for Annex I Parties, but also a seed for growing low-cost options in developing countries outside of the emissions trading framework. Furthermore, it is expected to trigger non-Annex I countries to take a pathway of less CO_2 intensive development through technology transfer, etc. In any case, the key issue is to design a regime that can fully realise the potential of the mechanism.

2.4 Sustainable Development of the Host Country

As for CDM, its role to assist the host developing country's sustainable development is explicitly stated in the provisions of the Protocol (Matsuo, 1998c). However, the concept of sustainable development may vary country by country. Article 6 of the Protocol does not specify this condition. However, through negotiations of JI in the FCCC, this condition may be implicitly considered for some cases of JI projects under Article 6 (although not for emissions trading). In other words, the project activities can (and should for CDM) make the most of ancillary benefits other than climate change mitigation.

Especially for CDM, it should be noted that it is a source of funds for adaptation measures in developing countries that are vulnerable to climate change. However, the cost burden from factors independent of (or opposite to) credit (CER) generation might reduce the merits of CDM as a whole through a decrease in the number of implemented projects. We must take care to balance the merits and burdens in total costs when designing the regime.

3. INTEGRATION OF THE MARKET AND POTENTIAL DEVELOPMENT OF TRADING SYSTEMS

3.1 Integration of the Emissions Allowance Market

It is expected that the emissions market will develop autonomously before the first commitment period, once details of each Kyoto mechanism is finalised at COP-6. Trading might take the form of over-the-counter (OTC) based transactions negotiated directly, or through brokers, and may also appear in secondary markets through commodity exchanges. The Chicago Board of Trade (CBOT), the International Petroleum Exchange (IPE) in London and the Sydney Futures Exchange (SFE) have expressed their intention to participate in the dealing of GHG (or CO_2) emission permits.

In addition to allowances, JI and CDM credits are also expected to be traded as equal-valued commodities with respect to both the environment

(Mg of CO_2-equivalent) and finance ($/unit), in the emissions market. Article 6 JI credits may not need additional procedures since they are part of assigned amount, while CDM CERs may have to be converted to the assigned amount of Annex I Parties with serial numbers attached. The Administrative Body for emissions trading established under the COP/ Meeting of the Parties (MOP) might be the appropriate institution to convert CDM credits to assigned amounts. In any case, the credits will increase the liquidity of the emissions market and promote cost efficient options globally through resales in the market. The voluntary participation of developing countries might be expected as well.

The Kyoto Protocol stipulates that CDM credits (CERs) may be generated from 2000. In reality, however, Kyoto Protocol is not likely to enter into force by 2000. Therefore, the COP 6 decision may take the form of recommendation to the first session of the COP/MOP to approve the credits retroactively the Protocol comes into force. In any case, before issuing the allowance (spot or advance) by the Administrative Body or COP/MOP (before official initiation of emissions trading), early trading of (potential) CDM CERs and contract-based forward/futures /options trading are possible. Spot/cash transactions are a straightforward trade of effective (true) permit or credit issued by the regulator. Advance is the permit issued by the regulator but becomes effective (becomes spot) after some vintage year. Forward and futures are not permits or credits but (business) contracts which guarantees the transfer of permits or credits (spot) at a set price on a future date. Forward contracts are OTC-based contracts and futures contracts are legal, standardized and anonymous established by an exchange. Options are contracts for choice options of transfer at a specified price on a specified date in the future. Standardized options like an Exchange establishes put/call (sell/purchase) options, while specialized (exotic) options are used through OTC. In the early stage of the market, project-based credits like CDM credits (spot) and forward/options contract of JI credits might be the principal commodities to be traded (Matsuo, 1998a). Hot-air (allowance) trading might dominate the market instead. However, investors might hesitate to have forward contracts with such country whose eligibility to trade and political risk is uncertain.

3.2 Integration of CDM and Article 6 Joint Implementation

There are some differences between CDM and Article 6 JI. However, partial harmonization of procedures is possible and might reduce transaction costs in some cases and influence the development of an integrated emissions market. Also, as discussed in sections 3.3 and 3.4, contract-based credit trading can reduce the difference effectively.

Upon the integration of the emissions market, CDM and Article 6 JI might take the same direction toward the reduction of transaction costs by eliminating various differences between the two. For example, sharing a standardized baseline setting, and approval procedure by the investor government may be useful. However, as Article 6 JI is less stringent in its procedures and also in fixing reductions, JI credit transfer does not affect the sum of assigned amounts of each country.

The participation of developing countries in the emission control framework (Annex I / B or C?), possibly on a voluntary basis, may be expected in the future. Participation of developing countries claims the amendment of the Protocol. Such countries may be categorized in Annex B or in another new category C established (with different type of quantitative commitments). Therefore, Article 6 JI may be applied instead of CDM. For this reason, it might be better not to differentiate between the procedures of CDM and JI.

3.3 International Joint Project-Based Trading

In the case of Article 6 JI, the transferred credits may constitute a part of the assigned amount (although the Protocol stipulates the additionality of reductions). Thus, in principle, it is possible to transfer credits through contract, independent of emission reduction attainable from the JI project (this is the flexibility mentioned in 3.1). In the case of CDM, it is possible to sign a contract to transfer the specified amount of CERs before project implementation based on the trial calculation of the emission reductions. However, reductions accrue only after they are certified. The investor may purchase/sell the permits/credits in the market in order to adjust the deficit/surplus to meet the emission reduction certification.

As mentioned above, contract-based credit trading is possible before the implementation of the project as in the case of allowance trading. In such cases, the contract is almost the same as futures/forward. However, credit transfer are completed officially only after the certification of GHG reductions. Credits may be shared based on absolute value, not the share of credits generated through negotiations. For example, a performance-based contract is a possibility similar to one processed by an energy service company (ESCO). In this case, the investor secures credits generated from the project. If the generated credits exceed the contracted performance, the investor earns the surplus in addition to the shared ratio of the credits projected.

As shown in the Prototype Carbon Fund (PCF) of the World Bank, a portfolio of several kinds of projects set forth by project-brokers may be contracted often in order to decrease transaction costs, to ease the risk to the investor and to utilise the know-how of project-brokers. In this case, the return for the investors may be in credits (and money). Performance-based contracts might be effective to reduce risks for investors in this base also.

3.4 Domestic Project-Based Trading

For CDM, it is possible that the non-Annex I country implements the project domestically with its own fund. For JI, it is impossible, but a similar idea (credit transfer within an Annex I country) is applicable for in-house projects to sell emission reductions attained in the international market as well.

For example, a company that plans to implement an option of switching an old lignite-fired power plant to a new gas-fired one, may sell the allowances to be avoided (or credits to be generated) in the future in order to raise funds for the project in advance. The gap between observed and projected reductions should be balanced using the investor's allowances after the effects of the project are certified (for non-Annex I CDM projects) or monitored (for Annex I projects).

In case of non-Annex I countries, the CDM CERs are generated and sold on the basis of the project activities implemented only by developing country entities. The form of the sale may be spot credit trading after certification, a contract to be transferred in the future (forward) or a contract for option to sell the credits at a fixed price in the future (options).

In the case of Annex I countries, the credits obtained are nothing but the allowances which were (or are expected to be) avoided. If they are sold, the procedure can be much simpler than that of Article 6 JI or CDM as different types of contracts are possible between domestic entities. An ESCO or an energy equipment manufacturer may use this method as a part of its business. In this regard, the method might provide new business opportunities for the Annex I companies.

4. INCENTIVES FOR THE PRIVATE SECTOR

4.1 The Role of the Private Sector and the Public Sector in the Market

Emissions trading and project-based emissions transfer mechanisms aim to utilise the market mechanisms as much as possible. Therefore, allowing profit-oriented economic entities to take a major role will likely bring out the full potential of the mechanisms. For that purpose, it is desirable to have as many diverse economic entities (in addition to those subject to regulations) as possible to participate in the market.

While speculators like hedge funds may target the emissions market, they are necessary entities in the market as risk-takers so regulated entities can hedge risks. As most of the commodities and securities can be targeted for the money game, this cannot be a reason to give up the emissions market. It is much more important to design the market so that each allowance Mg can correspond to each Mg of emissions reduction and so that it is effective to induce real reductions for regulated entities. The sulfur dioxide (SO_2) allowance market in the USA, which is an example of the emissions market, has succeeded in guaranteeing this condition through an accurate system of monitoring emissions and tracking allowances, as well as being a fully fluid market.

The role of the public sector is important as well. To begin with, the public sector must play the role of co-ordinator for allocating emissions and maintaining the framework for operating the (domestic) system properly. Also, it can be a participant in the market as a big trader, even though it is not as flexible as private sector participants. For example, in the case of domestic allocation, it is difficult to cover all of the emission sources, especially in the case of down-stream allocations. However, it is possible for the government to trade emissions from non-coverage sectors (e.g., residential and transportation sectors) instead. By covering all emission sources (directly and indirectly as well), it is possible to comply with the Kyoto Protocol. In other words, the excess emissions will be automatically covered by procurement from the international emissions market.

Deciding the role of the government in project-based mechanisms is a delicate issue. Principally, government approves the projects and sets the ground rules. Further, as a regulated Party, the government can be an investor in the projects. In the case of overseas development assistance (ODA) fund application, government participation may contradict the requirement for additionality of effects mentioned in the Protocol. However, there are consecutive forms of ODA or other public funds from grant to low-interest loans. It is true that a lot of projects, which need public funds, cannot be implemented only through economic principle. Moreover, public funds can play the role of catalyst, especially in the early stages. Therefore, it is rational to seek the best *use* of their features, and not exclude such options from the beginning. It must be noted that the public funds have been used extensively for AIJ, already.

4.2 How Should Incentives Be Set?

As mentioned above, the private sector may play a principal role in the market. For this reason, incentives to induce participation of the private sector in Annex I Parties (to reduce emissions) are needed.

One of these incentives would be to allow the selling of credits obtained by implementing/developing low-cost options abroad, in the international market. Trading companies, brokers and institutional investors may be the players in this case. However, this incentive is not enough to fully utilize the market functions. Although the participation of speculators and investors is anticipated, it is doubtful that the market can provide sufficient credits and emissions reductions to cover all the deficit of the OECD countries to comply with their quantified commitments. Current policy business-as-usual scenarios of IEA projects more than 20% emissions above Kyoto target in 2010 for Annex I as a whole. Market without entities who reduce emissions cannot fully draw the low-cost options (i.e., reductions) out.

Therefore, some more incentives are needed targeting for domestic emission sources (economic entities). The most straightforward and effective way is to introduce a domestic cap-and-trade framework and to allocate emissions (targets) to domestic entities. Each entity is expected to behave rationally to comply with its target by assessing the possibility of domestic/international low-cost options in comparison with the in-house emission reduction options and the trend of emissions allowance prices in the market.

Another incentive scheme is to install baseline-based credit-type trading schemes without covering all emission sources but to set some baselines for some sources. In this case, however, it is difficult to confine the domestic emissions under a ceiling; and the scheme will be somewhat complicated. So, this kind of trading scheme has been considered as an early-crediting scheme to promote early reductions prior to the first commitment period. Other indirect incentive schemes can include trading based on voluntary action/plan/agreement decided or agreed by the company itself, government purchase of the credit, and the introduction of tax benefits. Each government will choose its most suitable incentive scheme, unless inconsistent with international framework, to comply with the Protocol based on its sovereignty.

4.3 Options for Domestic Allocations

Among the incentives mentioned above, the cap-and-trade method has succeeded as the forerunner in the USA and can guarantee environmental benefits of complying with the Protocol if market works properly. The cap-and-trade method requires allocation. This issue is the most sensitive for policy-making and one of the most important for designing the scheme, while it may not influence economic efficiency very much. Therefore, I survey the particulars related to domestic allocation. I assume that tradeable gas is limited to CO_2 from fossil fuel combustion. It is good enough to consider only down-stream emissions for other GHGs.

There are two levels at which allocation could be made: Up-Stream and Down-Stream of energy flow. The up-stream sector consists of energy supply sectors that produce and/or import energy sources. The down-stream sectors include CO_2 emitters and/or energy consumers. Each country can choose its own definition for these categories. For example, power plant is defined as down-stream in the USA and up-stream in Japan. In general, the up-stream companies are few in number and easier to be regulated. If emissions are allocated to up-stream firms, the down-stream is involved in the scheme by adding the cost on the energy commodity consumed in down-stream sectors. In this case, it is not straightforward that the cost to buy extra permits is transferred to each product uniformly (based on C content). On the other hand, it might be natural and better to allocate emissions to down-stream firms because they know their own marginal costs. However, it is impossible to allocate emissions directly to all of the down-stream sources. An indirect method is used to cover all sources through trading by the government or up-stream firms instead of these sectors (Matsuo, 1998a).

We do not have to identify the sector monitoring emissions and comply to match them to the allowances (with an obligation to report to the government) with the sector to be allocated emission allowances. This is specifically true for carbon dioxide from fossil fuel combustion, which can be set by one-to-one correspondence to the fuel consumption. Especially interesting is the package of down-stream allocation and reporting from upstream. Even allocation to each individual may be possible. It may be easy to solve the equity in this case. For example, in the case of allocation to individuals, a person pays for allowances in addition to the price of gasoline, for example. It will be realistic to introduce an allowance trading system, which uses a pre-paid card or intelligent card.

In the case of a down-stream allocation, it is important to address sectors with allocation difficulties, such as small firms, residential/commercial and the transportation sectors. Theoretically, compliance with the Protocol can be automatic if the government or up-stream sector trades emissions from these sectors, instead of allowing the sectors to trade.

The criteria for allocation may include auctioning, grandfathering allocation based on past emission records and incentive based allocation. In the USA, SO_2 allowance trading scheme, some amount of allowances is rewarded for renewable energy users, opt-in entities, early implemented entities, etc. Equity is an issue that needs to be addressed in the distribution of allocation. In the case of auctioning, this issue will be significant while considering the redistribution of the auctioned revenues. Most of the existing or past tradeable permit schemes adopt the grandfathering rule. This reflects the need to account for vested interests in the introduction of any policies and measures, and is not characteristic of emissions trading only. For the grandfathering rule, there are many options available. These can include emissions-based or energy consumption-based, base year selection, a single-year or a multiple-year base (average or maximum), or a sector-by-sector criteria. These options can co-exist; so each country can select the best combination.

Tradable gases can be limited domestically to certain industrial gases for which sufficient monitoring accuracy can be attained. The Kyoto Protocol allows the trading of all designated GHGs. The subject of government allocation can be business entities, industrial organisations, and even individuals. Even the total amount allocated can exceed the level of quantified commitments given in the Kyoto Protocol. In such a case, the Government must purchase the excess part from the market. The frequency of allocation may be annual in order to be consistent with the domestic tax system. This scheme could be a government subsidy in a sense. This prototype has two aspects: equitable burden sharing in a country and international competitiveness. In general, subsidy itself can be used to soften distortions and promote energy efficiency as well. The former aspect is the issue of relative equity for allocation, so independent of total amount of allocations. The second aspects originated in Kyoto Protocol itself. If it is equitable and competitiveness is measured only by the burden of the country as a whole, there is no room for competitiveness problem. Otherwise, it is deeply connected to the political sovereignty of each country, not characteristic only for emissions trading.

5. DIRECTION OF FUTURE INTEGRATION

5.1 An Image of Market Maturity

The emissions market is expected to develop autonomously, driven by the private sector, mainly dealing with CERs (spot), JI credits (forward) and GHG allowance (forward/options) in the initial stage. The first milestone after COP-3 will come at COP-6 to be held toward the end of the year 2000 when the concrete design of these mechanisms will be settled. The next milestone will be the ratification of the Kyoto Protocol by the USA, which will largely secure the implementation of the Kyoto Protocol. If this comes in parallel with the official announcement by each government of the introduction of the domestic emissions trading scheme (including the early crediting scheme prior to 2008), it might be a big incentive for the private sector to participate in the market, although the allocation formula has not yet been fixed.

Of course, if the Administrative Body implementing the flexibility mechanisms issues the allowances (permits corresponds to each country's assigned amount) valid for the first commitment period (and beyond) as advance, this advance (with a vintage year to be valid) might form the basis of the deals in the emissions market.

If many non-Annex I Parties recognize the benefits of emissions trading, a new type of participation is expected such as opt-in. Setting of an attractive initial allocation (quantified commitment) for such developing countries is crucial for this participation. The concern of so-called tropical hot air should be taken into consideration balancing the merits to invite developing countries into the regulatory framework of GHGs emissions, which is under the environmental benefits of full market mechanism. An interesting idea of allocating emission budget (not limit) to some non-Annex I countries voluntarily is proposed (Philibert, 1999).

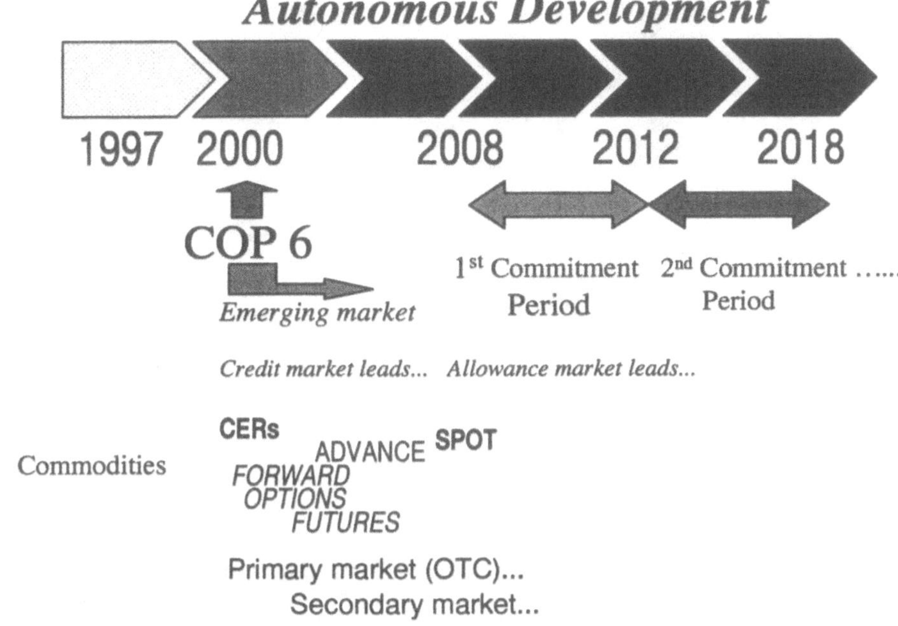

Figure 3. Schematic of prototype of international emissions market development.

The expected market will be international with a mixture of participants from the public sector and the private sector enterprises. Participants can be regulated or non-regulated entities of international or domestic domain. In order to develop a market of greater freedom and fluidity, it is preferable to allow the participation of governments and enterprises in non-Annex I countries through the CDM project or on their own as free traders.

Most deals in the emissions market will be done anonymously through the commodity exchanges; large volume trade will likely be based on OTC bilateral contracts in reference to the price information of secondary markets. Name of buyer and seller, date, serial numbers of transferred permits should be reported for any deal afterwards for record. On the other hand, for project-based credit transfer, trade at the initial stage will be bilateral, but trade with many small investors as a package—plural participants for single or multiple project(s) case—may dominate eventually.

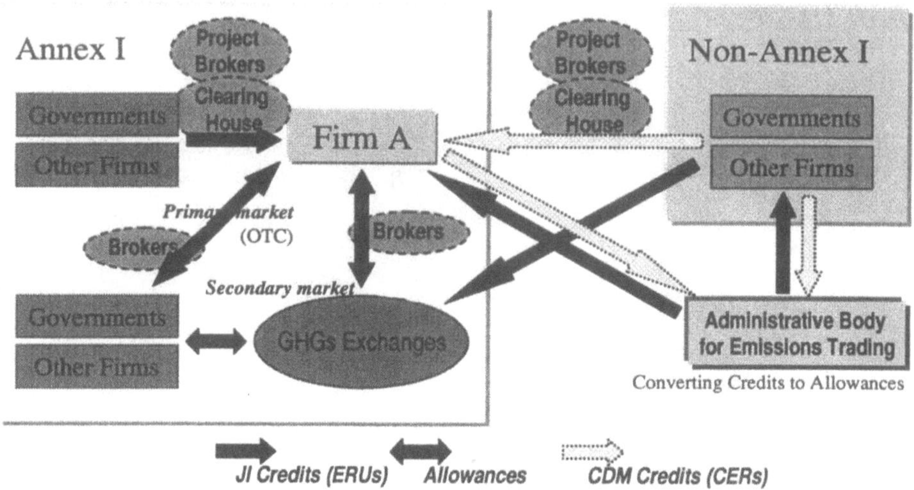

Figure 4. Schematic of prototype of international emissions market with an emphasis on flow of allowances and credits.

In the allowance market, most trade will be through brokers who are experts of derivatives. These derivatives provide tools to hedge risks. For credit transfer, a project market, where the project-brokers will play an active role, will develop with return paid in the form of cash and/or credits to investors.

Development of this kind of fluid market with many participants and a large amount of trade enable each Annex I Party and the Annex I Parties as a whole to comply with the quantified commitments of the Kyoto Protocol. If a government or a company would like to increase emissions, someone who could provide excess allowances/credits will appear in the market under the zero-sum condition of emissions. This means that the total amount of emissions will have to be confined within the ceiling through the market mechanism dynamically. This confinement effect is one of the most outstanding characters of the cap-and-trade type emissions trading in case that the enforcement mechanism is not strong. This terminology is inspired by the dynamic quark confinement effect in elementary particle physics. In contrast, the market may not be fluid, e.g., there may not be enough supply to meet the demand. This means that the market cannot function properly and will not be able to support compliance with the Kyoto Protocol. In other words, if the market works properly, it will be possible to reduce environment impacts in future with the market automatically adjusting to the lowering of the total emission limit step-by-step.

As for the size of the market, let us assume that 100% (turnover rate) of the annual CO_2 emissions target (around 4 Pg-C) in the Annex I countries for the 1st commitment period are traded as allowances including CDM CERs.

In the case of the SO_2 allowance market in the USA, the turnover rate is around 150%. If the mean market price of the allowance is \$30 Mg C, the market size will be around \$120 billion, which is five times as large as the world gold market and twice as large as the USA oil market. Development of the derivatives market may promote further expansion of the market. The emissions market will be a large market for a single commodity.

5.2 Application of Lessons Learned from the AIJ Pilot and Beyond

We have seen the CDM and Article 6 JI mechanisms through the viewpoints of the emissions market, which provides not only lower cost options but also environmental effectiveness through the confinement effect. AIJ pilot opens the door to this framework utilising market mechanism internationally. Seemingly, in the absence of AIJ pilot, it would be difficult to secure consensus on incorporating emissions trading framework to the Kyoto Protocol. The AIJ pilot helped us in learning the difference between and character of the credit-based trading and allowance-based trading as well.

The emissions market including credits is rather a promising scheme to promote compliance with the Kyoto Protocol in so far as it may work properly. In other words, a properly functioning emissions market regime can ensure Annex I Parties' compliance. The market can control the whole of the growing trend of emissions through all cost-effective activities (outside of the coverage of traditional policies and measures) in the Annex I region in addition to some cost-effective project activities in non-Annex I region. However, it is not clear whether the scheme functions properly as a compliance promoting scheme (an emission reduction scheme) prior to or after the commencement of the first commitment period.

As the Kyoto Protocol is an environmental treaty, it seems infeasible to have strong provisions for non-compliance like punishing the sovereign countries. In other words, the Protocol seems not be far from a gentlemen's agreement in spite of its legally binding character in the international community.

Furthermore, we have no experience of adopting international emissions market schemes, especially those affecting the majority of human activities, nor the experience of creating totally artificial commodities and developing a market for them (commodities are normally generated spontaneously). The SO_2 allowance scheme in the USA, which can be seen as a prototype of the GHG market, is a domestic scheme and regulates only a limited number of sources. Moreover, to regulate emissions through trade may not conform to the character and culture of non-Anglo-Saxon countries. On the other hand, the SO_2 allowance market in he USA has been successful environmentally with 100% compliance and is a mature market for a financial commodity, although the causality between these is not straightforward.

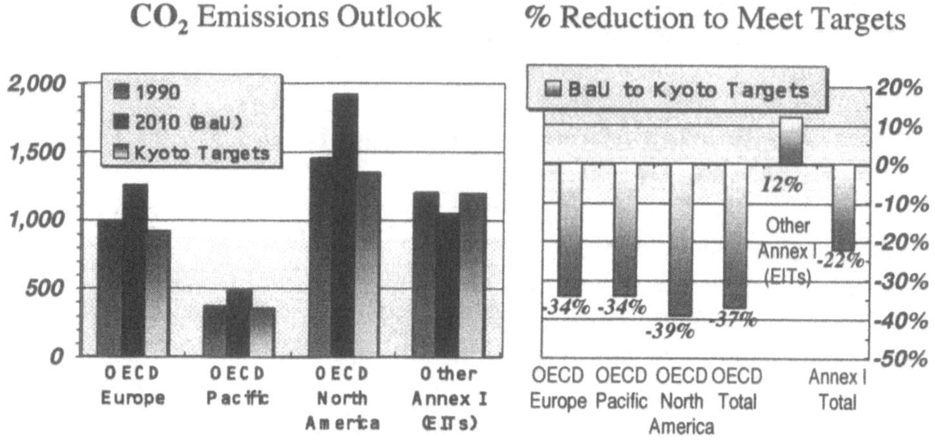

Figure 5. CO_2 emissions outlook for a business as usual scenario (IEA 1998).

Most of the successful trading schemes like SO_2 allowance trading schemes set initial targets at attainable levels (at least technically) and strengthened the target levels as markets matured. On the other hand, the targets of the Kyoto Protocol were set at a very ambitious level from the viewpoint of energy economics. Even if the hot-air is fully incorporated, the targets seem too far reached (at least for energy-related CO_2). For example, non-European Union (EU) OECD countries like the USA and Japan has already exceeded their CO_2 emissions by around 10% from the 1990 levels. The EU also does not have much reduction potential left after the reunion of the eastern states of Germany and the subsequent realisation of emissions reduction and dash-for-gas in the United Kingdom electricity sector. According to the IEA's World Energy Outlook of 1998, the actual emissions of OECD countries as a whole will exceed the quantified commitment of the Kyoto Protocol by 30–40% under the business-as-usual projection (no additional domestic measures and energy demand, supply and prices continues); and by 20% incorporating the hot-air in economy in transition countries.

Another concern to be solved is the supplementarity clause for emissions trading and JI. Based on interpretations of the Kyoto Protocol, emissions trading and JI shall be supplemental to domestic actions to meet the quantified commitment. EU has asserted to set some concrete ceiling for the tradable amount. This idea may suppress the merits of the market mechanism simultaneously. On the other hand, another idea to enhance domestic policies and measures by initiating the programme to establish the common performance indicators is proposed (Matsuo, 1998b). In this case, the indicators are not standards but the non-mandatory information for better understanding of each country's situation.

Consequently, what we must be aiming at under this difficult situation is the proper function of the emissions market so as to promote the compliance with the emission targets of the Kyoto Protocol through market mechanisms. However, it is uncertain whether the inexperienced market will work well for such confinement effects. In this regard, the efforts of:

- strengthening domestic policies and measures,
- procuring assigned amounts from non-Annex I regions through CDM, and
- correcting the imperfection of the emissions market in Annex I under Article 6 JI projects,

would be very important in reducing the 20% excess of emissions mentioned above for compliance of the Kyoto Protocol. In other words, if these efforts are insufficient, most of the Annex I Parties may not be able to comply with the Protocol.

As a technical and crucial problem, preparation of an accurate monitoring system is very important. This includes both the monitoring of assigned amount (inventory of GHGs) and the monitoring of effects (emission reductions) of projects. In economies in transition, a GHG inventory system with sufficient accuracy has not yet been prepared; this may be an obstacle for those countries to participate in the emissions trading regime. An international support system for the development of domestic monitoring systems in these countries should be provided. In the case of lower accuracy emission sources/sinks project-based uncertainty may be smaller than the uncertainties in assigned amount as a whole. We have accumulated a lot of experiences from the AIJ pilot for monitoring related issues, like intervals, how to deal with uncertainties, etc. However, the structural refinement of these regulating infrastructures incorporating the lessons learned is needed.

In relation to the monitoring issue, the baseline-setting for reference case is left as a crucial barrier for project-based mechanisms, especially in the case of CDM. The standardization of baseline-setting formula may hold down transaction costs as a whole, which sometimes were comparable to the implementation costs in some AIJ projects. In addition, standardization is necessary to keep consistency throughout the scheme (Matsuo, 1999). However, it is a trade-off with accuracy of the effects of the project.[1] We may need to seek some balance between them. For this issue, AIJ pilot experiences have provided a menu of options, like benchmarking, technology matrix, and top-down macro methods, in addition to time-dependence, spatial-dependence, etc. These are not yet harmonized but applications of the different methods are useful for designing credible credit-based mechanisms.

Throughout this chapter, I assume that one unit of credit is equivalent to one unit of allowance. On the other hand, we can design the allowance in a way that its value shall represent the possibility of an allowance-originating country to comply with the Protocol, by introducing concepts of the buyer liability. In the case of non-homogeneous permit scheme, we must carefully design the relation between allowance and Article 6 JI credit, and a way to deal with the value of CDM CER in the market. I am anxious about the possibility that any additional risk or complexity, additional transaction costs, may obstruct the development of a well-functioning fluid market in the early stages. On the other hand, we can design CERs to incorporate deductions reflecting uncertainties.

Other issues remain, for example, whether or not to set a formula for credit sharing, how to set criteria and guidelines for project approval and how to design domestic emissions trading schemes as an incentive for the private sector. The AIJ pilot experiences have provided useful information for addressing these design issues. For example, USIJI has rather rigorous criteria for these issues. There is an intensive dialogue between investors, hosts and regulatory authorities. The current understanding of JI or CDM is heavily based on these concrete dialogues.

Risk assessment associated with the project is important for decision-making to implement it. We have learned risks specific to the climate change targeting project activities through the AIJ pilot including identification of the risks and how to handle them.

The work left for us is how to design a regime overcoming these difficulties mentioned in this chapter and in this book. In this regard, it is very important to promote emission reduction effects as much as possible through market mechanisms. The emissions trading and project-based mechanisms are expected to form an integrated market to supplement each other. However, it is impossible to design a perfect regime from the beginning. It might be realistic to select a method with sufficient tolerance for adjusting the orbit, say step-by-step and learning-by-doing, fully utilising the experiences of AIJ pilot.

I am rather afraid that an irreparable situation may arise due to the delay of implementing mitigation options, especially by using such market-based instruments. COP-6 scheduled at the end of 2000 is anticipated to succeed in establishing a concrete scheme for designing the Kyoto mechanisms, thereby instituting the early autonomous development of the emissions market.

ACKNOWLEDGEMENTS

The author thanks Dr. Robert Dixon, Dr. Erik Haites, Mr. Richard Baron, Dr. Michael Dutschke and Ms. Maithili Iyer for comments. They are useful and productive to revise the draft.

REFERENCES

IEA (1998) World Energy Outlook of 1998.

N. Matsuo (1998a) Points and proposals for the emissions trading regime of climate change: For designing future system, IGES, Hayama.

N. Matsuo (1998b) A proposal on the supplementarity issue for emissions trading and joint implementation, IGES, Hayama.

N. Matsuo (1998c) How is the CDM compatible with sustainable development? —A view from project guidelines and adaptation measures, IGES, Hayama.

N. Matsuo (1999) Baseline as the critical issue of CDM: Possible pathway to standardization, in: CDM Workshop on Baseline for CDM, NEDO, Tokyo.

N. Matsuo, A. Maruyama, M. Hamamoto, M. Nakada and K. Enoki (1998) Issues and options in the design of the CDM, IGES, Hayama.

C. Philibert (1999) How could emissions trading benefit developing countries, UNEP, New York.

Chapter 16

IMPLICATIONS OF AIJ FOR CDM

I MINTZER, R DIXON
Pacific Institute for Studies in Development, Environment, and Security; Institute for Global Environmental Strategies

Key words: Joint Implementation, Activities Implemented Jointly, Clean Development Mechanism, Conference of the Parties, Kyoto Protocol, UN FCCC

Abstract: The Activities Implemented Jointly (AIJ) pilot provides the historical foundations for the Clean Development Mechanism (CDM) and the other cooperative mechanisms under the Kyoto Protocol. Despite high early expectations, the AIJ pilot produced only a small number of projects on the ground and the extent of learning by doing has been more limited than was originally anticipated. By reviewing the factors that constrained the implementation of AIJ projects, it is possible to draw some useful conclusions about what is needed to make the CDM more than a boutique market for greenhouse gas (GHG) emissions offsets.

1. THE BIRTH OF JOINT IMPLEMENTATION

In mid-1991, Norway introduced the concept of Joint Implementation (JI) into the UN negotiations leading to the Framework Convention on Climate Change (FCCC). Norway observed that, because national circumstances differed, it might be more expensive to reduce greenhouse gas (GHG) emissions in some countries than in others. This observation implied the idea that it might be cost-effective for two countries to form a partnership in these activities. In principle, they could share the costs of implementing an emissions reducing project in the country where costs were lowest. Thus the concept known today as JI was born (Carraro 1999).

The underlying motivation for introducing JI was to encourage industrial countries (or private entities within industrial countries) to invest in projects or activities that could employ low-cost emission reduction opportunities in developing countries and in countries with economies in transition (EITs). In this way, the benefits of global environmental protection might be achieved in a more cost-effective manner than if all of the necessary reductions were achieved in countries with the highest historical emissions (Jepma and van der Gaast 1999). The approach anticipated that credit for the resulting emissions reductions could be divided between the investing country and the host country for such projects.

In August 1993, developing country representatives began to raise a series of specific questions in the Intergovernmental Negotiating Committee (INC) about the operational aspects of JI (Parikh, 1994). Would credits for JI projects be made available before industrialized countries took on legally binding targets for domestic emissions reductions? Would credits be available for JI projects in developing countries or would the regime be limited to projects undertaken by two industrial countries? Would the availability of inexpensive JI credits arising from emissions reductions in developing countries or EITs reduce the rate of improvement in energy efficiency or the rate of innovation in new technology that would be achieved domestically in Annex 1 countries?

2. COP-1: FROM JOINT IMPLEMENTATION TO AIJ

Concerns about JI grew as the opening of the First Conference of the Parties (COP-1) in Berlin, Germany drew closer. During COP-1, a compromise was reached and incorporated into Decision 5, which established the AIJ pilot (United Nations, 1995). Decision 5 restructured the concept of JI, creating a pilot phase designed to promote learning by doing. This pre-operational period would last until an evaluation was completed before the end of the decade. The purpose of the pilot phase was to promote operational learning and to address methodological issues raised by this new class of joint ventures. To avoid confusion and residual ill will, the COP decided to change the name of the new regime from JI to AIJ. As part of this compromise, it was agreed that AIJ projects could be developed in either developing countries or in EITs. However, Decision 5 stipulated that there would be no internationally fungible credits awarded for projects that either achieved emissions reductions or expanded the uptake of GHGs by natural sinks during the AIJ pilot phase. Under these special and limited conditions, the Parties initiated a 5-year AIJ pilot phase to encourage cooperative international efforts to implement emissions reduction and carbon (C) sequestration projects.

Between 1995 and 1997, many governments and analysts expressed a sense of dissatisfaction with the results of the AIJ pilot phase. For some advocates of technology cooperation and GHG emissions reductions, the principal problem was that there were simply not enough projects under development (< 130 worldwide). No more than a dozen were actually operational and visible on the ground. As a consequence, little learning by doing was actually occurring.

For others, the primary issue was not so much the number as the geographic and sectoral distribution of AIJ projects. Almost half of the AIJ projects reported to the UN FCCC Secretariat by the end of 1997 were in the Baltic States. The second largest concentration occurred in Latin America. There were relatively few projects in Asia and only one under development in all of Africa (Dixon 1998). Nearly all the reported projects involved either the energy production sector or the forestry sector.

Among analysts, there were sharp concerns that agreement on fundamental AIJ methodological and institutional issues was nowhere in sight. Consensus among Parties about common approaches to developing project baselines, performance monitoring, evaluation, verification, and certification of emissions reductions still remained elusive (Dixon 1998).

3. THE CLEAN DEVELOPMENT FUND (CDF) AND THE CDM

At the UN FCCC COP-3, a number of governments were re-evaluating their approach to achieving technology cooperation and emissions reductions. One of the most promising ideas was a proposal by Brazil to establish a CDF (Government of Brazil, 1997). This proposal was viewed as an alternative to AIJ and contained the following novel elements:

- A target temperature level would be set for 2020 through negotiations among Parties or by consultation with a group of scientific experts, perhaps the Intergovernmental Panel on Climate Change (IPCC).
- Using this temperature target, a simple formula could be used to calculate an effective GHG emissions reduction target that all industrial countries must meet collectively so as to stay within the target temperature increase.
- Each industrial country would then be assigned a national emissions reduction target, based on assumptions of constant emissions at its 1990 level and consideration of its historic contribution to the cumulative buildup of GHGs.
- Industrial countries which fell out of compliance with their reduction targets could buy emissions reductions credits or contribute to a developing country CDF at the rate of US$3.33 per Mg C annually.

- The resources accumulated in the CDF would be disbursed by the financial mechanism of the FCCC, with 10% for adaptation projects and 90% for GHG emissions mitigation projects.

The Brazilian proposal for a Clean Development Fund gave way to a new proposal that emerged late in the COP-3 negotiations. This new proposal was forged in the hot crucible of a small contact group and never debated in the Committee of the Whole. It was referred to as the CDM. Accepted by the Parties as part of the compromise package that is now known as the Kyoto Protocol, the CDM was designed to promote sustainable development in non-Annex 1 Parties, stimulate cooperative investment ventures between Annex 1 and non-Annex 1 Parties, and reduce anthropogenic emissions of GHGs in a cost-effective fashion (United Nations, 1997a). In principle, it would replace the AIJ pilot and eliminate the need for a JI regime involving developing countries. Other chapters in this volume treat various proposals for implementation of the CDM in some detail. The following section highlights some of the key differences and similarities between AIJ and CDM. The final section suggests some lessons learned during the AIJ pilot phase that bear on the design and implementation of the CDM.

4. KEY DIFFERENCES AND SIMILARITIES BETWEEN AIJ AND THE CDM

Although similar in several ways, there are a number of key differences between the AIJ pilot phase and the CDM. The following paragraphs compare the two regimes in terms of the criteria for eligible participants and projects, the incentives and rewards for project developers, the primary obstacles and barriers to entry into the market, and the unresolved methodological issues that cloud the process of regime development.

4.1 Who can participate in AIJ and CDM Projects?

One of the distinctions between AIJ projects and CDM activities involves the criteria for participation. Decision 5 taken at COP-1 implies that participation in AIJ projects is open to all FCCC Parties, i.e., both Annex 1 Parties and non-Annex 1 Parties (United Nations, 1995). As a result, almost half of the AIJ projects reported to the FCCC Secretariat have involved partnerships between Annex 1 Parties and developing countries. The remainder has involved partnerships between OECD countries and EITs.

The eligibility rules for the CDM are different. Paragraph 12.2 of the Kyoto Protocol implies that CDM projects involve partnerships between Annex 1 Parties and developing countries (United Nations, 1997b). Annex 1

Parties that are also EITs are not eligible to participate in this flexibility mechanism under the Kyoto Protocol.

Some analysts have suggested that Article 12 is somewhat more permissive about participation. That is, they argue that the Protocol does not require the participation of an Annex 1 Party in a CDM project. In fact, they contend, a developing country could initiate and implement a CDM activity with no outside participation or external financing. The developing country participant could then sell the rights to any of the resulting emissions credits without having to share them with another Party representing the project's partners.

There is another similarity between the mechanisms. Both AIJ activities and CDM projects can involve private and/or public entities among the partners (Jepma and van der Gaast 1999). Indeed, the designers of the CDM intended that this new mechanism would have a significant long-term influence on the pattern and distribution of foreign direct investment worldwide.

4.2 What kinds of projects are eligible to become AIJ or CDM activities?

In the case of AIJ, Article 4 of the FCCC is broadly interpreted (United Nations, 1992). Most Parties and the FCCC Secretariat have understood Article 4 to imply that both GHG emissions reduction and carbon dioxide (CO_2) sink enhancement projects are eligible for consideration as AIJ.

By contrast, Article 12 of the Protocol does not specifically mention CO_2 sinks but nor does it exclude sink enhancement or land-use change and forestry (LUCF) projects from the regime (United Nations, 1992). However, since the concept of sink enhancement projects is not mentioned specifically in the text, some analysts and some Parties have argued that sink projects are not eligible for accreditation under the CDM.

A number of concerns have been raised concerning the risks of including CO_2 sink enhancement or forest conservation projects in the ambit of the CDM. To date, however, there is no credible evidence from the AIJ pilot phase to suggest that LUCF projects impede local economic and social development. Such projects do raise a number of technical and methodological issues, but these issues are not categorically different from the challenges associated with energy sector or industrial projects. In any event, a decision on inclusion of LUCF projects within the CDM is likely to be postponed until the completion in 2000 of a Special Report from the IPCC that has been requested by the FCCC's Subsidiary Body on Scientific and Technological Advice (SBSTA).

4.3 What are the incentives and rewards for participation by Parties or project developers in AIJ projects compared to CDM activities?

One of the key elements of COP-1 Decision 5 with respect to AIJ was the agreement that projects initiated under this regime would not earn internationally fungible credits during the AIJ pilot (United Nations, 1995). The impact of this was to severely limit the incentives for participation in AIJ projects by Annex 1 governments or private entities. Except for direct financial returns, the main payoffs from participation in these projects are limited to the intellectual benefits of learning by doing. For some companies, the socioeconomic benefits of knowing that the associated good works will help to reduce the risks of rapid climate change has significant intangible value. But Decision 5 is clear and unequivocal: Annex 1 Parties whose governments or private entities invest in AIJ projects will not be able to apply the resulting credits toward the Quantified Emissions Limitation and Reduction Obligations (QUELROs) that accrue under the Kyoto Protocol.

Decision 5 did not preclude Annex I Parties from implementing policies and measures that provide domestic incentives for participation in AIJ projects. A number of Annex 1 countries offer technical support and modest subsidies to firms seeking assistance in the preparation of AIJ project proposals. The scale of these domestic incentives has, to date, been relatively small and their impact on AIJ investments appears to be quite modest.

By contrast, certified emission reductions (CERs) under the CDM (Article 12 of the Protocol) may be bought and sold on the international market after year 2000. In addition, Annex 1 governments will have the option of applying their share of these credits to the fulfillment of their QUELROs during the first commitment period under the Protocol (2008-2012). In addition, CDM project participants can still capture the kinds of socioeconomic benefits that are expected to accrue from AIJ.

Paragraph 12.8 of the Protocol creates another important distinction between CDM and other flexibility mechanisms. This paragraph creates an important limitation to the rewards that can be captured by Parties and entities that invest in CDM projects. Article 12 specifies that the Executive Board or other operating entity of the regime will retain a share of the proceeds from CDM projects (United Nations, 1997b). This share of the proceeds will be allocated in part to cover the administrative costs of the CDM regime and in part to the funding of adaptation projects in developing countries. There remains, however, some continuing disagreement concerning whether the phrase share of the proceeds in Article 12 refers to:

- share of the investment in the project,
- share in the resulting stream of financial revenues from the project, or

– share of the CER units that accrue from the project activities.

5. OBSTACLES AND BARRIERS TO ENTRY: LESSONS FROM AIJ FOR CDM

In the four years since the beginning of the AIJ Pilot Phase, approximately 130 projects have been approved by Parties and reported to the FCCC Secretariat (Joint Implementation Quarterly, 1998). A number of reasons have been suggested for the small number of projects. The following paragraphs highlight some of the principal obstacles and barriers that have slowed or limited project development under the AIJ pilot phase.

The principal barrier to entry in this new C market has been uncertainty facing investors about the monetary value of their contribution to AIJ projects (van der Burg, 1994). Since emissions reductions resulting from AIJ projects can not earn internationally fungible credits, potential investors have very few benchmarks for estimating the worth of this future stream of value added. Absent the ability to monetize a value for the future emissions reductions, investors must make their decisions on the basis of the expected value of intangible goodwill that they may come to acquire. This goodwill could be in the form of increased likelihood of favorable treatment from domestic regulatory authorities in their own country or in the form of some increased likelihood of product acceptance by consumers in the country where the AIJ project is located. In either case, incorporating these non-financial returns into conventional project financial analyses (including calculations of net present value or internal rate of return) is difficult for developers to do with confidence.

The second obstacle or barrier to entry into the AIJ market has been persistent uncertainty about the incremental transaction costs associated with preparation of an AIJ project for approval by the governments of the host and investing countries. In many countries, these transaction costs have appeared to be quite high. Informal estimates suggest, for example, that the average cost of preparing project documents and fulfilling the process of acquiring approval from the US Initiative on Joint Implementation (USIJI) for AIJ projects has run as high as US$60,000 per project during some intervals in the last four years. Outside the USA, some of these costs have been defrayed for project developers by the support of government resources (e.g., Japan and the Netherlands). But the residual level of transaction costs even in these countries puts potential AIJ projects at a financial disadvantage when compared to more conventional joint venture alternatives.

The third obstacle or barrier to preparation and implementation of AIJ projects has involved the lack of host country capacity to absorb such undertakings. In many potential host countries there has been no single institution designated with the responsibility for working with project developers. The absence of a focal point has caused developers to wander between ministries and agencies, often giving up on the process before they are able to win project approval. In countries where there seems to be no inclination among policy makers to create an enabling environment, project developers tend to shy away from the entire process. However, when a host country government introduces this coordination capability, it lowers the barriers to entry for project developers and investors dramatically.

For example, at the initiation of the Decin municipal heating system project in the Czech Republic, there was very limited understanding of the climate problem or of the potential for AIJ pilot projects in the country. There was no agreement at the national level concerning which agency or ministry had jurisdiction over the project (Aslam, 1997). Throughout the project development cycle, leading Czech and USA based non-government organizations (NGOs), worked with the national government and the municipal authorities to illuminate the process. The NGOs also helped the government and municipal officials to explore the range of institutional and technical options available for structuring the project.

Following the Decin project development experience, the Czech government has delineated specific institutional responsibilities for development and approval of AIJ pilot projects. The Czech government has decided to partition these responsibilities among different bureaucracies. The national JI office is responsible for working with project developers on issues of approval, monitoring, and verification of AIJ pilot projects. The Ministry of Environment is responsible for establishing the criteria and guidelines under which the national JI office must operate. Aslam (1997) concludes that the evolution of clear institutional roles and responsibilities has contributed in important ways to the considerable subsequent achievements of the Czech Republic in attracting AIJ investments.

An even more powerful example of the importance to AIJ of strong institutions and national leadership may be observed in Costa Rica. Costa Rica made a national commitment to sustainable development and to the AIJ regime even before COP-1. The commitment was made at the highest levels of the Costa Rican government, reflecting the convictions of the President and the desires of the legislature to make sustainable development and constructive response to climate change a priority for strategic economic thinking in this small developing country.

Costa Rica created the first national JI program in 1994 when it established the Joint Implementation Program within the Ministry of Natural Resources, Energy and Mines (Aslam, 1997). In 1995, the Costa Rican government established the national Office of Joint Implementation (OCIC).

This new independent office was endowed with unusually broad ranging decision-making authority. OCIC could provide guidance to investors concerning host country priorities, project guidelines, and criteria for AIJ projects. The presence and skill of the OCIC staff increased investor confidence and led to approval of three USIJI projects as well as investments by Norway and the Netherlands.

OCIC also developed a new investment concept, the Certifiable Tradable Offset (CTO), and stimulated a second round of AIJ pilot projects. Each CTO instrument represents an offset equivalent to a ton of carbon emissions reduced or sequestered. The CTOs are backed by a reserve of unsold emissions credits and are guaranteed for 20 years by the Costa Rican government. OCIC first successfully traded CTOs internationally in 1998.

The strengthening of the legal and institutional framework in Costa Rica combined with the enthusiastic participation of public, private, and NGO actors in Costa Rican society to add substantial momentum to the AIJ pilot phase in this progressive country. The willingness of governmental, private sector and NGO stakeholders to take ownership of the AIJ concept and to commit financial and human resources to OCIC significantly increased the likelihood of success.

A fourth obstacle or barrier to entry for AIJ projects has been the perceived shortage of trained technical and managerial personnel in potential host countries. In many countries, professionals (including lawyers, managers, scientists, project finance and implementation specialists) and skilled laborers are in high demand. Attracting them to participation in AIJ projects is no easy feat in these competitive labor markets. In many cases, local staff must be trained on the job and, once trained, face a strong financial incentive to move on to higher paying work.

A number of additional market conditions that have nothing to do with the climate issue have created barriers to investment in AIJ projects. These include high inflation rates and scarce financing resources in host countries, perceived lack of credit-worthiness among municipal authorities and other potential project developers, and a generalized perception of high risk associated with investment projects in developing countries and EITs.

Unresolved methodological issues constitute a final example of obstacles to entry into the AIJ regime. The lack of clarity among the Parties about how best to address these methodological issues is one of the things which has contributed to the high transaction cost associated with planning and development of AIJ projects. The most knotty methodological issues in the debates about AIJ have included:

- How to establish credible and reliable project baselines;
- How to estimate the environmental benefits of AIJ projects;

- How to set minimal requirements for monitoring, evaluation, verification and certification of project performance; and
- How best to demonstrate financial additionality for AIJ projects.

Several approaches to addressing these issues are under discussion among the Parties and the FCCC Secretariat. No consensus exists on the best response to any of these issues. No decision-making process is yet agreed for resolving these issues in an orderly way. One might hope that they would be resolved prior to the formal evaluation of the AIJ pilot phase at the end of the decade, but there is no certainty that this will occur.

The obstacles and barriers to entry discussed above have slowed the development of the AIJ pilot into a full-blown market for carbon emissions reduction credits. Many of the same obstacles and barriers are likely to have a similar chilling effect on the evolution of the CDM regime.

CDM has the important advantage that internationally fungible credits should become available after the year 2000 if the COP can agree on the rules, methodologies and guidelines for operation of the new regime. Unfortunately, as of mid-1999, it is not clear that such an agreement can be reached in the next six months. Without such an agreement, investor enthusiasm will be dampened by the same uncertainty about the future value of emissions reductions that has bedeviled the AIJ pilot phase. If the FCCC Parties want the CDM and JI regimes under the Kyoto Protocol to be more than a boutique market with a handful of projects initiated each year, they will have to agree on an institutional structure for the CDM and on a practical approach to the issues listed above.

The following section briefly outlines some primary conclusions and lessons learned from the AIJ experience that can be applied to the development of the CDM regime.

6. CONCLUSIONS AND LESSONS LEARNED FROM AIJ

The CDM regime can evolve beyond the stage of boutique investments into a robust market for C emissions reduction credits, with thousands of projects initiated each year (Goldemberg 1998). In order to achieve this level of development through private sector partnerships, certain minimal conditions must apply. From the investor's or project developer's point of view, these include the following:

- Potential investors in CDM projects must face some credible likelihood of encountering emissions caps or emissions reduction requirements in their home markets. A domestic cap and trade system in each Annex 1 country will provide incentives for investment, even in the absence of tax relief, increased regulatory requirements or direct subsidies. Without such caps or requirements, CDM projects are unlikely to be as attractive as less complicated, conventional joint venture opportunities.

- As in the AIJ pilot phase, many of the early investments in CDM are likely to be made by small and medium scale enterprises. To expedite development of the CDM, national governments in Annex 1 countries must implement policies and measures to facilitate financing of CDM projects by these types of entities.
- Methodological and technical issues concerning baselines, project lifetimes, and appropriate system boundaries must be addressed in a way which provides clear guidance to project developers and limits the transaction costs associated with CDM projects.
- In order to ensure the replication and diffusion of technologies embodied in CDM projects, the international regime must ensure that a significant capacity-building component is incorporated into each CDM project design. In this context, capacity building activities are not intended to be limited to training of personnel in developing countries but should include institutional strengthening, expanding strategic resources of data and information and policy reform. To make the CDM more than a boutique activity, such capacity building is required in both industrial and developing countries.
- The CERs derived from CDM projects must have a financial value and be fungible with other forms of C offset credits in either a domestic or international market. Investors must have access to a credible and reliable protocol for estimating the environmental benefits of CDM projects. The uncertainty associated with the future value of the CERs must not dominate all other types of project costs.
- The transaction costs of initiating, developing, monitoring, reporting, and certifying emissions reductions from CDM activities must not be significantly higher than the costs associated with equivalent conventional joint ventures involving the same assets and producing similar financial results.
- The products or output of the CDM activity must have a viable potential for generating a self-sustaining commercial market in the region of project implementation or in countries facing similar circumstances.

From the host country perspective, the following conditions must apply:

- The host country must be able to recognize the proposed CDM project as being on the critical path associated with its development strategy, not just a vehicle for reducing its greenhouse gas emissions.
- The host country must have in place the institutional and regulatory infrastructure as well as the trained (or trainable) personnel needed to

adapt the necessary technologies and adopt the appropriate social organizations for implementing the CDM project.

- The project must incorporate capacity building components that are recognizable and valued by the host country and the receiving communities.
- The host country must see the project as enhancing its standing internationally and strengthening its prospects for long-term economic development.

Whatever it's final outline, the CDM regime must build on the lessons learned from the AIJ pilot phase. If the conditions listed above can be met, the potential benefits of CDM projects in terms of both economic development and global sustainability may make this new mechanism an important contributor to international economic cooperation for decades to come.

REFERENCES

Aslam, M.A. (1997) Endogenous Capacity Building for AIJ: Developing Country Needs, ENVORK, Islamabad.

Carraro, C. (1999) International Environmental Agreements on Climate Change, Kluwer Academic Publishers, Dordrecht, in press.

Dixon, R.K. (1998) The U.S. Initiative on Joint Implementation: An Asian-Pacific Perspective. Asian Perspective 22:5-19.

Goldemberg, J. (Ed.) (1998) The Clean Development Mechanism: Issues and Options, UN Development Program, New York, 180p.

Government of Brazil (1997) Proposal for a Clean Development Fund, Ministry of Foreign Affairs, Brasilia.

Jackson, T. and Begg, K. (1999) Accounting and Accreditation of Activities Implemented Jointly, European Commission, Brussels, 410p.

Jepma, C. J. and van der Gaast, W. (1999) On the Compatibility of Flexible Instruments, Kluwer Academic Publishers, Dordrecht, in press.

Joint Implementation Quarterly (1998) Volume 4, no. 4, Foundation JIN, Paterswolde.

Parikh, J. (1994) North-South Cooperation in Climate Change through Joint Implementation, Indira Ghandi Institute for Development Research, Bombay.

United Nations (1992) Framework Convention on Climate Change, Article 4.2, United Nations, Geneva.

United Nations (1995) Decision 5/CP.1, Climate Change Secretariat, United Nations, New York.

United Nations (1997a) Kyoto Protocol to the UN Framework Convention on Climate Change, United Nations, New York.

United Nations (1997b) Kyoto Protocol to the UN Framework Convention on Climate Change, Article 12, United Nations, New York.

van der Burg, T. (1994) Economic Aspects of Joint Implementation, in Onno Kuik, P. Peters, and N. Schrijver, (Ed.) Joint Implementation to Curb Climate Change: Legal and Economic Aspects, Kluwer Academic Publishers, Dordrecht.

Index (by Chapter)

Zeitfracht Medien GmbH
Ferdinand-Jühlke-Straße 7
99095 Erfurt, Deutschland
produktsicherheit@kolibri360.de